Fixed Point Theory and Related Topics

Fixed Point Theory and Related Topics

Special Issue Editor
Hsien-Chung Wu

MDPI • Basel • Beijing • Wuhan • Barcelona • Belgrade • Manchester • Tokyo • Cluj • Tianjin

Special Issue Editor
Hsien-Chung Wu
Department of Mathematics,
National Kaohsiung Normal University
Taiwan

Editorial Office
MDPI
St. Alban-Anlage 66
4052 Basel, Switzerland

This is a reprint of articles from the Special Issue published online in the open access journal *Axioms* (ISSN 2075-1680) (available at: https://www.mdpi.com/journal/axioms/special_issues/fixed_point_theory_related).

For citation purposes, cite each article independently as indicated on the article page online and as indicated below:

LastName, A.A.; LastName, B.B.; LastName, C.C. Article Title. *Journal Name* **Year**, *Article Number*, Page Range.

ISBN 978-3-03928-432-0 (Pbk)
ISBN 978-3-03928-433-7 (PDF)

© 2020 by the authors. Articles in this book are Open Access and distributed under the Creative Commons Attribution (CC BY) license, which allows users to download, copy and build upon published articles, as long as the author and publisher are properly credited, which ensures maximum dissemination and a wider impact of our publications.

The book as a whole is distributed by MDPI under the terms and conditions of the Creative Commons license CC BY-NC-ND.

Contents

About the Special Issue Editor vii

Preface to "Fixed Point Theory and Related Topics" ix

Azadeh Ghanifard, Hashem Parvaneh Masiha, Manuel De La Sen and Maryam Ramezani
Viscosity Approximation Methods for $*$−Nonexpansive Multi-Valued Mappings in Convex Metric Spaces
Reprinted from: *Axioms* **2020**, *9*, 10, doi:10.3390/axioms9010010 1

Mohamed Amine Farid, Karim Chaira, ElMiloudi Marhrani and Mohamed Aamri
Measure of Weak Noncompactness and Fixed Point Theorems in Banach Algebras with Applications
Reprinted from: *Axioms* **2019**, *8*, 130, doi:10.3390/axioms8040130 9

Thenmozhi Shanmugam, Marudai Muthiah and Stojan Radenović
Existence of Positive Solution for the Eighth-Order Boundary Value Problem Using Classical Version of Leray–Schauder Alternative Fixed Point Theorem
Reprinted from: *Axioms* **2019**, *8*, 129, doi:10.3390/axioms8040129 31

Hsien-Chung Wu
Informal Complete Metric Space and Fixed Point Theorems
Reprinted from: *Axioms* **2019**, *8*, 126, doi:10.3390/axioms8040126 45

Yaé Ulrich Gaba and Erdal Karapınar
A New Approach to the Interpolative Contractions
Reprinted from: *Axioms* **2019**, *8*, 110, doi:10.3390/axioms8040110 59

Ram Prasad Daripally, Naveen Venkata Kishore Gajula, Hüseyin Işık, Srinuvasa Rao Bagathi and Adi Lakshmi Gorantla
C^*-Algebra Valued Fuzzy Soft Metric Spaces and Results for Hybrid Pair of Mappings
Reprinted from: *Axioms* **2019**, *8*, 99, doi:10.3390/axioms8030099 63

Edraoui Mohamed, Aamri Mohamed and Lazaiz Samih
Relatively Cyclic and Noncyclic P-Contractions in Locally $\mathbb{K}$-Convex Space
Reprinted from: *Axioms* **2019**, *8*, 96, doi:10.3390/axioms8030096 81

Vahid Parvaneh, Nawab Hussain, Aiman Mukheimer and Hassen Aydi
On Fixed Point Results for Modified JS-Contractions with Applications
Reprinted from: *Axioms* **2019**, *8*, 84, doi:10.3390/axioms8030084 89

Hüseyin Işık, Hassen Aydi, Nabil Mlaiki and Stojan Radenović
Best Proximity Point Results for Geraghty Type $\mathcal{Z}$-Proximal Contractions with an Application
Reprinted from: *Axioms* **2019**, *8*, 81, doi:10.3390/axioms8030081 103

Erdal Karapınar
Recent Advances on the Results for Nonunique Fixed in Various Spaces
Reprinted from: *Axioms* **2019**, *8*, 72, doi:10.3390/axioms8020072 115

Badshah-e-Rome, Muhammad Sarwar and Poom Kumam
Fixed Point Theorems via α-ϱ-Fuzzy Contraction
Reprinted from: *Axioms* **2019**, *8*, 69, doi:10.3390/axioms8020069 151

Taoufik Sabar, Abdelhafid Bassou and Mohamed Aamri
Best Approximation Results in Various Frameworks
Reprinted from: *Axioms* **2019**, *8*, 67, doi:10.3390/axioms8020067 . . . . **161**

Atiya Perveen, Idrees A. Khan and Mohammad Imdad
Relation Theoretic Common Fixed Point Results for Generalized Weak Nonlinear Contractions with an Application
Reprinted from: *Axioms* **2019**, *8*, 49, doi:10.3390/axioms8020049 . . . . **171**

Yoshihiro Sugimoto
Applications of Square Roots of Diffeomorphisms
Reprinted from: *Axioms* **2019**, *8*, 43, doi:10.3390/axioms8020043 . . . . **191**

Hamid Faraji, Dragana Savić and Stojan Radenović
Fixed Point Theorems for Geraghty Contraction Type Mappings in *b*-Metric Spaces and Applications
Reprinted from: *Axioms* **2019**, *8*, 34, doi:10.3390/axioms8010034 . . . . **199**

Nizar Souayah, Hassen Aydi, Thabet Abdeljawad and Nabil Mlaiki
Best Proximity Point Theorems on Rectangular Metric Spaces Endowed with a Graph
Reprinted from: *Axioms* **2019**, *8*, 17, doi:10.3390/axioms8010017 . . . . **211**

About the Special Issue Editor

Hsien-Chung Wu is a Professor of the Department of Mathematics, National Kaohsiung Normal University, Taiwan. He is the sole author of more than 110 scientific papers published in international journals. He is an area editor of the *International Journal of Uncertainty, Fuzziness and Knowledge-Based Systems* and an associate editor of *Fuzzy Optimization and Decision Making*. His current research includes the nonlinear analysis in mathematics and the applications of fuzzy sets theory in operations research.

Preface to "Fixed Point Theory and Related Topics"

This book contains the successful submissions to a Special Issue of *Axioms* on the subject area of "Fixed Point Theory and Related Topics". Fixed point theory arose from the Banach contraction principle and has been studied for a long time. Its application mostly relies on the existence of solutions to mathematical problems that are formulated from economics and engineering. Fixed points of functions depend heavily on the considered spaces that are defined using the intuitive axioms. Different spaces will result in different types of fixed point theorems. The articles in this Special Issue are summarized below.

Three articles study the best proximity point under different settings. H. Isik, H. Aydi, N. Mlaiki, and S. Radenovic study the best proximity point for Geraghty-type z-proximal contractions. N. Souayah, H. Aydi, T. Abdeljawad, and N. Mlaiki study the best proximity point on rectangular metric spaces endowed with a graph. T.Sabar, A. Bassou, and M. Aamri also study the best proximity point in the framework of newly introduced metric space.

Two articles study the fixed point in fuzzy metric space. D. Ram Prasad, G. Kishore, G, H. Isik, B. Srinuvasa Rao, and G. Adi Lakshmi study the fixed point in c^*-algebra valued fuzzy soft metric spaces. B. Rome, M. Sarwar, and P. Kumam study the fixed point theorems considering fuzzy contraction.

Two articles study the common fixed points. A. Ghanifard, H. Masiha, M. De La Sen, and M. Ramezani study the common fixed points for nonexpansive multi-valued mappings in convex metric spaces. A. Perveen, I. Khan, and M. Imdad also study the common fixed points for generalized weak nonlinear contractions.

E. Mohamed, A. Mohamed, and L. Samih study the fixed point theorems for relatively cyclic and noncyclic p-contractions in locally k-convex space. V. Parvaneh, N. Hussain, A. Mukheimer, and H. Aydi study the fixed points for modified JS-contractions. H. Faraji, D. Savic, and S. Radenovic study the fixed point theorems for Geraghty-type contraction type mappings in b-complete b-metric spaces.

Y. Gaba and E. Karapinar study the common fixed points for Kannan-type contractions. E. Karapinar also provides a short survey for the non-unique fixed point results in various abstract spaces.

T. Shanmugam, M. Muthiah, and S. Radenovic study the existence of positive solutions for the eighth-order boundary value problem using a classical version of Leray-Schauder alternative fixed point theorem.

M. Farid, K. Chaira, E. Marhrani, and M. Aamri study the fixed point theorems in Banach algebras. H.-C. Wu studies the fixed point theorem in a newly proposed informal complete metric space. Y. Sugimoto studies the square roots of diffeomorphisms.

Hsien-Chung Wu
Special Issue Editor

Article

Viscosity Approximation Methods for ∗−Nonexpansive Multi-Valued Mappings in Convex Metric Spaces

Azadeh Ghanifard [1], Hashem Parvaneh Masiha [1], Manuel De La Sen [2],* and Maryam Ramezani [3]

[1] Faculty of Mathematics, K. N. Toosi University of Technology, Tehran 16569, Iran; a.ghanifard@email.kntu.ac.ir (A.G.); masiha@kntu.ac.ir (H.P.M.)
[2] Institute of Research and Development of Processes University of the Basque Country, 48940 Leioa, Spain
[3] Department of Mathematics, University of Bojnord, Bojnord 94531, Iran; m.ramezani@ub.ac.ir
* Correspondence: manuel.delasen@ehu.eus

Received: 12 December 2019; Accepted: 12 January 2020; Published: 17 January 2020

Abstract: In this paper, we prove convergence theorems for viscosity approximation processes involving ∗−nonexpansive multi-valued mappings in complete convex metric spaces. We also consider finite and infinite families of such mappings and prove convergence of the proposed iteration schemes to common fixed points of them. Our results improve and extend some corresponding results.

Keywords: ∗−nonexpansive multi-valued mapping; viscosity approximation methods; fixed point; convex metric space

MSC: 47H10; 26A51

---

1. Introduction

Many of the real world known problems that scientists are looking to solve are nonlinear. Therefore, translating linear version of such problems into their equivalent nonlinear version has a great importance. Mathematicians have tried to transfer the structure of covexity to spaces that are not linear spaces. Takahashi [1], Kirk [2,3], and Penot [4], for example, presented this notion in metric spaces. Takahashi [1] introduced the following notion of convexity in metric spaces:

Definition 1. ([1]) Let (X, d) be a metric space and $I = [0, 1]$. A mapping $W : X \times X \times I \to X$ is said to be a convex structure on X if for each $x, y, u \in X$ and all $t \in I$,

$$d(u, W(x, y, t)) \leq t d(u, x) + (1 - t) d(u, y).$$

A metric space (X, d) together with a convex structure W is called a convex metric space and is denoted by (X, W, d).

A subset C of X is called convex if $W(x, y, t) \in C$, for all $x, y \in C$ and all $t \in I$.

Example 1. Let $X = M_2(\mathbb{R})$. For any $A = \begin{bmatrix} a_1 & a_2 \\ a_3 & a_4 \end{bmatrix}$ and $B = \begin{bmatrix} b_1 & b_2 \\ b_3 & b_4 \end{bmatrix}$ and $t \in I = [0, 1]$, we define the mapping $W : X \times X \times I \to X$ by

$$W(A, B, t) = \begin{bmatrix} ta_1 + (1-t)b_1 & ta_2 + (1-t)b_2 \\ ta_3 + (1-t)b_3 & ta_4 + (1-t)b_4 \end{bmatrix}$$

and the metric $d : X \times X \to [0, +\infty)$ by

$$d(A, B) = \Sigma_{i=1}^{4} |a_i - b_i|.$$

Then (X, W, d) is a convex metric space.

Example 2. Let $X = \mathbb{R}^2$ with the metric

$$d((x_1, x_2), (y_1, y_2)) = \max\{|x_1 - y_1|, |x_2 - y_2|\},$$

for any $(x_1, x_2), (y_1, y_2) \in X$ and define the mapping $W : X \times X \times [0, 1] \to X$ by

$$W((x_1, x_2), (y_1, y_2), t) = (tx_1 + (1-t)y_1, tx_2 + (1-t)y_2),$$

for each $(x_1, x_2), (y_1, y_2) \in X$ and $t \in [0, 1]$. Then (X, W, d) is a convex metric space.

Example 3. Let $X = C([0, 1])$ be the metric space with the metric $d(f, g) = \int_0^1 |f(x) - g(x)| dx$ and define $W : X \times X \times [0, 1] \to X$ by $W(f, g, t) = tf + (1-t)g$, for all $f, g \in X$ and $t \in [0, 1]$. Then (X, W, d) is a convex metric space.

This notion of convex structure is a generalization of convexity in normed spaces and allows us to obtain results that seem to be possible only in linear spaces. One of its useful applications is the iterative approximation of fixed points in metric spaces. All of the sequences that are used in fixed point problems require linearity or convexity of the space. So, this concept of convexity helps us to define various iteration schemes and to solve fixed point problems in metric spaces. In recent years, many authors have established several results on the covergence of some iterative schemes using different contractive conditions in convex metric spaces. For more details, refer to [5–14].

Now, let us recall some definitions and concepts that will be needed to state our results:

Definition 2. ([15]) Let (X, d) be a metric. A subset D is called proximinal if for each $x \in X$ there exists an element $y \in D$ such that $d(x, y) = d(x, D)$, where $d(x, D) = \inf\{d(x, z) : z \in D\}$.

We denote the family nonempty proximinal and bounded subsets of D by $P(D)$ and the family of all nonempty closed and bounded subsets of X by $CB(X)$.

For two bounded subsets A and B of a metric space (X, d), the Pompeiu–Hausdorff metric between A and B is defined by

$$H(A, B) = \max\{\sup_{x \in A} d(x, B), \sup_{y \in B} d(A, y)\}.$$

Definition 3. ([16]) Let (X, d) be a metric space. A multi-valued mapping $T : X \to CB(X)$ is said to be nonexpansive if $H(Tx, Ty) \le d(x, y)$, for all $x, y \in X$.
An element $p \in X$ is called a fixed point of T if $p \in T(p)$. The set of all fixed points of T are denoted by $F(T)$.

Definition 4. ([17]) Let (X, d) be a metric space and D be a nonempty subset of X. A multi-valued mapping $T : D \to CB(D)$ is called $*$−nonexpansive if for all $x, y \in D$ and $u_x \in T(x)$ with $d(x, u_x) = \inf\{d(x, z) : z \in T(x)\}$, there exists $u_y \in T(y)$ with $d(y, u_y) = \inf\{d(y, w) : w \in T(y)\}$ such that

$$d(u_x, u_y) \le d(x, y).$$

It is clear that if T is a $*$−nonexpansive map, then P_T is a nonexpansive map, where P_T for $T : D \to P(D)$ is defined by

$$P_T(x) = \{y \in T(x) : d(x, y) = d(x, T(x))\},$$

for all $x \in D$.

Definition 5. ([16]) *Let (X,d) be a metric space. A multi-valued mapping $T : X \to CB(X)$ is said to satisfy condition (I) if there is a nondecreasing function $f : [0,\infty) \to [0,\infty)$ with $f(0) = 0$, $f(r) > 0$ for $r \in (0,\infty)$ such that $d(x,T(x)) \geq f(d(x,F(T)))$, for all $x \in X$.*

First of all, Moudafi [18] introduced the viscosity approximation method for approximating the fixed point of nonexpansive mappings in Hilbert spaces. Since then, many authors have been extending and generalizing this result by using different contractive conditions on several spaces. For some new works in these fields, we can refer to [19–27]. Inspired and motivated by the research work going on in these fields, in this paper we investigate the convergence of some viscosity approximation processes for $*$−nonexpansive multi-valued mappings in a complete convex metric spaces. The convergence theorems for finite and infinite family of such mappings are also presented. Our results can improve and extend the corresponding main theorems in the literature.

2. Main Results

At first, we present two lemmas that are used to prove our main result. Since the idea is similar to the one given in Lemmas 2.1 and 2.2 in [28], we only state the results without the proof:

Lemma 1. *Let $\{u_n\}$ and $\{v_n\}$ be sequences in a convex metric space (X,W,d) and $\{a_n\}$ be a sequence in $[0,1]$ such that $\limsup_n a_n < 1$. Set*

$$d = \limsup_{n \to \infty} d(u_n,v_n) \text{ or } d = \liminf_{n \to \infty} d(u_n,v_n).$$

Let $u_{n+1} = W(v_n, u_n, a_n)$ for all $n \in \mathbb{N}$. Suppose that

$$\limsup_{n \to \infty}(d(v_{n+1},v_n) - d(u_{n+1},u_n)) \leq 0,$$

and $d < \infty$. Then

$$\liminf_{n \to \infty} |d(v_{n+k}, u_n) - (1 + a_n + a_{n+1} + \ldots + a_{n+k-1})d| = 0,$$

for all $k \in \mathbb{N}$.

Lemma 2. *Let $\{u_n\}$ and $\{v_n\}$ be bounded sequences in a convex metric space (X,W,d) and $\{a_n\}$ be a sequence in $[0,1]$ with $0 < \liminf_n a_n \leq \limsup_n a_n < 1$. Suppose that $u_{n+1} = W(v_n, u_n, a_n)$ and*

$$\limsup_{n \to \infty}(d(v_{n+1},v_n) - d(u_{n+1},u_n)) \leq 0.$$

Then $\lim_{n \to \infty} d(v_n, u_n) = 0$

Now, we state and prove the main theorem of this paper:

Theorem 1. *Let D be a nonempty, closed and convex subset of a complete convex metric space (X,W,d) and $T : D \to P(D)$ be a $*$−nonexpansive multi-valued mapping with $F(T) \neq \emptyset$, such that T satisfies condition (I). Suppose that $a_n \in [0,1]$ such that $0 < \liminf_n a_n \leq \limsup_n a_n < 1$ and $c_n \in (0,+\infty)$ such that $\lim_{n \to \infty} c_n = 0$. Let $\{x_n\}$ be the Mann type iterative scheme defined by*

$$x_{n+1} = W(z_n, x_n, a_n), \qquad (1)$$

where $d(z_{n+1}, z_n) \leq H(P_T(x_{n+1}), P_T(x_n)) + c_n$ for $z_n \in P_T(x_n)$. Then $\{x_n\}$ converges to a fixed point of T.

Proof. Take $p \in F(T)$. Then $p \in P_T(p) = \{p\}$ and we have

$$\begin{aligned}
d(x_{n+1}, p) &= d(W(z_n, x_n, a_n), p) \\
&\leq a_n d(z_n, p) + (1 - a_n) d(x_n, p) \\
&\leq a_n H(P_T(x_n), P_T(p)) + (1 - a_n) d(x_n, p) \\
&\leq a_n d(x_n, p) + (1 - a_n) d(x_n, p) = d(x_n, p).
\end{aligned}$$

Hence, $\{d(x_n, p)\}$ is a decreasing and bounded below sequence and thus $\lim_{n \to \infty} d(x_n, p)$ exists for any $p \in F(T)$. Therefore $\{x_n\}$ is bounded and so $\{z_n\}$ is bounded. On the other hand,

$$d(z_{n+1}, z_n) \leq H(P_T(x_{n+1}), P_T(x_n)) + c_n \leq d(x_{n+1}, x_n) + c_n.$$

Thus

$$\limsup_{n \to \infty} (d(z_{n+1}, z_n) - d(x_{n+1}, x_n)) \leq 0.$$

Applying Lemma 2, we get

$$\lim_{n \to \infty} d(z_n, x_n) = 0.$$

Hence, we have $\lim_{n \to \infty} d(x_n, T(x_n)) = 0$. Since T satisfies condition (I), we conclude that $\lim_{n \to \infty} d(x_n, F(T)) = 0$. Next, we show that $\{x_n\}$ is a Cauchy sequence. Since $\lim_{n \to \infty} d(x_n, F(T)) = 0$, thus for $\varepsilon_1 > 0$, there exists $n_1 \in \mathbb{N}$ such that for all $n \geq n_1$

$$d(x_n, F(T)) \leq \frac{\varepsilon_1}{3}.$$

Thus, there exists $p_1 \in F(T)$ such that for all $n \geq n_1$,

$$d(x_n, p_1) \leq \frac{\varepsilon_1}{2}.$$

It follows that

$$d(x_{n+m}, x_n) \leq d(x_{n+m}, p_1) + d(p_1, x_n) \leq d(x_n, p_1) + d(p_1, x_n)$$
$$\leq \frac{\varepsilon_1}{2} + \frac{\varepsilon_1}{2} = \varepsilon_1,$$

for all $m, n \geq n_1$. Therefore $\{x_n\}$ is a Cauchy sequence and hence it is convergent. Let $\lim_{n \to \infty} x_n = p^*$. We will show that p^* is a fixed point of T.

Since $\lim_{n \to \infty} x_n = p^*$, thus for given $\varepsilon_2 > 0$, there exists $n_2 \in \mathbb{N}$ such that for all $n \geq n_2$,

$$d(x_n, p^*) \leq \frac{\varepsilon_2}{4}.$$

Moreover, $\lim_{n \to \infty} d(x_n, F(T)) = 0$ implies that there exists a natural number $n_3 \geq n_2$ such that for all $n \geq n_3$,

$$d(x_n, F(T)) \leq \frac{\varepsilon_2}{12},$$

and thus there exists $p_2 \in F(T)$ such that for all $n \geq n_3$,

$$d(x_n, p_2) \leq \frac{\varepsilon_2}{8}.$$

Therefore

$$\begin{aligned}
d(T(p^*), p^*) &\leq d(T(p^*), p_2) + d(p_2, T(x_{n_3})) + d(T(x_{n_3}), p_2) + d(p_2, x_{n_3}) + d(x_{n_3}, p^*) \\
&\leq H(P_T(p^*), P_T(p_2)) + 2H(P_T(p_2), P_T(x_{n_3})) + d(p_2, x_{n_3}) + d(x_{n_3}, p^*) \\
&\leq d(p^*, p_2) + 2d(p_2, x_{n_3}) + d(p_2, x_{n_3}) + d(x_{n_3}, p^*) \\
&\leq d(p^*, x_{n_3}) + d(x_{n_3}, p_2) + 2d(p_2, x_{n_3}) + d(p_2, x_{n_3}) + d(x_{n_3}, p^*) \\
&= 2d(x_{n_3}, p^*) + 4d(x_{n_3}, p_2) \leq \frac{\varepsilon_2}{2} + \frac{\varepsilon_2}{2} = \varepsilon_2.
\end{aligned}$$

Thus, $p^* \in T(p^*)$ and therefore p^* is a fixed point of T. □

As a result of Theorem 1, Corollaries 1 and 2 are obtained:

Corollary 1. *Let D be a nonempty, closed and convex subset of a complete convex metric space (X, W, d), $T : D \to P(D)$ be $*-$nonexpansive multi-valued mapping with $F(T) \neq \emptyset$ such that T satisfies condition (I) and $f : D \to D$ be a contractive mapping with a contractive constant $k \in (0, 1)$. Then the iterative sequence $\{x_n\}$ defined by*

$$x_{n+1} = W(z_n, f(x_n), a_n)$$

where $z_n \in P_T(x_n)$ and $0 < \liminf_n a_n \leq \limsup_n a_n < 1$, converges to a fixed point of T.

Corollary 2. *Let D be a nonempty, closed, and convex subset of a complete convex metric space (X, W, d) and $T : D \to P(D)$ be $*-$nonexpansive multi-valued mapping with $F(T) \neq \emptyset$. Let $\{x_n\}$ be the Ishikawa type iterative scheme defined by*

$$\begin{aligned}
x_{n+1} &= W(z_n', x_n, a_n) \\
y_n &= W(z_n, x_n, b_n)
\end{aligned}$$

where $z_n' \in P_T(y_n)$, $z_n \in P_T(x_n)$, and $\{a_n\}, \{b_n\} \in [0, 1]$. Then $\{x_n\}$ converges to a fixed point of T if and only if $\lim_{n \to \infty} d(x_n, F(T)) = 0$.

The above result can be generalized to the finite and infinite family of $*-$nonexpansive multi-valued mappings:

Theorem 2. *Let D be a nonempty, closed, and convex subset of a complete convex metric space (X, W, d) and $\{T_i : D \to P(D) : i = 1, \ldots, k\}$ be a finite family of $*-$nonexpansive multi-valued mappings such that $F := \cap_{i=1}^k F(T_i) \neq \emptyset$. Consider the iterative process defined by*

$$\begin{aligned}
y_{1n} &= W(z_{1n}, x_n, a_{1n}), \\
y_{2n} &= W(z_{2n}, x_n, a_{2n}), \\
&\cdots \\
y_{(k-1)n} &= W(z_{(k-1)n}, x_n, a_{(k-1)n}), \\
x_{n+1} &= W(z_{kn}, x_n, a_{kn}),
\end{aligned}$$

where $a_{in} \in [0, 1]$ and $z_{in} \in P_{T_i}(y_{(i-1)n})$ ($y_{0n} = x_n$), for all $n \in \mathbb{N}$ and $i = 1, 2, \ldots, k$. Then $\{x_n\}$ converges to a point in F if and only if $\lim_{n \to \infty} d(x_n, F) = 0$.

Proof. The necessity of conditions is obvious and we will only prove the sufficiency. Let $p \in F$. we have

$$\begin{aligned}
d(y_{1n}, p) &= d(W(z_{1n}, x_n, a_{1n}), p) \\
&\leq a_{1n} d(z_{1n}, p) + (1 - a_{1n}) d(x_n, p) \\
&\leq a_{1n} H(P_{T_1}(x_n), P_{T_1}(p)) + (1 - a_{1n}) d(x_n, p) \\
&\leq a_{1n} d(x_n, p) + (1 - a_{1n}) d(x_n, p) = d(x_n, p), \\
d(y_{2n}, p) &= d(W(z_{2n}, x_n, a_{2n}), p) \\
&\leq a_{2n} d(z_{2n}, p) + (1 - a_{2n}) d(x_n, p) \\
&\leq a_{2n} H(P_{T_2}(y_{1n}), P_{T_2}(p)) + (1 - a_{2n}) d(x_n, p) \\
&\leq a_{2n} d(y_{1n}, p) + (1 - a_{2n}) d(x_n, p) \\
&\leq a_{2n} d(x_n, p) + (1 - a_{2n}) d(x_n, p) = d(x_n, p), \\
&\vdots \\
d(y_{(k-1)n}, p) &= d(W(z_{(k-1)n}, x_n, a_{(k-1)n}), p) \\
&\leq a_{(k-1)n} d(z_{(k-1)n}, p) + (1 - a_{(k-1)n}) d(x_n, p) \\
&\leq a_{(k-1)n} H(P_{T_{k-1}}(y_{(k-2)n}), P_{T_{k-1}}(p)) + (1 - a_{(k-1)n}) d(x_n, p) \\
&\leq a_{(k-1)n} d(y_{(k-2)n}, p) + (1 - a_{(k-1)n}) d(x_n, p) \\
&\leq a_{(k-1)n} d(x_n, p) + (1 - a_{(k-1)n}) d(x_n, p) = d(x_n, p).
\end{aligned}$$

Thus

$$\begin{aligned}
d(x_{n+1}, p) &= d(W(z_{kn}, x_n, a_{kn}), p) \\
&\leq a_{kn} d(z_{kn}, p) + (1 - a_{kn}) d(x_n, p) \\
&\leq a_{kn} H(P_{T_k}(y_{(k-1)n}), P_{T_k}(p)) + (1 - a_{kn}) d(x_n, p) \\
&\leq a_{kn} d(y_{(k-1)n}, p) + (1 - a_{kn}) d(x_n, p) \\
&\leq a_{kn} d(x_n, p) + (1 - a_{kn}) d(x_n, p) = d(x_n, p).
\end{aligned}$$

Therefore, $\{d(x_n, p)\}$ is a decreasing sequence and so $d(x_{n+m}, p) \leq d(x_n, p)$, for all $n, m \in \mathbb{N}$. As in the proof of Theorem 1, $\{x_n\}$ is a Cauchy sequence and thus $\lim_{n \to \infty} x_n$ exists and equals to some $p^* \in D$. Again, with a similar process as in the proof of Theorem 1, we conclude that $p^* \in P_{T_i}(q)$ for all $i = 1, \ldots, k$. Hence $p^* \in F$ and this completes the proof of theorem. □

Theorem 3. *Let D be a nonempty, closed, and convex subset of a complete convex metric space (X, W, d) and $\{T_i : D \to P(D) : i = 1, \ldots\}$ be an infinite family of $*$−nonexpansive multi-valued mappings such that $F := \cap_{i=1}^{\infty} F(T_i) \neq \emptyset$. Consider the iterative process defined by*

$$\begin{aligned}
x_{n+1} &= W(z'_n, x_n, a_n) \\
y_n &= W(z_n, x_n, b_n)
\end{aligned}$$

where $z'_n \in P_{T_n}(y_n)$, $z_n \in P_{T_n}(x_n)$ and $\{a_n\}, \{b_n\} \in [0, 1]$. Then $\{x_n\}$ converges to a point in F if and only if $\lim_{n \to \infty} d(x_n, F) = 0$.

Author Contributions: Data curation, A.G.; Formal analysis, A.G.; Software, A.G.; Writing—original draft, A.G.; Conceptualization, H.P.M.; Project administration, H.P.M.; Supervision, M.D.L.S.; Funding acquisition, M.D.L.S.; Writing—review and editing, M.D.L.S. and M.R.; Validation, M.R. All authors have read and agreed to the published version of the manuscript.

Funding: This research was funded by Basque Government through grant IT1207-19.

Acknowledgments: The authors are grateful to the referees for valuable suggestions and to the Basque Government for Grant IT1207-19.

Conflicts of Interest: The authors declare no conflict of interest.

References

1. Takahashi, W. A convexity in metric spaces and nonexpansive mappings. *Kodai Math. Sem. Rep.* **1970**, *22*, 142–149. [CrossRef]
2. Kirk, W.A. An abstract fixed point theorem for nonexpansive mappings. *Proc. Am. Math. Soc.* **1981**, *82*, 640–642. [CrossRef]
3. Kirk, W.A. Fixed point theory for nonexpansive mappings II. *Contemp. Math.* **1983**, *18*, 121–140.
4. Penot, J.P. Fixed point theorems without convexity. *Bull. Soc. Math. France Mem.* **1979**, *60*, 129–152. [CrossRef]
5. Chang, S.S.; Kim, J.K. Convergence theorems of the Ishikawa type iterative sequences with errors for generalized quasi-contractive mappings in convex metric spaces. *Appl. Math. Lett.* **2003**, *16*, 535–542. [CrossRef]
6. Chang, S.S.; Kim, J.K.; Jin, D.S. Iterative sequences with errors for asymptotically quasi-nonexpansive type mappings in convex metric spaces. *Arch. Inequal. Appl.* **2004**, *2*, 365–374.
7. Ding, X.P. Iteration processes for nonlinear mappings in convex metric spaces. *J. Math. Anal. Appl.* **1988**, *132*, 114–122. [CrossRef]
8. Khan, A.R.; Ahmed, M.A. Convergence of a general iterative scheme for a finite family of asymptotically quasi-nonexpansive mappings in convex metric spaces and applications. *Comput. Math. Appl.* **2010**, *59*, 2990–2995. [CrossRef]
9. Kim, J.K.; Kim, K.H.; Kim, K.S. Three-step iterative sequences with errors for asymptotically quasi-nonexpansive mappings in convex metric spaces. *Nonlinear Anal. Convex Anal.* **2004**, *1365*, 156–165.
10. Rafiq, A. Fixed point of Ciric quasi-contractive operators in generalized convex metric spaces. *Gen. Math.* **2006**, *14*, 79–90.
11. Saluja, G.S.; Nashine, H.K. Convergence of implicit iteration process for a finite family of asymptotically Quasi-nonexpansive mappings in convex metric spaces. *Opuscula Math.* **2010**, *30*, 331–340. [CrossRef]
12. Tian, Y.X. Convergence of an Ishikawa type Iterative scheme for asymptotically quasi- nonexpansive mappings. *Comput. Math. Appl.* **2005**, *49*, 1905–1912. [CrossRef]
13. Wang, C.; Zhu, J.H.; Damjanovic, B.; Hu, L.G. Approximating fixed points of a pair of contractive type mappings in generalized convex metric spaces. *Appl. Math. Comput.* **2009**, *215*, 1522–1525. [CrossRef]
14. Wang, C.; Liu, L.W. Convergence theorems of fixed points of uniformly quasi-Lipschitzian mappings in convex metric spaces. *Nonlinear Anal.* **2009**, *70*, 2067–2071. [CrossRef]
15. Roshdi, K. Best approximation in metric spaces. *Proc. Amer. Math. Soc.* **1988**, *103*, 579–586.
16. Shahzad, N.; Zegeye, H. On Mann and Ishikawa iteration schemes for multi-valued maps in Banach spaces. *Nonlinear Anal.* **2009**, *71*, 838–844. [CrossRef]
17. Hussain, T.; Latif, A. Fixed points of multivalued nonexpansive maps. *Math. Japon.* **1988**, *33*, 385–391.
18. Moudafi, A. Viscosity approximation methods for fixed-points problems. *J. Math. Anal. Appl.* **2000**, *241*, 46–55. [CrossRef]
19. Deng, W.Q. A new viscosity approximation method for common fxed points of a sequence of nonexpansive mappings with weakly contractive mappings in Banach spaces. *J. Nonlinear Sci. Appl.* **2016**, *9*, 3920–3930. [CrossRef]
20. Khan, A.R.; Yasmin, N.; Fukhar-ud-din, H.; Shukri, S.A. Viscosity approximation method for generalized asymptotically quasi-nonexpansive mappings in a convex metric space. *Fixed Point Theory Appl.* **2015**, *2015*, 196. [CrossRef]
21. Lin, Y.C.; Sharma, B.K.; Kumar, A.; Gurudwan, N. Viscosity approximation method for common fixed point problems of a finite family of nonexpansive mappings. *J. Nonlinear Convex Anal.* **2017**, *18*, 949–966.
22. Liu, X.; Chen, Z.; Xiao, Y. General viscosity approximation methods for quasi-nonexpansive mappings with applications. *J. Inequal. Appl.* **2019**, *2019*, 71. [CrossRef]
23. Liu, C.; Song, M. The new viscosity approximation methods for nonexpansive nonself-mappings. *Int. J. Mod. Nonlinear Theory Appl.* **2016**, *5*, 104–113. [CrossRef]
24. Naqvi, S.F.A.; Khan, M.S. On the viscosity rule for common fixed points of two nonexpansive mappings in Hilbert spaces. *Open J. Math. Sci.* **2017**, *1*, 111–125. [CrossRef]
25. Thong, D.V. Viscosity approximation methods for solving fixed-point problems and split common fixed-point problems. *J. Fixed Point Theory Appl.* **2016**. [CrossRef]

26. Xiong, T.; Lan, H. Strong convergence of new two-step viscosity iterative approximation methods for set-valued nonexpansive mappings in CAT(0) spaces. *J. Funct. Spaces* **2018**, *2018*. [CrossRef]
27. Khan, S.H.; Fukhar-ud-din, H. Approximating fixed points of ρ-nonexpansive mappings by RK-iterative process in modular function spaces. *J. Nonlinear Var. Anal.* **2019**, *3*, 107–114.
28. Suzuki, T. Strong convergence theorems for infinite families of nonexpansive mappings in general Banach spaces. *Fixed Point Theory Appl.* **2005**, *1*, 103–123. [CrossRef]

© 2020 by the authors. Licensee MDPI, Basel, Switzerland. This article is an open access article distributed under the terms and conditions of the Creative Commons Attribution (CC BY) license (http://creativecommons.org/licenses/by/4.0/).

Article

Measure of Weak Noncompactness and Fixed Point Theorems in Banach Algebras with Applications

Mohamed Amine Farid [1], Karim Chaira [2], El Miloudi Marhrani [1,*] and Mohamed Aamri [1]

[1] Laboratory of Algebra, Analysis and Applications (L3A), Faculty of Sciences Ben M'Sik, Hassan II University of Casablanca, B.P 7955, Sidi Othmane, Casablanca 20700, Morocco; amine.farid17@gmail.com (M.A.F.); aamrimohamed82@gmail.com (M.A.)
[2] CRMEF, Avenue Allal El Fassi, Madinat Al Irfan, B.P 6210, Rabat 10000, Morocco; chaira_karim@yahoo.fr
* Correspondence: marhrani@gmail.com

Received: 12 September 2019; Accepted: 5 November 2019; Published: 14 November 2019

Abstract: In this paper, we prove some fixed point theorems for the nonlinear operator $A \cdot B + C$ in Banach algebra. Our fixed point results are obtained under a weak topology and measure of weak noncompactness; and we give an example of the application of our results to a nonlinear integral equation in Banach algebra.

Keywords: Banach algebras; fixed point theorems; measure of weak noncompactness; weak topology; integral equations

MSC: 47H09; 47H10; 47H30

---

1. Introduction

Integral equations are involved in various scientific problems such as transport theory, the theory of radiative transfer, biomathematics, etc (see [1–6]). The use of these equations dates back to 1730 with Bernoulli in the study of oscillatory problems. With the development of functional analysis, more general results were obtained by L. Schwartz, H. Poincaré, I. Fredholm, and others.

The problems of the existence of solutions for an integral equation can then be resolved by searching fixed points for nonlinear operators in a Banach algebra. For this, many researchers have been interested in the case where the Banach algebra is endowed with its strong topology; however, few of them were interested to the existence of a fixed point for mappings acting on a Banach algebra equipped with its weak topology [7–11]; such a topology allows obtaining some generalizations of these results.

The history of fixed point theory in Banach algebra started in 1977 with R.W. Legget [12], who considered the existence of solutions for the equation:

$$x = x_0 + x \cdot Bx, \ (x_0, x) \in X \times \Omega \qquad (1)$$

where Ω is a nonempty, bounded, closed, and convex subset of a Banach algebra X and B is a compact operator from Ω into X. Many authors [10,11,13,14] generalized Equation (1) to the equation:

$$x = Ax \cdot Bx + Cx, \ x \in \Omega, \qquad (2)$$

where Ω is a nonempty, bounded, closed, and convex subset of a Banach algebra and $A, C : X \longrightarrow X$, $B : \Omega \longrightarrow X$ are nonlinear operators. Most of these authors have obtained the desired results through the study of the operator $\left(\frac{I-C}{A}\right)^{-1} B$.

This study was based mainly on the properties of operators A, B, C, and $\frac{I-C}{A}$ (cf. condensing, relatively weakly compact, etc.).

The study of nonlinear integral equations in Banach algebra via fixed point theory was in initiated by B.C. Dhage [15]. In 2005, B.C. Dhage [14] studied the existence of solutions for the equation:

$$x = Ax \cdot Bx$$

The results were obtained in the case of the norm topology on Banach algebra. In 2014, Banas et al. [8] proved some existence results of operator equations under the weak topology using the measure of weak noncompactness. In 2015, Ben Amar et al. used the De Blasi measure of non-compactness to obtain some generalizations of these results. In 2019, A.B. Amar et al. [16] established new fixed point theorems for the sum of two mappings in Banach space and showed that the condition «weakly condensing» can by relaxed by the assumption «countably weakly condensing».

In this paper, we use the measure of noncompactness to prove some fixed point results for a nonlinear operator of type $AB + C$ in a Banach algebra. We note that the condition «relatively weakly compact», which is not easy to verify, is not required in most results in [16]. Our results are formulated using the operator $I - \frac{I-C}{A}$ under the weak topology in a Banach algebra.

As an application, we discuss the existence of solutions for an abstract nonlinear integral equation in the Banach algebra $C([0,1], X)$; and an example of a nonlinear integral equation in the Banach algebra $C([0,1], \mathbb{R})$.

2. Preliminaries

Let $(X, \|\ \|)$ be a Banach space with zero element θ. We denote respectively $P(X)$, $P_{cv}(X)$, $P_{bd}(X)$ and $P_{cl,cv}(X)$ the family of all nonempty subsets, nonempty and convex subsets, nonempty and bounded subsets, nonempty closed and convex subsets of X.

For any $\varepsilon > 0$, we denote B_ε the closed ball of X centered at origin with radius ε. Moreover, we write $x_n \to x$ and $x_n \rightharpoonup x$ respectively to denote the strong convergence and the weak convergence of a sequence $\{x_n\}_n$ to x.

For a subset K of X, we write $\overline{K}$, $\overline{K}^w$, conv$K$, and $\overline{\text{conv}}K$, to denote the closure, the weak closure, the convex hull, and the closed convex hull of the subset K, respectively; and by $\mathcal{R}(T)$, the range of the operator T.

Definition 1. *Let Ω be a nonempty subset of X. We say that a multivalued map $H : \Omega \to P(\Omega)$ has a weakly closed graph if the following property holds: if for every net $\{x_\delta\}_\delta$, with $x_\delta, x \in \Omega$ such that $x_\delta \rightharpoonup x$ and $\{y_\delta\}_\delta$ such that $y_\delta \in Hx_\delta$, $y_\delta \rightharpoonup y$, then $Hx \cap S(x,y) \neq \emptyset$; here, $S(x,y) := \{\lambda y + (1-\lambda)x \,;\, \lambda \in [0,1]\}$.*

We say that a map $H : \Omega \to P(\Omega)$ has a w-weakly closed graph in $\Omega \times X$ if it has a weakly closed graph in $\Omega \times X$ with respect to the weak topology.

Definition 2 ([9]). *Let X be a Banach space. An operator $T : X \to X$ is said to be weakly sequentially continuous on X if for every sequence $\{x_n\}_n$ with $x_n \rightharpoonup x$, we have $Tx_n \rightharpoonup Tx$.*

Note that T is weakly sequentially continuous if and only if $I - T$ is weakly sequentially continuous.

Definition 3. *Let X be a Banach space. An operator $T : X \longrightarrow X$ is said to be weakly compact if $T(M)$ is relatively weakly compact for every bounded subset $M \subset X$.*

Definition 4 ([17]). *Let Ω be a nonempty weakly closed set of a Banach space X and $T : \Omega \to X$ a weakly sequentially continuous operator. T is said to be a weakly semi-closed operator at θ if the conditions $\{x_n\}_n \subset \Omega$, $x_n - Tx_n \rightharpoonup \theta$ imply that there exists $x \in \Omega$ such that $Tx = x$.*

We recall that a function $\omega : P_{bd}(X) \to [0, +\infty)$ is said to be a measure of weak noncompactness (MWNC) on X if it satisfies the following properties.

1. For any bounded subsets Ω_1, Ω_2 of X, we have $\Omega_1 \subseteq \Omega_2$ implies $\omega(\Omega_1) \leq \omega(\Omega_2)$.
2. $\omega(\overline{conv}(\Omega)) = \omega(\Omega)$, for all bounded subsets $\Omega \subset X$.
3. $\omega(\Omega \cup \{a\}) = \omega(\Omega)$ for all $a \in X$, $\Omega \in P_{bd}(X)$.
4. $\omega(\Omega) = 0$ if and only if Ω is relatively weakly compact in X.

The MWNC ω is said to be:

1. Positive homogeneous, if $\omega(\lambda \Omega) = \lambda \omega(\Omega)$, for all $\lambda > 0$ and $\Omega \in P_{bd}(X)$.
2. Subadditive, if $\omega(\Omega_1 + \Omega_2) \leq \omega(\Omega_1) + \omega(\Omega_2)$, for all $\Omega_1, \Omega_2 \in P_{bd}(X)$.

As an example of MWNC, we have the De Blasi measure of weak noncompactness [18], defined on $P_{bd}(X)$ by:

$$\mu(M) = \inf\{\varepsilon > 0;\ \text{there exists } K \text{ weakly compact such that}: M \subset K + B_\varepsilon\},$$

it is well known that μ is homogenous, subadditive, and satisfies the set additivity property:

$$\mu(M \cup N) = \max\{\mu(M), \mu(N)\}, \quad \text{for all } M, N \in P_{bd}(X).$$

For more properties of the MWNC, we refer to [19].

Let us formulate some other definitions needed in this paper.

Definition 5. *Let Ω be a subset of a Banach space X, ω be an MWNC on X, and $0 \leq k < 1$. Let T be a mapping from Ω into X; we say that:*

1. *T is k-ω-contractive if $\omega(T(M)) \leq k\omega(M)$ for any bounded set $M \subset \Omega$;*
2. *T is ω-condensing if $\omega(T(M)) < \omega(M)$ for any bounded set $M \subset \Omega$ with $\omega(M) > 0$;*
3. *T is countably k-ω-contractive, if $\omega(T(M)) \leq k\omega(M)$ for any countable bounded set $M \subset \Omega$;*
4. *T is countably ω-condensing if $\omega(T(M)) < \omega(M)$ for any countable bounded set $M \subset \Omega$ with $\omega(M) > 0$;*
5. *T is weakly countable one-set-contractive if $\omega(T(M)) \leq \omega(M)$ for any bounded set $M \subset \Omega$.*

Clearly, every k-ω-contractive is countably k-ω-contractive, but the converse is not always true.

Definition 6. *A mapping $T : \Omega \subset X \longrightarrow X$ is said to be:*

1. *Lipschitzian with the Lipschitz constant $k > 0$:*

$$\|Tx - Ty\| \leq k\|x - y\|, \quad \text{for all } x, y \in \Omega.$$

If $k = 1$, T is called nonexpansive, and if $k \in [0, 1[$, T is called a contraction.

2. *Pseudocontractive if for each $r > 0$, we have:*

$$\|x - y\| \leq \|r(Ty - Tx) + (1 + r)(x - y)\|, \quad \text{for all } x, y \in \Omega.$$

3. *Accretive if for each $\lambda \geq 0$, we have:*

$$\|x - y\| \leq \|x - y + \lambda(Tx - Ty)\|, \quad \text{for all } x, y \in \Omega.$$

In addition, if $\mathcal{R}(I + \lambda T) = X$ for every $\lambda > 0$, then T is called m-accretive.

Note that T is pseudocontractive if and only if $I - T$ is accretive.

Definition 7. *An operator $T : \Omega \subseteq X \to X$ is called $\mathcal{D}$-Lipschitzian if there exists a continuous and nondecreasing function $\Phi_T : [0, +\infty) \to [0, +\infty)$ with $\Phi_T(0) = 0$ such that:*

$$\|Tx - Ty\| \leq \Phi_T(\|x - y\|), \text{ for all } x, y \in \Omega.$$

Sometimes, Φ_T is called a $\mathcal{D}$-function of T on X. Moreover, if $\Phi_T(r) < r$ for all $r > 0$, then the operator T is called a nonlinear contraction with a contraction function Φ_T.

Definition 8. *An operator $T : \Omega \subseteq X \to X$ is said to be ψ-expansive if there exists a function $\psi : [0, \infty) \to [0, \infty)$ such that $\psi(0) = 0$, $\psi(r) > r$ for any $r > 0$, ψ is either continuous or nondecreasing, and $\|Tx - Ty\| \geq \psi(\|x - y\|)$ for all $x, y \in \Omega$.*

Definition 9. *We say that $H : \Omega \subseteq X \to P(X)$ is countably ω-condensing if $H(\Omega)$ is bounded on X and $\omega(H(M)) < \omega(M)$ for all countable bounded subsets M of Ω with $\omega(M) > 0$.*

The following result is crucial:

Theorem 1 ([20]). *Let X be a Banach space.*

(i) *Let H be a bounded subset of $\mathcal{C}([0, T], X)$. Then:*

$$\sup_{t \in [0,T]} \mu(H(t)) \leq \mu(H),$$

where $H(t) = \{x(t); x \in H\}$.

(ii) *Let $H \subset \mathcal{C}([0, T], X)$ be bounded and equicontinuous. Then:*

$$\mu(H) = \sup_{t \in [0,T]} \mu(H(t)) = \mu(H([0, T])),$$

where $H([0, T]) = \cup_{t \in [0,T]} H(t)$.

Here, μ is the De Blasi measure of weak noncompactness.

Lemma 1 ([21]). *Let X be a Banach space and $T : X \longrightarrow X$ a k-Lipschitzian map and weakly sequentially continuous. Then, for each bounded subset S of X, we have:*

$$\mu(T(S)) \leq k\mu(S), \text{ for all } x, y \in X;$$

here, μ is the De Blasi measure of weak noncompactness.

We recall that an algebra X is a vector space endowed with an internal composition law denoted by «·», which is associative and bilinear. A normed algebra is an algebra endowed with a norm $\|.\|$ satisfying the following property:

$$\|x \cdot y\| \leq \|x\| \|y\|, \text{ for all } x, y \in X.$$

A complete normed algebra is called a Banach algebra. For basic properties of Banach algebra, refer to [22].

In general, the product of two weakly sequentially continuous mappings on a Banach algebra is not necessarily weakly sequentially continuous.

Definition 10 ([9]). *We will say that the Banach algebra X satisfies condition* ($\mathcal{P}$) *if:*

$$(\mathcal{P}) \begin{cases} \text{For any sequences } \{x_n\}_n \text{ and } \{y_n\}_n \text{ in } X \text{ such that } x_n \rightharpoonup x \text{ and } y_n \rightharpoonup y, \\ \text{we have } x_n y_n \rightharpoonup xy. \end{cases}$$

Note that, every finite dimensional Banach algebra satisfies condition ($\mathcal{P}$). If X satisfies condition ($\mathcal{P}$), then the space $C(K; X)$ of all continuous functions from a compact Hausdorff space K into X is also a Banach algebra satisfying condition ($\mathcal{P}$) (see [9]).

Definition 11. *Let X be a Banach algebra. An operator* $T : X \to X$ *is called regular on X if it maps X into the set of all invertible elements of X.*

In [16] (Theorem 3.1), Afif Ben Amar et al. proved the following result:

Theorem 2 ([16], Theorem 3.1). *Let Ω be a nonempty closed convex subset of a Banach space X and ω be an MWNC on X. Assume that $T : \Omega \to \Omega$ is a weakly sequentially continuous and countably ω-condensing mapping with a bounded range. Then, T has a fixed point.*

Theorem 3 ([16], Theorem 3.3). *Let Ω be a nonempty closed convex subset of a Banach space X, ω be a positive homogeneous MWNC on X, and $T : \Omega \to \Omega$ be weakly sequentially continuous, weakly countably one-set-contractive. In addition, assume that T is weakly semi-closed at 0 with a bounded range. Then, T has a fixed point.*

Theorem 4 ([16], Theorem 3.2). *Let Ω be a nonempty convex closed subset of a Banach space E, $U \subset E$ be a weakly open subset of Ω with $0 \in U$, and ω be a subadditive MWNC on E. Assume $T : \overline{U}^w \to X$ is a weakly sequentially continuous countably ω-condensing map with $T(\overline{U}^w)$ bounded. Then, either T has a fixed point or there exists $u \in \partial_\Omega U$ and $\lambda \in]0, 1[$ such that $u = \lambda T(u)$ ($\partial_\Omega U$ denotes the weak boundary of U in Ω).*

The following lemma is useful for the sequel.

Lemma 2. *Let X be a Banach algebra satisfying condition* ($\mathcal{P}$). *Then, for any bounded subset M of X and relatively weakly compact subset K of X, we have $w(MK) \leq \|K\| w(M)$.*

3. Results

Our first main result is a new version of Theorem 3.2 proven by Jeribi et al. in [23].

Theorem 5. *Let Ω be a nonempty, bounded, closed, and convex subset of a Banach algebra X and ω be a subadditive MWNC on X. Let $A, C : X \longrightarrow X$, and $B : \Omega \longrightarrow X$ be three operators that satisfy the following conditions:*

(i) A is regular on X, and $\left(\frac{I-C}{A}\right)^{-1}$ exists on $B(\Omega)$,
(ii) B and $\frac{I-C}{A}$ are weakly sequentially continuous,
(iii) $I - \frac{I-C}{A}$ is countably α-ω-contractive on Ω,
(iv) B is countably β-ω-contractive,
(v) $x = Ax \cdot By + Cx$, for all $y \in \Omega$ implies $x \in \Omega$.

Then, there exists $x \in \Omega$ such that $x = Ax \cdot Bx + Cx$, whenever $\frac{\beta}{1-\alpha} < 1$.

Proof. Note that $x = Ax \cdot Bx + Cx$, $x \in \Omega$ if and only if x is a fixed point for the operator $T := \left(\frac{I-C}{A}\right)^{-1} B$.

Let $y \in \Omega$; from Assumption (i), there is a unique $x_y \in X$ such that:

$$\left(\frac{I-C}{A}\right) x_y = By,$$

then:

$$x_y = Ax_y \cdot By + Cx_y;$$

by Condition (v), we have $x_y \in \Omega$, and then, T is well defined on Ω.

By Theorem 2, it suffices to prove that the map T is weakly sequentially continuous and countably ω-condensing.

Let $\{x_n\}_n$ be a sequence in Ω such that $x_n \rightharpoonup x$; the set $\{x_n : n \in \mathbb{N}\}$ is relatively weakly compact; and since B is weakly sequentially continuous, the set $\{Bx_n : n \in \mathbb{N}\}$ is relatively weakly compact. Assume that $\omega(\{Tx_n : n \in \mathbb{N}\}) > 0$. Since:

$$T = B + \left(I - \frac{I-C}{A}\right) T,$$

and $I - \frac{I-C}{A}$ is countably α-ω-contractive, we obtain:

$$\begin{aligned} \omega\left(\{Tx_n : n \in \mathbb{N}\}\right) &\leq \omega\left(\{Bx_n : n \in \mathbb{N}\}\right) + \omega\left(\left(I - \frac{I-C}{A}\right)(\{Tx_n : n \in \mathbb{N}\})\right) \\ &\leq \alpha \omega\left(\{Tx_n : n \in \mathbb{N}\}\right) \\ &< \omega\left(\{Tx_n : n \in \mathbb{N}\}\right), \end{aligned}$$

which is absurd. It follows that $\{Tx_n : n \in \mathbb{N}\}$ is weakly relatively compact; hence, there exists a subsequence $\{x_{\sigma(n)}\}_n$ of $\{x_n\}_n$ such that $Tx_{\sigma(n)} \rightharpoonup y$ for some $y \in \Omega$. Moreover, $\frac{I-C}{A}$ is weakly sequentially continuous; then, $I - \frac{I-C}{A}$ is weakly sequentially continuous, and then:

$$\left(I - \frac{I-C}{A}\right) Tx_{\sigma(n)} \rightharpoonup \left(I - \frac{I-C}{A}\right) y,$$

As we have $\left(I - \frac{I-C}{A}\right) T = -B + T$ and $-Bx_{\sigma(n)} + Tx_{\sigma(n)} \rightharpoonup -Bx + y$, we obtain:

$$-Bx + y = y - \left(\frac{I-C}{A}\right) y$$

which gives $Tx = y$, and therefore, $Tx_{\sigma(n)} \rightharpoonup Tx$.

We claim that $Tx_n \rightharpoonup Tx$. Assume that there exists a subsequence $\{x_{\sigma_1(n)}\}_n$ of $\{x_n\}_n$ and a weak neighborhood V^w of Tx such that $Tx_{\sigma_1(n)} \notin V^w$ for all $n \in \mathbb{N}$. Since $\{x_{\sigma_1(n)}\}_n$ converge weakly to x, we may extract a subsequence $\{x_{\sigma_1\sigma_2(n)}\}_n$ of $\{x_{\sigma_1(n)}\}_n$ such that $Tx_{\sigma_1\sigma_2(n)} \rightharpoonup Tx$ and $Tx_{\sigma_1\sigma_2(n)} \notin V^w$, which is absurd. Hence, $Tx_n \rightharpoonup Tx$; it follows that T is weakly sequentially continuous.

T is countably ω-condensing. Indeed, let M be a countably subset of Ω with $\omega(M) > 0$; we have:

$$\begin{aligned} \omega(T(M)) &\leq \omega(B(M)) + \omega\left(\left(I - \frac{I-C}{A}\right)(T(M))\right) \\ &\leq \beta \omega(M) + \alpha \omega(T(M)), \end{aligned}$$

then $\omega(T(M)) \leq \frac{\beta}{1-\alpha}\omega(M) < \omega(M)$, which ends the proof. $\square$

Corollary 1. *Let Ω be a nonempty, bounded, closed, and convex subset of a Banach algebra X and ω be a subadditive MWNC on X. Let $C : X \longrightarrow X$ and $B : \Omega \longrightarrow X$ be two operators that satisfy the following conditions:*

(i) $(I-C)^{-1}$ exists on $B(\Omega)$,
(ii) B and $I-C$ are weakly sequentially continuous,
(iii) C is countably α-ω-contractive on Ω,
(iv) B is countably β-ω-contractive,
(v) $x = By + Cx$, for all $y \in \Omega$ implies $x \in \Omega$.

Then, there exists $x \in \Omega$ such that $x = Bx + Cx$, whenever $\frac{\beta}{1-\alpha} < 1$.

Remark 1.

1. Note that Hypothesis (ii) in Theorem 5 may be replaced by "A, B, and C are weakly sequentially continuous", but the Banach algebra X must satisfy condition $(\mathcal{P})$.
2. In Theorem 5, we do not require the conditions "A satisfies condition $(\mathcal{H}1)$" and "$A(\Omega)$ is relatively weakly compact", but in Theorem 3.2 in [23], these conditions are necessary.
3. In Theorem 5, Condition (i) may be replaced by
   $(\widetilde{ii})$ A is regular on X and, A and C are nonlinear contractions on X with contraction functions Φ_A and Φ_C, respectively, and $L\Phi_A(r) + \Phi_C(r) < r$, for $r > 0$ and $L = \|B(\Omega)\|$.

In the following result, we will use the notion of $\mathcal{D}$-Lipschitzian operators.

Theorem 6. *Let Ω be a nonempty, bounded, closed, and convex subset of a Banach algebra X satisfying condition $(\mathcal{P})$ and ω a subadditive MWNC on X. Let $A, C : X \longrightarrow X$, and $B : \Omega \longrightarrow X$ be three weakly sequentially continuous operators with the following conditions:*

(i) A is regular on X,
(ii) $I - \frac{I-C}{A}$ is countably α-ω-contractive on Ω,
(iii) B is countably β-ω-contractive,
(iv) A and C are $\mathcal{D}$-Lipschitzian with the $\mathcal{D}$-function ϕ_A and ϕ_C, respectively, and $L\phi_A(r) + \phi_C(r) < r$ for $r > 0$ and $L = \|B(\Omega)\|$,
(v) $x = Ax \cdot By + Cx$, for all $y \in \Omega$ implies $x \in \Omega$.

Then, there exists $x \in \Omega$ such that $x = Ax \cdot Bx + Cx$, whenever $\frac{\beta}{1-\alpha} < 1$.

Proof. Let $y \in \Omega$ and $F_y : X \longrightarrow X$ by $F_y(x) = Ax \cdot By + Cx$.
For each $x, z \in X$, (iv) gives:

$$\begin{aligned} \|F_y(x) - F_y(z)\| &\leq \|Ax \cdot By - Az \cdot By\| + \|Cx - Cz\| \\ &\leq \|Ax - Az\|\|By\| + \|Cx - Cz\| \\ &\leq L\phi_A(\|x-z\|) + \phi_C(\|x-z\|). \end{aligned}$$

By the Boyd–Wong fixed point theorem ([24]), the mapping F_y has a unique fixed point x_y. Hence, the operator $T = \left(\frac{I-C}{A}\right)^{-1} B : \Omega \longrightarrow X$ is well defined; and by (v), we have $T(\Omega) \subset \Omega$.

Let $\{x_n\}_n$ be a sequence in Ω such that $x_n \rightharpoonup x$; as seen in the proof of Theorem 5, there exists a subsequence $\{x_{\sigma_1(n)}\}_n$ of $\{x_n\}_n$ such that $Tx_{\sigma_1(n)} \rightharpoonup y$ for some $y \in \Omega$. Since:

$$T = AT \cdot B + CT,$$

and A, B, and C are weakly sequentially continuous, we obtain:

$$Tx_{\sigma_1(n)} = A(Tx_{\sigma_1(n)}) \cdot Bx_{\sigma_1(n)} + C(Tx_{\sigma_1(n)}) \rightharpoonup y = Ay \cdot Bx + Cy$$

Thus, $y = Tx$, and then, $T_{\sigma_1(n)} \rightharpoonup Tx$. As above, we can prove that $Tx_n \rightharpoonup Tx$; and then, T is weakly sequentially continuous. By Theorems 2 and 5, T is countably ω-condensing. □

Remark 2. *Note that the hypothesis "A and C are weakly sequentially continuous" in Theorem 6 can be replaced by "$\frac{I-C}{A}$ is weakly sequentially continuous", and in this case, the condition $(\mathcal{P})$ is not required.*

Theorem 7. *Let Ω be a nonempty, closed, convex, and bounded subset of a Banach algebra X and ω be a subadditive MWNC on X. Let $A, C : X \longrightarrow X$, and $B : \Omega \longrightarrow X$ be three operators satisfying the following conditions:*

(i)   *A is regular on X, and B is weakly sequentially continuous,*
(ii) *$\frac{I-C}{A}$ is ψ-expansive, accretive, and continuous,*
(iii) *$I - \frac{I-C}{A}$ is countably α-ω-contractive on Ω,*
(iv) *B is countably β-ω-contractive,*
(v)   *$x = Ax \cdot By + Cx$, for all $y \in \Omega$ implies $x \in \Omega$.*

Then, there exists $x \in \Omega$ such that $x = Ax \cdot Bx + Cx$, whenever $\frac{\beta}{1-\alpha} < 1$.

Proof. For $y \in \Omega$, we define the mapping $F_y : X \longrightarrow X$ by:

$$F_y(x) = \left(I - \frac{I-C}{A}\right)x + By$$

Since $\frac{I-C}{A}$ is continuous and accretive, $I - \frac{I-C}{A}$ is continuous and pseudocontractive, and F_y is continuous and pseudocontractive.

Moreover, we have:

$$\|(I - F_y)x - (I - F_y)z\| = \left\|\left(\frac{I-C}{A}\right)x - \left(\frac{I-C}{A}\right)z\right\|,$$

for all $x, z \in \Omega$, and $\frac{I-C}{A}$ is ψ-expansive. Then, $I - F_y$ is ψ-expansive, continuous, and accretive. It follows that $I - F_y$ is m-accretive (see [25], Corollary 3.2). By [26], Theorem 8, we deduce that $I - F_y$ is surjective. Then, there exists an $x \in X$ such that $\theta = (I - F_y)x$. It follows that:

$$x = F_y(x) = \left(I - \frac{I-C}{A}\right)x + By$$

which implies $By = \left(\frac{I-C}{A}\right)x \in \left(\frac{I-C}{A}\right)(X)$. We conclude by Theorem 5. □

In the following result, we present a nonlinear alternative of the Leary–Schauder type in Banach algebra.

Theorem 8. *Let Ω be a nonempty, bounded, closed, and convex subset of a Banach algebra X, U be a weakly open subset of Ω with $\theta \in U$, and ω be a subadditive MWNC on X. Let $A, C : X \longrightarrow X$, and $B : \overline{U}^w \longrightarrow X$ be three operators satisfying the following conditions:*

(i)   *A is regular on X, and $\left(\frac{I-C}{A}\right)^{-1}$ exists on $B(\Omega)$,*
(ii) *B and $\frac{I-C}{A}$ are weakly sequentially continuous,*
(iii) *$I - \frac{I-C}{A}$ is countably α-ω-contractive on Ω,*
(iv) *B is countably β-ω-contractive,*
(v)   *$x = Ax \cdot By + Cx$, for all $y \in \overline{U}^w$ implies $x \in \Omega$.*

Then, either:

(i) *there exists $x \in U$ such that $x = Ax \cdot Bx + Cx$, or*
(ii) *there exists $u \in \partial_\Omega U$ and $\lambda \in]0, 1[$ such that $u = \lambda A\left(\frac{u}{\lambda}\right) \cdot Bu + \lambda C\left(\frac{u}{\lambda}\right)$,*

where $\partial_\Omega U$ denotes the weak boundary of U in Ω and $\frac{\alpha}{1-\beta} < 1$.

Proof. Let $T := \left(\frac{I-C}{A}\right)^{-1} B$; Condition (vi) implies $T(\overline{U}^w) \subset \Omega$, and T is weakly sequentially continuous and countably ω-condensing. Theorem 4 implies that T has a fixed point in U, or there exists $u \in \partial_\Omega U$ and $\lambda \in]0,1[$ such that $u = \lambda T(u)$, then either there exists $x \in U$ such that $x = Ax \cdot Bx + Cx$, or there exists $u \in \partial_\Omega U$ and $\lambda \in]0,1[$ such that $u = \lambda A\left(\frac{u}{\lambda}\right) \cdot Bu + \lambda C\left(\frac{u}{\lambda}\right)$. $\square$

Corollary 2. *Let Ω be a nonempty, bounded, closed, and convex subset of a Banach algebra X, U be a weakly open subset of Ω with $\theta \in U$, and ω be a subadditive MWNC on X. Let $C : X \longrightarrow X$ and $B : \overline{U}^w \longrightarrow X$ be two operators that satisfy the following conditions:*

(i) $(I - C)^{-1}$ exists on $B(\Omega)$,
(ii) B and $I - C$ are weakly sequentially continuous,
(iii) C is countably α-ω-contractive on Ω,
(iv) B is countably β-ω-contractive,
(v) $x = By + Cx$, for all $y \in \overline{U}^w$ implies $x \in \Omega$.

Then,

(i)   *either there exists $x \in U$ such that $x = Bx + Cx$, or*
(ii)   *there exists $u \in \partial_\Omega U$ and $\lambda \in]0,1[$ such that $u = \lambda Bu + \lambda C\left(\frac{u}{\lambda}\right)$,*

where $\partial_\Omega U$ denotes the weak boundary of U in Ω, and $\frac{\alpha}{1-\beta} < 1$.

Remark 3. *In Theorem 8, Condition (i) may be replaced by*

$(\tilde{i})$ $\frac{I-C}{A}$ *is ψ-expansive and $Bx \in \left(\frac{I-C}{A}\right)(X)$, for all $x \in \Omega$.*

Theorem 9. *Let Ω be a nonempty, closed, convex, and bounded subset of a Banach algebra X, U be a weakly open subset of Ω with $\theta \in U$ and ω be a subadditive MWNC on X. Let $A, C : X \longrightarrow X$, and $B : \overline{U}^w \longrightarrow X$ be three operators satisfying the following conditions:*

(i) A is regular on X,
(ii) B and $\frac{I-C}{A}$ are weakly sequentially continuous,
(iii) $\frac{I-C}{A}$ is ψ-expansive, accretive, and continuous,
(iv) $I - \frac{I-C}{A}$ is countably α-ω-contractive on Ω,
(v) B is countably β-ω-contractive,
(vi) $x = Ax \cdot By + Cx$, for all $y \in \overline{U}^w$ implies $x \in \Omega$.

Then, either:

(i)   *there exists $x \in \Omega$ such that $x = Ax \cdot Bx + Cx$, or*
(ii)   *there exists $u \in \partial_\Omega U$ and $\lambda \in]0,1[$ such that $u = \lambda A\left(\frac{1}{\lambda}u\right) \cdot Bu + C\left(\frac{1}{\lambda}u\right)$.*

where $\partial_\Omega U$ denotes the weak boundary of U in Ω, and $\frac{\alpha}{1-\beta} < 1$.

Proof. Define $T : \Omega \longrightarrow \Omega$ by $Tx = \left(\frac{I-C}{A}\right)^{-1} Bx$. As seen in the proof of Theorem 7, the operator T is well defined; moreover, T is weakly sequentially continuous and countably ω-condensing, and by (vi), we have $T(\overline{U}^w) \subset \Omega$; we conclude by Theorem 4. $\square$

Remark 4. *If we take A is the unit element in the Banach algebra X, we obtain Theorem 3.9 in [16].*

In the following result, the operator $\frac{I-C}{A}$ is not invertible.

Theorem 10. *Let Ω be a nonempty, bounded, closed, and convex subset of a Banach algebra X and ω be a subadditive MWNC on X. Let $A, C : X \longrightarrow X$, and $B : \Omega \longrightarrow X$ be three operators that satisfy the following conditions:*

(i) A is regular,
(ii) $I - \frac{I-C}{A}$ is countably α-ω-contractive on Ω,
(iii) B is countably β-ω-contractive,
(iv) for every net $\{x_\delta\}_\delta$, $x_\delta \in \Omega$, if $x_\delta \rightharpoonup x$, $x \in \Omega$, then, $Bx_\delta \rightharpoonup Bx$ and $\left(\frac{I-C}{A}\right) x_\delta \rightharpoonup \left(\frac{I-C}{A}\right) x$,
(v) for every net $\{x_\delta\}_\delta$, $x_\delta \in \Omega$, if $\left(\frac{I-C}{A}\right) x_\delta \rightharpoonup y$, $y \in \Omega$, then there exists a weakly convergent subset of $\{x_\delta\}_\delta$,
(vi) $\left(\frac{I-C}{A}\right)^{-1} Bx$ is convex, for all $x \in \Omega$;
(vii) $Bx \in \left(\frac{I-C}{A}\right)(X)$ for all $x \in \Omega$ and $x = Ax \cdot By + Cx$, for all $y \in \Omega$ implies $x \in \Omega$.

Then, there exists $x \in \Omega$ such that $x = Ax \cdot Bx + Cx$, whenever $\frac{\beta}{1-\alpha} < 1$.

Proof. By (vii), the multivalued mapping:

$$H : \Omega \longrightarrow P(\Omega)$$
$$x \longmapsto Hx = \left(\frac{I-C}{A}\right)^{-1} Bx,$$

is well defined.

Step 1. H has a ω-weakly closed graph in $\Omega \times \Omega$.

Let $\{x_\delta\}_\delta$ and $\{y_\delta\}_\delta$ be nets in Ω such that $x_\delta \rightharpoonup x \in \Omega$, $y_\delta \rightharpoonup y \in \Omega$ and $y_\delta \in Hx_\delta$.

Since $\left(\frac{I-C}{A}\right) y_\delta = Bx_\delta$, we obtain $\left(\frac{I-C}{A}\right) y_\delta \rightharpoonup \left(\frac{I-C}{A}\right) y$ and $Bx_\delta \rightharpoonup Bx$; it follows that $\left(\frac{I-C}{A}\right) y = Bx$ and then $y \in Hx$; which gives:

$$y \in S(x,y) = \{\lambda y + (1-\lambda)x : \lambda \in [0,1]\}$$

then, $Hx \cap S(x,y) \neq \emptyset$, and H has a ω-weakly closed graph.

Step 2. By Step 1, Hx is closed, for all $x \in \Omega$, and by (vi), $H(\Omega) \subset P_{cl,cv}(\Omega)$.

Step 3. H maps weakly compact sets into relatively weakly compact sets.

Let K be a weakly compact set in Ω, and let $\{y_n\}_n$ be a sequence in $H(K)$; choose $\{x_n\}_n$ in K such that $y_n \in Hx_n$ for all $n \in \mathbb{N}$ and $\{x_{\sigma_1(n)}\}_n$ a subsequence of $\{x_n\}_n$ such that $x_{\sigma_1(n)} \rightharpoonup x$. By (iv), $\left(\frac{I-C}{A}\right) y_{\sigma_1(n)} = Bx_{\sigma_1(n)} \rightharpoonup Bx$, and (v) implies that $\{y_n\}_n$ has a weakly convergent subsequence. Then, by the Eberlein–Šmulian theorem [27], $H(K)$ is relatively weakly compact.

Step 4. H is countably ω-condensing.

Let M be a countable subset of Ω with $\omega(M) > 0$; we have:

$$\left(\frac{I-C}{A}\right)(Hx) = \{Bx\}, \text{ for all } x \in M,$$

then, for all $y \in Hx$ we have:

$$\left(\frac{I-C}{A}\right) y = Bx;$$

hence:

$$y = Bx + \left(I - \frac{I-C}{A}\right) y;$$

consequently:

$$Hx \subset Bx + \left(I - \frac{I-C}{A}\right)(Hx), \text{ for all } x \in M,$$

then:

$$H(M) \subset B(M) + \left(I - \frac{I-C}{A}\right)(H(M)),$$

and:

$$\omega(H(M)) \leq \omega(B(M)) + \omega\left(\left(I - \frac{I-C}{A}\right)(H(M))\right)$$
$$\leq \beta\omega(M) + \alpha\omega(H(M)),$$

It follows that $\omega((H(M)) \leq \frac{\beta}{1-\alpha}\omega(M) < \omega(M)$; and then, H is countably ω-condensing. By Theorem 3.18 in [16], we conclude that H has a fixed point in Ω. □

The following result requires the condition "relatively weakly compact" and where $\frac{\beta}{1-\alpha} \leq 1$.

Theorem 11. *Let Ω be a nonempty, bounded, closed, and convex subset of a Banach algebra X and ω be a positive homogenous MWNC on X. Let $A, C : X \longrightarrow X$, and $B : \Omega \longrightarrow X$ be three operators that satisfy the following conditions:*

(i) A is regular on X, and $\left(\frac{I-C}{A}\right)^{-1}$ exists on $B(\Omega)$,
(ii) B and $\frac{I-C}{A}$ are weakly sequentially continuous,
(iii) $\left(I - \frac{I-C}{A}\right)(\Omega)$ is relatively weakly compact,
(iv) B is countably β-ω-contractive,
(v) If $\{x_n\}_n$ is a sequence in Ω such that $(I - B)x_n \rightharpoonup x$, then $\{x_n\}_n$ has a weakly convergent subsequence,
(vi) $I - \frac{I-C}{A}$ is countably α-ω-contractive on Ω,
(vii) $x = Ax \cdot By + Cx$, for all $y \in \Omega$ implies $x \in \Omega$.

Then, there exists $x \in \Omega$ such that $x = Ax \cdot Bx + Cx$, whenever $\frac{\beta}{1-\alpha} \leq 1$.

Proof. Let $x \in \Omega$, and consider:

$$T : \Omega \longrightarrow \Omega$$
$$x \longmapsto Tx = \left(\frac{I-C}{A}\right)^{-1} Bx;$$

by (i) and (vii), it is clear that T is well defined.

We will show that T satisfies the conditions of Theorem 3. From the proof of Theorem 5, we can see that T is weakly sequentially continuous, and then, it suffices to prove that T is weakly countably one-set-contractive and semi-closed at 0.

Let M be a countably subset of Ω; we have:

$$T = B + \left(I - \frac{I-C}{A}\right)T,$$

then:

$$\omega(T(M)) \leq \omega(B(M)) + \omega\left(\left(I - \frac{I-C}{A}\right)(T(M))\right)$$
$$\leq \beta\omega(M) + \alpha\omega(T(M)),$$

and so:

$$\omega(T(M)) \leq \frac{\beta}{1-\alpha} \leq \omega(M);$$

therefore, T is weakly countably one-set-contractive.

Now, let $\{x_n\}_n$ be a sequence in Ω such that $(I - T)x_n \to 0$.

$$y_n = (I - T)x_n = x_n - Bx_n - \left(I - \frac{I-C}{A}\right)Tx_n$$

By (iii), there exists a subsequence $\{x_{\sigma_1(n)}\}_n$ of $\{x_n\}_n$ such that $\left(I - \frac{I-C}{A}\right)Tx_{\sigma_1(n)} \rightharpoonup y$; and then, $(I - B)x_{\sigma_1(n)} = y_{\sigma_1(n)} + \left(I - \frac{I-C}{A}\right)Tx_{\sigma_1(n)} \rightharpoonup y$. By (v), we conclude that there exists a subsequence $\{x_{\sigma_1\sigma_2(n)}\}_n$ of $\{x_{\sigma_1(n)}\}_n$, which converges to some element x. Since $(I - T)x_{\sigma_1\sigma_2(n)} \to 0$ and T is weakly sequentially continuous, we obtain $Tx = x$, and then, T is weakly semi-closed at θ. □

Let Ω be a nonempty closed and convex subset of a Banach algebra X, and let $A, C : X \longrightarrow X$, and $B : \Omega \longrightarrow X$ be three operators. For any $D \subseteq \Omega$, we set (see [28]):

$$\mathcal{F}(A, C, B, D) = \{x \in X : x = Ax \cdot By + Cx, \, y \in D\}.$$

If $A = 1_X$ and $C = 0$, we obtain $\mathcal{F}(1_X, 0, B, D) = B(D)$.

Theorem 12. *Let X be a Banach algebra satisfying condition $(\mathcal{P})$ and Ω be a nonempty, closed, convex, and bounded subset of X; ω is an MWNC on X. Let $A, C : X \longrightarrow X$, and $B : \Omega \longrightarrow X$ be three operators satisfying the following conditions:*

(i)   *A is regular on X, and B is weakly sequentially continuous,*
(ii)   *$I - \frac{I-C}{A}$ is a contraction on Ω,*
(iii)   *$\omega(\mathcal{F}(A, C, B, D)) < \omega(D)$, for any countably subset D of Ω with $\omega(D) > 0$,*
(iv)   *$\mathcal{F}(A, C, B, \Omega) \subset \Omega$,*
(v)   *If $\{x_n\} \subset \mathcal{F}(A, C, B, \Omega)$, then $\{Ax_n\}_n$ and $\{Cx_n\}_n$ have weakly convergent subsequences (converging respectively to y and z), and if $x_n \rightharpoonup x$, we have $y = Ax$ and $z = Cx$.*

Then, there exists $x \in \Omega$ such that $x = Ax \cdot Bx + Cx$.

Proof. For $y \in \Omega$, we define the mapping:

$$\begin{aligned} F_y : X &\longrightarrow X \\ x &\longmapsto F_y(x) = \left(I - \frac{I-C}{A}\right)x + By, \end{aligned}$$

(ii) implies that F_y is a contraction; then, F_y has a unique fixed point $\tau(y) \in X$; we have $\tau(y) = \left(I - \frac{I-C}{A}\right)\tau(y) + By$ or equivalently $\tau(y) = A\tau(y) \cdot By + C\tau(y)$; which shows that $\tau(y) \in \mathcal{F}(A, C, B, \Omega)$. It follows that $\tau(\Omega) \subset \Omega$.

Let M be a countable subset of Ω such that $\omega(M) > 0$; we have:

$$\begin{aligned} \tau(M) &= \{\tau(x) : x \in M\} \\ &= \{\tau(x) = A(\tau(x)) \cdot Bx + C(\tau(x)) : x \in M\} \\ &= \{\tau(x) : \tau(x) \in \mathcal{F}(A, C, B, M)\} \\ &\subseteq \mathcal{F}(A, C, B, M), \end{aligned}$$

Hence, $\omega(\tau(M)) \leq \omega(\mathcal{F}(A, C, B, M)) < \omega(M)$; then, τ is countably ω-condensing.

Moreover, τ is weakly sequentially continuous. Indeed, let $\{x_n\}_n$ be a sequence in Ω such that $x_n \rightharpoonup x$; since B is weakly sequentially continuous, we have $Bx_n \rightharpoonup Bx$, and since $\{\tau(x_n)\}_n \subset \mathcal{F}(A, C, B, \Omega)$, there exists a subsequence $\{\tau(x_{\sigma_1(n)})\}_n$ and $\{\tau(x_{\sigma_2(n)})\}_n$ of $\{\tau(x_n)\}_n$ such that $A\tau(x_{\sigma_1(n)}) \rightharpoonup y$ and $C\tau(x_{\sigma_2(n)}) \rightharpoonup z$. It follows that:

$$\tau(x_{\sigma_1\sigma_2(n)}) = A\tau(x_{\sigma_1\sigma_2(n)}) \cdot Bx_{\sigma_1\sigma_2(n)} + C\tau(x_{\sigma_1\sigma_2(n)}) \rightharpoonup y \cdot Bx + z.$$

With (v), we obtain $A(y \cdot Bx + z) = y$ and $C(y \cdot Bx + z) = z$; and then, $y \cdot Bx + z = A(y \cdot Bx + z) \cdot Bx + C(y \cdot Bx + z)$.

The uniqueness of the fixed point implies that $\tau(x) = y \cdot Bx + z$; and therefore, $\tau(x_{\sigma_1 \sigma_2(n)}) \rightharpoonup \tau(x)$. We claim that $\tau(x_n) \rightharpoonup \tau(x)$. For this, assume that there exists a weak neighborhood V of $\tau(x)$ and a subsequence $\{x_{\varphi_1(n)}\}_n$ of $\{x_n\}_n$ such that $x_{\varphi_1(n)} \notin V$ for all $n \in \mathbb{N}$. Since $x_{\varphi_1(n)} \rightharpoonup x$, we can extract a subsequence $\{x_{\varphi_1 \varphi_2(n)}\}_n$ of $\{x_{\varphi_1(n)}\}_n$ such that $\tau(x_{\varphi_1 \varphi_2(n)}) \rightharpoonup \tau(x)$. This is not possible, since $x_{\varphi_1 \varphi_2(n)} \notin V$ for all $n \in \mathbb{N}$. We conclude that τ is weakly sequentially continuous. By Theorem 2, there exists $x \in \Omega$ such that $x = \tau(x) = Ax \cdot Bx + Cx$. □

If $A = 1_X$ in Theorem 12, we obtain Theorem 3.11 in [16].

Theorem 13. *Let Ω be a nonempty, closed, convex, and bounded subset of a Banach algebra X; ω is an MWNC on X. Let $A, C : X \longrightarrow X$, and $B : \Omega \longrightarrow X$ be three operators that satisfy the following conditions:*

(i) *A is regular on X, and $\frac{I-C}{A}$ is one-to-one,*
(ii) *$I - \frac{I-C}{A}$ is nonexpansive,*
(iii) *B and $\frac{I-C}{A}$ are weakly sequentially continuous,*
(iv) *$\omega\left(\mathcal{F}(A, C, B, D)\right) < \omega(D)$, for any countably subset D of Ω with $\omega(D) > 0$,*
(v) *$\left(I - \frac{I-C}{A}\right) x + By \in \Omega$ for all $x, y \in \Omega$,*
(vi) *If $\{x_n\}_n \subset \Omega$ such that $\left\{\left(\frac{I-C}{A}\right) x_n\right\}_n$ is weakly convergent, then the sequence $\{x_n\}_n$ has a weakly convergent subsequence.*

Then, there exists $x \in \Omega$ such that $x = Ax \cdot Bx + Cx$.

Proof. Let $y \in \Omega$, and define $F_y : \Omega \longrightarrow X$ by:

$$F_y(x) = \left(I - \frac{I-C}{A}\right) x + By$$

By (ii), F_y is nonexpansive, and by (v), we have $F(\Omega) \subset \Omega$. Then, by ([29], Theorem 2.15), there exists a sequence $\{x_n\}_n$ in Ω such that $\|x_n - F_y(x_n)\| \longrightarrow 0$, and then, $\left(\frac{I-C}{A}\right) x_n \longrightarrow By$. Using (vi), we can extract a subsequence $\{x_{\sigma_1(n)}\}_n$ of $\{x_n\}_n$ such that $x_{\sigma_1(n)} \rightharpoonup x \in \Omega$, and then, $\left(\frac{I-C}{A}\right) x_{\sigma_1(n)} \rightharpoonup \left(\frac{I-C}{A}\right) x$; then:

$$By = \left(\frac{I-C}{A}\right) x \in \left(\frac{I-C}{A}\right) (\Omega)$$

which implies $B(\Omega) \subset \left(\frac{I-C}{A}\right) (\Omega)$.

Define $T : \Omega \longrightarrow \Omega$ by $Tx = \left(\frac{I-C}{A}\right)^{-1} Bx$. Let $D \subseteq \Omega$ and $x \in D$; the equality $Tx = A(Tx) \cdot Bx + C(Tx)$ implies that $Tx \in \mathcal{F}(A, C, B, D)$; then:

$$T(D) \subset \mathcal{F}(A, C, B, D)$$

for any subset D of Ω.

The assumption (iv) implies that T is countably ω-condensing. Moreover, T is weakly sequentially continuous. Indeed, let $\{x_n\}_n$ be a sequence such that $x_n \rightharpoonup x$; we have $Bx_n \rightharpoonup Bx$; then, $\left(\frac{I-C}{A}\right) Tx_n \rightharpoonup Bx$. By (vi), there exists a subsequence $\{x_{\sigma_2(n)}\}_n$ such that $Tx_{\sigma_2(n)} \rightharpoonup y \in \Omega$; thus, $\left(\frac{I-C}{A}\right) Tx_{\sigma_2(n)} \rightharpoonup \left(\frac{I-C}{A}\right) y$, which leads to $\left(\frac{I-C}{A}\right) y = Bx$, and so, $Tx_{\sigma_2(n)} \rightharpoonup Tx$. As in the proof of Theorem 5, we can prove that $Tx_n \rightharpoonup Tx$, and we apply Theorem 2 to end the proof. □

Remark 5. *If we take $A = 1_X$ in Theorem 13, we obtain Theorem 3.13 in [16].*

4. Application

Let X be a real Banach algebra satisfying condition $(\mathcal{P})$; we denote $E = C([0,1], X)$ the Banach space of all X-valued continuous functions defined on $[0,1]$, endowed with the norm $\|x\|_\infty = \sup_{t \in [0,1]} \|x(t)\|$. In this section, we discuss the following abstract nonlinear quadratic integral equation $((FIE))$ (see [30]):

$$x(t) = K(t, x(\xi(t))) + Tx(t) \left(q(t) + \int_0^{\sigma(t)} g(s, x(\eta(s))) ds \right), \; t \in J = [0,1],$$

where $q : J \longrightarrow X$, $g, K : J \times X \longrightarrow X$, $\xi, \sigma, \eta : J \longrightarrow J$, and $T : E \longrightarrow E$. Note that E is a Banach algebra satisfying condition $(\mathcal{P})$ and the integral in (FIE) is the Pettis integral, while the solutions of (FIE) are in E. Make the following assumptions for (FIE):

Hypothesis 1 (H1).

(i) *The functions $\xi, \sigma, \eta : J \longrightarrow J$ are continuous, and σ is nondecreasing,*
(ii) *the function $q : J \longrightarrow X$ is continuous,*

Hypothesis 2 (H2).

(i) *for all $t \in [0,1]$, $K(t,.) : X \longrightarrow X$ is weakly sequentially continuous,*
(ii) *for each $u \in X$, $K(.,u) : J \longrightarrow X$ is continuous,*
(iii) *there is a continuous function $\delta : J \longrightarrow [0, +\infty)$ with bound $\Delta = \sup_{t \in J} |\delta(t)|$ such that $\|K(t, x(t)) - K(t, y(t))\| \leq \delta(t) \|x(t) - y(t)\|$ for all $x, y \in E$ and $t \in [0,1]$,*

Hypothesis 3 (H3). *The operator $T : E \longrightarrow E$ satisfies:*

(i) *there is a continuous function $\gamma : J \longrightarrow [0, +\infty)$ with bound $\Gamma = \sup_{t \in J} |\gamma(t)|$ such that $\|Tx(t) - Ty(t)\| \leq \gamma(t) \|x(t) - y(t)\|$, for all $x, y \in E$ and $t \in [0,1]$,*
(ii) *T is weakly sequentially continuous on E,*
(iii) *T is regular on E; $\frac{1_E}{T}$ is well defined on E; $\frac{1_E}{T}$ is weakly compact; and there exists $m_0 \in [0,1)$ such that $\sup_{x \in E} \left\| 1_E - \frac{1_E}{Tx} \right\|_\infty \leq m_0$, where 1_E represents the unit element in the Banach algebra E,*

Hypothesis 4 (H4).

(i) *for each continuous $x : [0,1] \longrightarrow X$, the function $s \longmapsto g(s, x(s))$ is weakly measurable on $[0,1]$, and for almost every $t \in [0,1]$, the map $u \longmapsto g(t, u)$ is weakly sequentially continuous on X,*
(ii) *there are a function $\phi \in L^1([0,1], \mathbb{R}^+)$ and a continuous nondecreasing function $\vartheta : [0, +\infty) \longrightarrow [0, +\infty)$ such that:*

$$\|g(s,u)\| \leq \phi(s) \vartheta(\|u\|) \; a.e \text{ for all } s \in [0,1], \text{ and all } u \in X,$$

(iii) *there is a constant $0 \leq \beta < 1$ such that:*

$$\mu(g([0,1] \times W)) \leq \beta \mu(W),$$

for any countably bounded subset W of X,

Hypothesis 5 (H5). *There is a constant $r > 0$ such that $Q\Gamma + \Delta < 1$, where:*

$$Q = Q_1 + \vartheta(r) \int_0^1 \phi(s) ds \quad \text{with } Q_1 = \sup_{t \in J} \|q(t)\|$$

Now, we are in a position to state our main result of this section:

Theorem 14. *Assume that Hypotheses (H1)–(H5) hold and $r_0 = \frac{L+Q\|T0\|_\infty}{1-\Delta-Q\Gamma} \leq r$ with $L = \sup_{t\in[0,1]} \|K(t,0)\|$; then (FIE) has a solution in $C([0,1],X)$ whenever $m_0 + \Delta(1+m_0) < 1$ and $\frac{\beta}{1-(m_0+\Delta(1+m_0))} < 1$.*

Proof. The integral equation (FIE) may be written in the form:

$$x(t) = Ax(t) \cdot Bx(t) + Cx(t),$$

where:

$$Bx(t) = q(t) + \int_0^{\sigma(t)} g(s, x(\eta(s)))ds,$$
$$Ax(t) = Tx(t),$$
$$Cx(t) = K(t, x(\xi(t))).$$

Let $\Omega = \{x \in C([0,1],X) : \|x\|_\infty \leq r_0\}$; note that Ω is a closed, convex, and bounded subset of E. We will show that the mappings A, B, and C verify all the conditions of Theorem 6.

Step 1. We show that A and C are Lipschitzian. First, we verify that the mapping C is well defined. Let $x \in E$, and let $\{t_n\}_n$ be a sequence in J such that $t_n \to t \in J$. We have:

$$\begin{aligned}
\|Cx(t_n) - Cx(t)\| &= \|K(t_n, x(\xi(t_n))) - K(t, x(\xi(t)))\| \\
&\leq \|K(t_n, x(\xi(t_n))) - K(t_n, x(\xi(t)))\| + \|K(t_n, x(\xi(t))) - K(t, x(\xi(t)))\| \\
&\leq \delta(t)\|x(\xi(t_n)) - x(\xi(t))\| + \|K(t_n, x(\xi(t))) - K(t, x(\xi(t)))\| \\
&\leq \Delta\|x(\xi(t_n)) - x(\xi(t))\| + \|K(t_n, x(\xi(t))) - K(t, x(\xi(t)))\|.
\end{aligned}$$

Since $K(.,x)$ is continuous and ξ is continuous, then $\|Cx(t_n) - Cx(t)\| \to 0$; we conclude that $Cx \in E$. Now, let $x, y \in E$ and $t \in J$; we have:

$$\|Cx(t) - Cy(t)\| \leq \delta(t)\|x(\xi(t)) - y(\xi(t))\|,$$

then:

$$\|Cx - Cy\|_\infty \leq \Delta\|x - y\|_\infty,$$

and we have:

$$\begin{aligned}
\|Ax(t) - Ay(t)\| &= \|Tx(t) - Ty(t)\| \\
&\leq \gamma(t)\|x(t) - y(t)\|,
\end{aligned}$$

then:

$$\|Ax - Ay\|_\infty \leq \Gamma\|x - y\|_\infty.$$

Thus, A and C are Lipschitzians with the Lipschitz constants Δ and Γ, respectively.

Step 2. From the assumption (H3)(ii), the mapping A is weakly sequentially continuous on E. Now, we show that C is weakly sequentially continuous on E; for this, let $\{x_n\}_n$ in E such that $x_n \rightharpoonup x \in E$, then $\{x_n\}_n$ is bounded on E; from Dobrokov's theorem ([31], p. 36), we get for all $t \in [0,1]$, $x_n(t) \rightharpoonup x(t)$. Since $K(t,.)$ is weakly sequentially continuous for all $t \in [0,1]$, we get $Cx_n(t) \rightharpoonup Cx(t)$. Again, from Dobrokov's theorem, we deduce that $Cx_n \rightharpoonup Cx$, then C is weakly sequentially continuous on E. Now, we prove that B is weakly sequentially continuous. Firstly, we verify that if $x \in \Omega$, then $Bx \in E$. Let $x \in \Omega$ and $t, t' \in [0,1]$, such that $t \leq t'$; without loss of generality, we may assume that $Bx(t) - Bx(t') \neq 0$. Using the Hahn–Banach theorem, we get that there exists $x^* \in X^*$ such that $x^*(Bx(t) - Bx(t')) = \|Bx(t) - Bx(t')\|$ and $\|x^*\|_* = 1$; hence:

$$\begin{aligned} \|Bx(t) - Bx(t')\| &= x^*(Bx(t) - Bx(t')) \\ &= x^*(q(t) - q(t')) + \int_{\sigma(t)}^{\sigma(t')} x^*(g(s, x(\eta(s)))) ds \\ &\leq \sup_{t \in J} \|q(t) - q(t')\| + \vartheta(\|x\|_\infty) \int_{\sigma(t)}^{\sigma(t')} \phi(s) ds \\ &\leq \sup_{t \in J} \|q(t) - q(t')\| + \vartheta(r_0) \int_{\sigma(t)}^{\sigma(t')} \phi(s) ds; \end{aligned}$$

consequently, $Bx \in E$. As q and σ are uniformly continuous on the compact $[0, 1]$, we get that $B(\Omega)$ is an equicontinuous family of functions. Now, we show that B is weakly sequentially continuous on Ω. Let $\{x_n\}_n$ be a sequence in Ω such that $x_n \rightharpoonup x \in \Omega$, then we get for all $t \in [0, 1]$, $x_n(t) \rightharpoonup x(t)$. Furthermore, for $n \in \mathbb{N}$ and $x^* \in X^*$:

$$x^*(Bx_n(t)) = x^*(q(t)) + \int_0^{\sigma(t)} x^*(g(s, x_n(\eta(s)))) ds, \text{ for all } t \in J,$$

From (H1)(i) and (H4)(i), we have $x^*(g(s, x_n(\eta(s)))) \to x^*(g(s, x(\eta(s))))$ for all $s \in [0, 1]$. The Lebesgue dominated convergence theorem yields:

$$\int_0^{\sigma(t)} x^*(g(s, x_n(\eta(s)))) ds \longrightarrow \int_0^{\sigma(t)} x^*(g(s, x(\eta(s)))) ds,$$

then $Bx_n(t) \rightharpoonup Bx(t)$; by Dobrokov's theorem ([31], p. 36), we get $Bx_n \rightharpoonup Bx$.

Step 3. B i countably β-ω-contractive. First, we show that $B(\Omega)$ is bounded. Let $x \in \Omega$ and $t \in [0, 1]$. Without loss of generality, we may assume that $Bx(t) \neq 0$. Using the Hahn–Banach theorem, we deduce that there exists $x^* \in X^*$ such that $x^*(Bx(t)) = \|Bx(t)\|$ and $\|x^*\|_* = 1$. Hence,

$$\begin{aligned} \|Bx(t)\| &= x^*(Bx(t)) \\ &= x^*(q(t)) + \int_0^{\sigma(t)} x^*(g(s, x(\eta(s)))) ds \\ &\leq \sup_{t \in J} \|q(t)\| + \int_0^1 \|g(s, x(\eta(s)))\| ds \\ &\leq Q_1 + \vartheta(r) \int_0^1 \phi(s) ds = Q, \end{aligned}$$

then $B(\Omega)$ is bounded.

Now, let V be a countably bounded subset of Ω; for each $t \in [0, 1]$, we have by ([32], Theorem 3):

$$\begin{aligned} \mu(B(V)(t)) &\leq \mu\left(\left\{\int_0^{\sigma(t)} g(s, x(\eta(s))) ds : x \in V\right\}\right) \\ &\leq \mu\left(\sigma(t) \overline{co}\{g(s, x(\eta(s))) : x \in V, s \in [0, 1]\}\right) \\ &\leq \mu\left(g\left([0, 1] \times V([0, 1])\right)\right) \\ &\leq \beta \mu\left(V([0, 1])\right) \\ &\leq \beta \sup_{t \in J} \mu(V(t)) \\ &\leq \beta \mu(V), \end{aligned}$$

because V is bounded, then we can apply Theorem 1.

Since $V(B)$ is bounded and equicontinuous, again, by Theorem 1, we get:

$$\mu(B(V)) \leq \beta \mu(V).$$

Consequently, B is countably β-μ-contractive.

Step 4. Now, we prove that $I - \frac{I-C}{A}$ is countably α'-μ-contractive where $\alpha' = m_0 + \Delta(1 + m_0)$. Firstly, for all $x \in E$ by Step 1, we have $Cx \in E$, and by (H3)(iii), we have $\frac{1_E}{Ax} \in E$; hence, $\left(I - \frac{I-C}{A}\right) x \in E$. Now, let $x \in \Omega$; we have:

$$\left\| \left(I - \frac{I-C}{A}\right) x \right\|_\infty = \left\| x - \frac{x - Cx}{Ax} \right\|_\infty$$

$$\leq \left\| 1_E - \frac{1_E}{Ax} \right\|_\infty \|x\|_\infty + \left\| \frac{1_E}{Ax} \right\|_\infty \|Cx\|_\infty$$

$$\leq m_0 r_0 + (1 + m_0)(\Delta r_0 + L),$$

then $\left(I - \frac{I-C}{A}\right)(\Omega)$ is bounded. Now, let V be a bounded subset of Ω such that $\mu(V) > 0$; note that for all $x \in V$, we have:

$$\left(I - \frac{I-C}{A}\right) x = x - \frac{x - Cx}{Ax}$$

$$= \left(1_E - \frac{1_E}{Ax}\right) \cdot x + \frac{1_E}{Ax} \cdot Cx,$$

then:

$$\left(I - \frac{I-C}{A}\right)(V) \subset \left(1_E - \frac{1_E}{A(V)}\right) \cdot V + \frac{1_E}{A(V)} \cdot C(V),$$

because $\frac{1_E}{A}$ is weakly compact; then, by the assumption (H3)(iii), we get:

$$\mu\left(\left(I - \frac{I-C}{A}\right)(V)\right) \leq \mu\left(\left(1_E - \frac{1_E}{A(V)}\right) \cdot V\right) + \mu\left(\frac{1_E}{A(V)} \cdot C(V)\right)$$

$$\leq \left\| 1_E - \frac{1_E}{A(V)} \right\|_\infty \mu(V) + \left\| \frac{1_E}{A(V)} \right\|_\infty \mu(C(V)).$$

because C is Δ-Lipschitzian and weakly sequentially continuous; by Lemma 1, we get $\mu(C(V)) \leq \Delta \mu(V)$, then

$$\mu\left(\left(I - \frac{I-C}{A}\right)(V)\right) \leq \left\| 1_E - \frac{1_E}{A(V)} \right\|_\infty \mu(V) + \left\| \frac{1_E}{A(V)} \right\|_\infty \Delta \mu(V),$$

then:

$$\mu\left(\left(I - \frac{I-C}{A}\right)(V)\right) \leq m_0 \mu(V) + (1 + m_0) \Delta \mu(V)$$

$$\leq \alpha' \mu(V),$$

where $\alpha' = m_0 + \Delta(1 + m_0) < 1$; then, $I - \frac{I-C}{A}$ is countably α'-μ-contractive.

Step 5. We show that for all $x \in E$ and $y \in \Omega$, if $x = Ax \cdot By + Cx$, then $x \in \Omega$. We have for all $t \in [0,1]$:

$$x(t) = Ax(t) \cdot By(t) + Cx(t),$$

then,
$$\begin{aligned}\|x(t)\| &\leq \|Cx(t)\| + \|Ax(t) \cdot By(t)\| \\ &\leq \|Cx(t)\| + \|Ax(t)\|\|By(t)\| \\ &\leq \Delta\|x(t)\| + L + Q\left(\Gamma\|x(t)\| + \|A0\|\right),\end{aligned}$$

then,
$$\|x(t)\| \leq \frac{L + Q\|A0\|}{1 - \Delta - Q\Gamma} = r_0,$$

hence,
$$\|x\|_\infty \leq r_0,$$

consequently, $x \in \Omega$.

Applying Theorem 6, we get a fixed point for $A \cdot B + C$ and hence a solution to (FIE) in E. □

5. Example

Consider the Banach algebra $E = \mathcal{C}([0,1], \mathbb{R})$ of all continuous real-valued functions on $J = [0,1]$, with norm $\|x\|_\infty = \sup_{t\in[0,1]} |x(t)|$. In this case, $X = \mathbb{R}$, and E is a Banach algebra satisfying condition $(\mathcal{P})$ and reflexive. Let $b : [0,1] \longrightarrow X$ be a continuous and nonnegative function such that $\sup_{t\in J} |b(t)| = \frac{1}{4}$. We consider the following nonlinear integral equation:

$$x(t) = \frac{1}{4}t^3 x\left(\frac{t^2}{2}\right) + \left(1 + \int_0^t \frac{b(s)}{1+|x(s)|} ds\right) \cdot \left(\sqrt{t} + \int_0^t \frac{s^2}{20} \frac{|x(s)| \cdot x(s)}{e^{|x(s)|}} ds\right), \quad t \in J. \quad (3)$$

To show that (3) has a solution in E, we will verify that all conditions of Theorem 14 are satisfied.

Define $K : [0,1] \times \mathbb{R} \longrightarrow \mathbb{R}$, by $K(t, x(t)) = \frac{1}{4}t^3 x\left(\frac{t^2}{2}\right)$ (in this case $\zeta(t) = \frac{t^2}{2}$). For all $t \in [0,1]$, the function $K(t, .) : X \longrightarrow X$ is continuous (then weakly sequentially continuous, because $X = \mathbb{R}$), and for all $x \in X$, the function $K(., x) : J \longrightarrow X$ is continuous. Now, let $x, y \in E$ and $t \in [0,1]$; we have:

$$|K(t, x(t)) - K(t, y(t))| \leq \delta(t)|x(t) - y(t)|$$

where the function $\delta : t \mapsto \frac{1}{4}t^3$ is continuous with bound $\Delta = \sup_{t\in J} |\delta(t)| = \frac{1}{4}$.

Next, we introduce the function $T : E \longrightarrow E$ such that $Tx(t) = 1 + \int_0^t \frac{b(s)}{1+|x(s)|} ds$ for all $t \in J$. As seen in Step 2 in the proof of Theorem 14, the operator T is weakly sequentially continuous, regular on E, and $\frac{1_E}{T}$ is well defined on E. Let $x \in E$ and $t \in [0,1]$; we have:

$$\left|1 - \frac{1}{Tx(t)}\right| = \frac{\int_0^t \frac{b(s)}{1+|x(s)|} ds}{1 + \int_0^t \frac{b(s)}{1+|x(s)|} ds} \leq \int_0^1 b(s) \, ds \leq \frac{1}{4};$$

thus, $\sup_{x \in X} \|1_E - \frac{1_E}{Tx}\|_\infty \leq m_0$, where $m_0 = \frac{1}{4}$.

Moreover, $\frac{1_E}{T}$ is weakly compact on E; indeed, let $x \in E$, and let $t, t' \in J$ such that $t \leq t'$. Without loss of generality, we may assume that $\left(\frac{1_E}{T}\right)x(t) - \left(\frac{1_E}{T}\right)x(t') \neq 0$. Using the Hahn–Banach theorem, we deduce that there exists $x^* \in X^*$ such that $x^*\left(\left(\frac{1_E}{T}\right)x(t) - \left(\frac{1_E}{T}\right)x(t')\right) = \left|\left(\frac{1_E}{T}\right)x(t) - \left(\frac{1_E}{T}\right)x(t')\right|$ and $\|x^*\|_* = 1$, hence,

$$\left|\left(\frac{1_E}{T}\right)x(t) - \left(\frac{1_E}{T}\right)x(t')\right| \leq \frac{1}{4}|t - t'|,$$

then $\left(\frac{1_E}{T}\right)(E)$ is weakly equicontinuous. Now, let $\{x_n\}_n$ be a sequence in E, and fix $t \in J$; we have:

$$\left|\left(\frac{1_E}{T}\right)x_n(t)\right| = \left|\frac{1}{1+\int_0^t \frac{b(s)}{1+|x_n(t)|}ds}\right| \leq 1;$$

therefore, $\{\left(\frac{1_E}{T}\right)x_n(t)\}_n$ is weakly equi-bounded. Let $t \in J$; since $X = \mathbb{R}$ is reflexive, then by [33], the set $\{\left(\frac{1_E}{T}\right)x_n(t) : n \in \mathbb{N}\}$ is weakly relatively sequentially compact. The Arzela–Ascoli theorem implies that there exists a subsequence $\{\left(\frac{1_E}{T}\right)x_{\sigma(n)}\}_n$ such that $\left(\frac{1_E}{T}\right)x_{\sigma(n)} \to \left(\frac{1_E}{T}\right)x \in E$; then, $\left(\frac{1_E}{T}\right)(E)$ is relatively weakly compact. Therefore, $\frac{1_E}{T}$ is weakly compact.

Let $x, y \in E$ and $t \in [0,1]$; we have:

$$\begin{aligned} |Tx(t) - Ty(t)| &\leq \int_0^t b(s)(|x(s) - y(s)|)ds \\ &\leq \gamma(t)\|x - y\|_\infty, \end{aligned}$$

where $\gamma : t \mapsto \frac{t}{4}$ is continuous with bound $\Gamma = \sup_{t \in J} |\gamma(t)| = \frac{1}{4}$.

Finally, we define $g : [0,1] \times X \longrightarrow X$, by $g(s, x(s)) = \frac{s^2}{20} \frac{|x(s)| \cdot x(s)}{e^{|x(s)|}}$. For each $u \in X$, the function $g(., u) : [0,1] \longrightarrow X$ is weakly measurable on $[0,1]$, and for almost every $t \in [0,1]$, the function $g(t,.) : X \longrightarrow X$ is continuous (then weakly sequentially continuous). Furthermore, we have:

$$|g(s, u)| \leq \vartheta(|u|)\phi(s) \text{ a.e for all } s \in [0,1], \text{ and all } u \in X,$$

where $\phi(s) = s^2$ and $\vartheta(v) = \frac{v}{20}$ for all $v \in [0, +\infty)$ since $e^{|z|} \geq |z|$ for all $z \in X$.
Moreover, if W is a countably bounded subset of X, we have:

$$\begin{aligned} \mu\left(g([0,1] \times W)\right) &= \mu(\{g(s, u) : u \in W \text{ and } s \in [0,1]\}) \\ &\leq \mu\left(\left\{\left(\frac{1}{20}\frac{s^2 \cdot |u|}{e^{|u|}}\right).u : u \in W \text{ and } s \in [0,1]\right\}\right) \\ &\leq \mu\left([0, \frac{1}{5}].W\right) \\ &\leq \frac{1}{5}\mu(W), \end{aligned}$$

Then:

$$\mu(g([0,1] \times W)) \leq \beta\mu(W), \text{ where } \beta = \frac{1}{5}.$$

We set $q : [0,1] \longrightarrow [0, +\infty)$, such that $q(t) = \sqrt{t}$; we have that q is continuous and $Q_1 = \sup_{t \in J} |q(t)| = 1$.

If we take $r = 4$, we get $Q = Q_1 + \vartheta(4) \int_0^1 \phi(s)ds = \frac{16}{15}$ and $Q\Gamma + \Delta = \frac{31}{60} < 1$ (then, for all $s \in \mathbb{R}^+$, $Q\phi_A(s) + \phi_C(s) = Q\Gamma s + \Delta s < s$ where $\phi_A(s) = \Gamma s$ and $\phi_C(s) = \Delta s$).
Now, we have $\|T0\|_\infty = \sup_{t \in J} |1 + \int_0^t b(s) ds| \leq \frac{5}{4}$ and $r_0 = \frac{L+Q\|T0\|_\infty}{1-\Delta-Q\Gamma} \leq \frac{80}{29}$, then $r_0 \leq r$, $m_0 + (1+m_0) \cdot \Delta = \frac{9}{16} < 1$ and $\frac{\beta}{1-(m_0+(1+m_0)\cdot \Delta)} = \frac{16}{35} < 1$.
Theorem 14 proves the existence of a solution to Equation (3).

6. Conclusions

In this paper, we proved some fixed point theorems for the nonlinear operator $A \cdot B + C$ in a Banach algebra under a weak topology and with the help of the measure of weak noncompactness. Our results improved and generalized some interesting fixed point theorems in the literature. Our examples showed that the results in this paper can be applied to prove the existence of the solution of a nonlinear integral equation in Banach algebra.

Author Contributions: All authors contributed equally and significantly to writing this article.

Funding: We have no funding for this article.

Acknowledgments: The authors are thankful to the Editors and the anonymous referees for their valuable comments, which reasonably improved the presentation of the manuscript.

Conflicts of Interest: The authors declare that they have no competing interests.

References

1. Chandrasekhar, S. *Radiative Transfer*; Dover Publications Inc.: New York, NY, USA, 1960.
2. Gripenberg, G. On some epidemic models. *Q. Appl. Math.* **1981**, *39*, 317–327. [CrossRef]
3. Hu, S.; Khavanin, M.; Zhuang, W. Integral equations arising in the kinetic theory of gases. *Appl. Anal.* **1989**, *34*, 261–266. [CrossRef]
4. Kurz, L.; Nowosad, P.; Saltzberg, B.R. On the solution of a quadratic integral equation arising in signal design. *J. Franklin Inst.* **1966**, *281*, 437–454. [CrossRef]
5. Spiga, G.; Bowden, R.L.; Boffi, V.C. On the solutions to a class of nonlinear integral equations arising in transport theory. *J. Math. Phys.* **1984**, *25*, 3444–3450. [CrossRef]
6. George, H.; Pimbley, J.R. Positive solutions of a quadratic integral equation. *Arch. Ration. Mech. Anal.* **1967**, *24*, 107–127.
7. Ali, A.A.; Ben Amar, A. Measures of weak noncompactness, nonlinear Leray-Schauder alternatives in Banach algebras satisfying condition (P) and an application. *Quaest. Math.* **2015**, *39*, 319–340. [CrossRef]
8. Banaś, J.; Taoudi, M.A. Fixed points and solutions of operator equations for the weak topology in Banach algebras. *Taiwan. J. Math.* **2014**, *18*, 871–893. [CrossRef]
9. Ben Amar, A.; Chouayekh, S.; Jeribi, A. New fixed point theorems in Banach algebras under weak topology features and applications to nonlinear integral equations. *J. Funct. Anal.* **2010**, *259*, 2215–2237. [CrossRef]
10. Ben Amar, A.; O'Regan, D. Measures of weak noncompactness and new fixed point theory in Banach algebras satisfying condition (*P*). *Fixed Point Theory* **2017**, *18*, 37–46. [CrossRef]
11. Ben Amar, A.; Chouayekh, S.; Jerbi, A. Fixed point theory in a new class of Banach algebras and application. *Afr. Mat.* **2013**, *24*, 705–724. [CrossRef]
12. Leggett, R.W. On certain nonlinear integral equations. *J. Math. Anal. Appl.* **1977**, *57*, 462–468. [CrossRef]
13. Banans, J.; Lecko, M. Fixed points of the product of operators in Banach algebra. *Panam. Math. J.* **2002**, *12*, 101–109.
14. Dhage, B.C. On a fixed point theorem in Banach algebras with applications. *Appl. Math. Lett.* **2005**, *18*, 273–280. [CrossRef]
15. Dhage, B.C. On some variants of Schauder's fixed point principle and applications to nonlinear integral equations. *J. Math. Sci.* **1988**, *22*, 603–611.
16. Ben Amar, A.; Derbel, S.; O'Regan, D.; Xiang, T. Fixed point theory for countably weakly condensing maps and multimaps in non-separable Banach spaces. *J. Fixed Point Theory Appl.* **2019**, *21*, 8. [CrossRef]
17. Ben Amar, A., Xu, S.: Measures of weak noncompactness and fixed point theory for 1-set weakly contractive operators on unbounded domains. *Anal. Theory Appl.* **2019**, *27*, 224–238.
18. De Blasi, F.S. On a property of the unit sphere in Banach spaces. *Bull. Math. Soc. Sci. Math. Roum.* **1977**, *21*, 259–262.
19. Banaś, J.; Rivero, J. On measures of weak noncompactness. *Ann. Mat. Pura Appl.* **1988**, *151*, 213–224. [CrossRef]
20. O'Regan, D. Operator equations in Banach spaces relative to the weak topology. *Arch. Math.* **1998**, *71*, 123–136. [CrossRef]
21. Agarwal, R.P.; O'Regan, D.; Taoudi, M.A. Browder-Krasnoselskii-type fixed point theorems in Banach spaces. *Fixed Point Theory Appl.* **2010**, *2010*, 243716. [CrossRef]
22. Rudin, W. *Functional Analysis*, 2nd ed.; McGraw-Hill: New York, NY, USA, 1991.
23. Jeribi, A.; Krichen, B.; Mefteh, B. Fixed point theory in WC–Banach algebras. *Turk. J. Math.* **2016**, *40*, 283–291. [CrossRef]

24. Boyd, D.W.; Wong, J.S.W. On nonlinear contractions. *Proc. Am. Math. Soc.* **1969**, *20*, 458–464. [CrossRef]
25. Barbu, V. *Nonlinear Semigroups and Differential Equations in Banach Spaces*; Noordhoff International Publishing: Leyden, IL, USA, 1976.
26. Garcia-Falset, J.; Morales, C.H. Existence theorems for m-accretive operators in Banach spaces. *J. Math. Anal. Appl.* **2005**, *309*, 453–461. [CrossRef]
27. Edwards, R.E. Theory and Applications, Functional Analysis; *Holt, Rinehart and Winston*: New York, NY, USA, 1965.
28. Chlebowicz, A.; Taoudi, M.A. Measures of Weak Noncompactness and Fixed Points. In *Advances in Nonlinear Analysis via the Concept of Measure of Noncompactness*; Banaś, J., Jleli, M., Mursaleen, M., Samet, B., Vetro, C., Eds.; Springer: Singapore, 2017.
29. Smart, D.R. *Fixed Point Theorems*; Cambridge University Press: Cambridge, UK, 1980.
30. Khchine, A.; Maniar, L.; Taoudi, M.A. Leray-Schauder-type fixed point theorems in Banach algebras and application to quadratic integral equations. *Fixed Point Theory Appl.* **2016**, *2016*, 88. [CrossRef]
31. Dobrakov, I. On representation of linear operators on $C_0(T, X)$. *Czechoslov. Math. J.* **1971**, *21*, 13–30.
32. Cichon, M.; Kubiaczyk, I.; Sikorska, A. The Henstock-Kurzweil-Pettis integrals and existence theorems for the Cauchy problem. *Czechoslov. Math. J.* **2004**, *54*, 279–289. [CrossRef]
33. Zeidler, E. *Nonlinear Functional Analysis and Its Applications*; Springer: New York, NY, USA, 1986; Volume I.

© 2019 by the authors. Licensee MDPI, Basel, Switzerland. This article is an open access article distributed under the terms and conditions of the Creative Commons Attribution (CC BY) license (http://creativecommons.org/licenses/by/4.0/).

Article

Existence of Positive Solution for the Eighth-Order Boundary Value Problem Using Classical Version of Leray–Schauder Alternative Fixed Point Theorem

Thenmozhi Shanmugam [1,*], Marudai Muthiah [1] and Stojan Radenović [2,3,*]

1. Department of Mathematics, Bharathidasan University, Tiruchirappalli 620 024, Tamilnadu, India; mmarudai@yahoo.co.in
2. Nonlinear Analysis Research Group, Ton Duc Thang University, Ho Chi Minh City 758 307, Vietnam
3. Faculty of Mathematics and Statistics, Ton Duc Thang University, Ho Chi Minh City 758 307, Vietnam
* Correspondence: sthenu85@gmail.com (T.S.); stojan.radenovic@tdtu.edu.vn (S.R.)

Received: 13 September 2019; Accepted: 20 October 2019; Published: 14 November 2019

Abstract: In this work, we investigate the existence of solutions for the particular type of the eighth-order boundary value problem. We prove our results using classical version of Leray–Schauder nonlinear alternative fixed point theorem. Also we produce a few examples to illustrate our results.

Keywords: eighth-order boundary value problem; Green's function; Leray–Schauder nonlinear alternative; nontrivial solution; fixed points

PACS: 34B10; 34B15; 34K10

1. Introduction

Eighth-order differential equations govern the physics of some hydrodynamic stability problems. Chandrasekhar [1] proved that when an infinite horizontal layer of fluid is heated from below and under the action of rotation, instability sets in. When the instability sets in as overstability, the problem is modeled by an eighth-order ordinary differential equation for which the existence and uniqueness of the solution can be found in the book [2]. Many authors used different numerical methods to study higher order boundary value problems. For example, Reddy [3] presented a finite element method involving the Petrov–Galerkin method with quintic B-splines as basis functions and septic B-splines as weight functions to solve a general eighth-order boundary value problem with a particular case of boundary conditions. Prorshouhi et al. [4] presented a variational iteration method for the solution of a special case of eighth- order boundary value problems. Ballem and Kasi Viswanadham [5] presented a simple finite element method which involves the Galerkin approach with septic B-splines as basis functions to solve the eighth- order two-point boundary value problems. Graef et al. [6] applied the Guo–Krasnosel'skii fixed point theorem to solve the higher-order nonlinear boundary value problem. Graef et al. [7] used various fixed point theorems to give some existence results for a nonlinear nth-order boundary value problem with nonlocal conditions. Hussin and Mandangan [8] solved linear and nonlinear eighth-order boundary value problems using a differential transformation method. Kasi Viswanadham and Ballem [9] presented a finite element method involving the Galerkin method with quintic B-splines as basis functions to solve a general eighth-order two-point boundary value problem. Liu et al. [10] used the Leggett–Williams fixed point theorem to establish existence results for solutions to the m-point boundary value problem for a second- order differential equation under multipoint boundary conditions. Napoli and Abd-Elhameed [11] analyzed a numerical algorithm for the solution of eighth-order boundary value problems. Noor and Mohyud-Din [12] implemented a relatively new analytical technique—the variational iteration decomposition method for solving the eighth-order boundary value problems. Xiaoyong and Fengying [13] used the collocation method based on the second kind Chebyshev wavelets to find the numerical solutions for the eighth-order initial and boundary value problems. Some basic fixed point theorems on altering distance functions and on G-metric spaces were discussed in [14], and also some fixed point results in cone metric spaces were collectively given in [15]. Metric fixed point theory and metrical fixed point theory results were discussed in [16,17]. Deng et al. [18] generalized some results using measure of noncompactness. Omid et al. [19] studied differential equations with the conformable derivatives. Todorčević [20] presented harmonic quasiconformal mappings and hyperbolic type metrics defined on planar and multidimensional domains. Recently Zouaoui Bekri [21] studied sixth-order nonlinear boundary value problem using the Leray–Schauder alternative theorem. Ma [22] has given the existence and uniqueness theorems based on the Leray–Schauder fixed point theorem for some fourth-order nonlinear boundary value problems. Zvyagin and Baranovskii [23] have constructed a topological characteristic to investigate a class of controllable systems. Ahmad and Ntouyas [24] conferred some existence results based on some standard fixed point theorems and Leray–Schauder degree theory for an nth-order nonlinear differential equation with four-point nonlocal integral boundary conditions. Motivated by these study, we investigate the existence of solutions for the eighth-order boundary value problem.

$$\begin{cases} y^{(8)}(x) = \phi(x, y(x), y''(x)), & 0 < x < 1, \\ y(0) = y'(0) = y''(0) = y'''(0) = y^{(4)}(1) = y^{(5)}(1) = y^{(6)}(1) = y^{(7)}(1) = 0, \end{cases} \quad (1)$$

where $\phi \in C([0,1] \times \mathbb{R} \times \mathbb{R}, \mathbb{R})$ and $\mathbb{R} = (-\infty, \infty)$.

2. Preliminaries

We consider the following eighth-order boundary value problem under the assumption that $\phi \in C([0,1] \times \mathbb{R} \times \mathbb{R}, \mathbb{R})$. $E = C([0,1])$ with the norm

$$\|y\| = \max\{|y|_\infty, |y''|_\infty\} \text{ where } |y|_\infty = \max_{0 \leq x \leq 1} |y(x)| \text{ for any } y \in E.$$

The following Lemma is used to prove our main theorem.

Lemma 1. *(By Lemma 1 in [25]) Let $f \in C[0,1]$. Then the following eighth-order boundary value problem*

$$\begin{cases} y^{(8)}(x) = f(x), & 0 < x < 1 \\ y(0) = y'(0) = y''(0) = y'''(0) = y^{(4)}(1) = y^{(5)}(1) = y^{(6)}(1) = y^{(7)}(1) = 0, \end{cases} \quad (2)$$

has the integral formulation

$$y(x) = \int_0^1 G(x,s) f(s) ds$$

where $G : [0,1] \times [0,1] \longrightarrow [0, \infty)$ is the Green's function given by

$$G(x,s) = \frac{1}{5040} \begin{cases} x^4[(s-x)^3 + 4s(s-x)^2 + 10s^2(3s-x)], & 0 \leq x < s \leq 1, \\ s^4[(x-s)^3 + 4x(x-s)^2 + 10x^2(3x-s)], & 0 \leq s < x \leq 1. \end{cases} \quad (3)$$

Proof. Consider $y^{(8)}(x) = 0$ for $0 \leq x \leq 1$. Then,

$$y(x) = A + Bx + Cx^2 + Dx^3 + Ex^4 + Fx^5 + Gx^6 + Hx^7,$$

so that the Green's function is of the form

$$G(x,s) = \frac{1}{5040} \begin{cases} \alpha_1 + \alpha_2 x + \alpha_3 x^2 + \alpha_4 x^3 + \alpha_5 x^4 + \alpha_6 t^5 + \alpha_7 t^6 + \alpha_8 t^7, & 0 \leq x < s \leq 1, \\ \beta_1 + \beta_2(1-x) + \beta_3(1-x)^2 + \beta_4(1-x)^3 + \beta_5(1-x)^4 \\ + \beta_6(1-x)^5 + \beta_7(1-x)^6 + \beta_8(1-x)^7, & 0 \leq s < x \leq 1. \end{cases} \quad (4)$$

where α_i and β_i are continuous functions for $i = 1, \ldots, 8$.

From the boundary conditions we have,

$$G(0,s) = \frac{\partial G(0,s)}{\partial x} = \frac{\partial^2 G(0,s)}{\partial x^2} = \frac{\partial^3 G(0,s)}{\partial x^3} = 0$$

i.e.,

$$\alpha_1 = \alpha_2 = \alpha_3 = \alpha_4 = 0$$

and

$$\frac{\partial^4 G(1,s)}{\partial x^4} = \frac{\partial^5 G(1,s)}{\partial x^5} = \frac{\partial^6 G(1,s)}{\partial x^6} = \frac{\partial^7 G(1,s)}{\partial x^7} = 0$$

i.e.,

$$\beta_5 = \beta_6 = \beta_7 = \beta_8 = 0.$$

We deduce the Green's function for the problem is,

$$G(x,s) = \frac{1}{5040} \begin{cases} \alpha_5 x^4 + \alpha_6 x^5 + \alpha_7 x^6 + \alpha_8 x^7, & 0 \leq x < s \leq 1, \\ \beta_1 + \beta_2(1-x) + \beta_3(1-x)^2 + \beta_4(1-x)^3, & 0 \leq s < x \leq 1. \end{cases} \quad (5)$$

Since G satisfied continuity conditions up to the sixth-order and jump discontinuity at the seventh-order by -1, we get,

$$\begin{cases} \beta_1 + \beta_2(1-s) + \beta_3(1-s)^2 + \beta_4(1-s)^3 - \alpha_5 s^4 - \alpha_6 s^5 - \alpha_7 s^6 - \alpha_8 s^7 = 0, \\ -\beta_2 - 2\beta(1-s) - 3\beta(1-s)^2 - 4\alpha_5 s^3 - 5\alpha_6 s^4 - 6\alpha_7 s^5 - 7\alpha_8 s^6 = 0, \\ 2\beta_3 + 6\beta_4(1-s) - 12\alpha_5 s^2 - 20\alpha_6 s^3 - 30\alpha_7 s^4 - 42\alpha_8 s^5 = 0, \\ -6\beta_4 - 24\alpha_5 s - 60\alpha_6 s^2 - 120\alpha_7 s^3 - 210\alpha_8 s^4 = 0, \\ -24\alpha_5 - 120\alpha_6 s - 360\alpha_7 s^2 - 840\alpha_8 s^3 = 0, \\ -120\alpha_6 - 720\alpha_7 s - 2520\alpha_8 s^2 = 0, \\ -720\alpha_7 - 5040\alpha_8 s = 0, \\ -5040\alpha_8 = 1. \end{cases} \quad (6)$$

By solving the above system, we can find the coefficients $\beta_1, \beta_2, \beta_3, \beta_4, \alpha_5, \alpha_6, \alpha_7, \alpha_8$,

i.e., $\beta_1 = -\frac{s^7}{5040} + \frac{s^6}{720} - \frac{s^5}{240} + \frac{s^4}{144}$, $\beta_2 = -\frac{s^6}{720} + \frac{s^5}{120} - \frac{s^4}{48}$, $\beta_3 = -\frac{s^5}{240} + \frac{s^4}{48}$, $\beta_4 = -\frac{s^4}{144}$,

$\alpha_5 = \frac{s^3}{144}$, $\alpha_6 = -\frac{s^2}{240}$, $\alpha_7 = \frac{s}{720}$, $\alpha_8 = -\frac{1}{5040}$.

And finally, substituting these coefficients in Equation (5) we arrive to the expression of a Green's function

$$G(x,s) = \frac{1}{5040} \begin{cases} x^4[(s-x)^3 + 4s(s-x)^2 + 10s^2(3s-x)], & 0 \le x < s \le 1, \\ s^4[(x-s)^3 + 4x(x-s)^2 + 10x^2(3x-s)], & 0 \le s < x \le 1. \end{cases} \quad (7)$$

□

Lemma 2. *For all $(x,s) \in [0,1] \times [0,1]$, we have*

$$0 \le G(x,s) \le G(s,s).$$

Proof. The proof is obvious, so we leave it. □

Define the integral operator $T : E \longrightarrow E$ by

$$T(y(x)) = \frac{1}{5040} \int_0^x s^4[(x-s)^3 + 4x(x-s)^2 + 10x^2(3x-s)]f(s)\,ds +$$

$$\frac{1}{5040} \int_x^1 x^4[(s-x)^3 + 4s(s-x)^2 + 10s^2(3s-x)]f(s)\,ds$$

By Lemma 1, the boundary value problem (Equation (1)) has a solution iff the operator T has a fixed point in E. Hence to find the solution of a given boundary value problem, it is enough to find the fixed point for the operator T in E. Since T is compact and hence T is completely continuous.

Theorem 1. *[26,27] Let $(E, \|.\|)$ be a Banach space, $U \subset E$ be an open bounded subset such that $0 \in U$ and $T : \overline{U} \longrightarrow E$ be a completely continuous operator. Then*
(1) either T has a fixed point in $\overline{U}$, or
(2) there exist an element $x \in \partial U$ and a real number $\lambda > 1$ such that $\lambda x = T(x)$.

3. Main Results

In this section, we prove some important results which will help to prove the existence of a nontrivial solution for the eighth-order boundary value problem in Equation (1). Consider $\phi \in C([0,1] \times \mathbb{R} \times \mathbb{R}, \mathbb{R})$

Theorem 2. *Suppose that $\phi(x,0,0) \ne 0$ and there exist nonnegative functions $p, q, r \in L^1[0,1]$ such that*

$$|\phi(x,y,z)| \le p(x)|y| + q(x)|z| + r(x), \quad a.e.\ (x,y,z) \in [0,1] \times \mathbb{R} \times \mathbb{R},$$

and

$$\frac{1}{720} \int_0^1 [5s^7 + s^6 + 5s^4][p(s) + q(s)]\,ds < 1.$$

Then the boundary value problem (Equation (1)) has at least one nontrivial solution $y^* \in C([0,1])$.

Proof. Let

$$A = \frac{1}{720} \int_0^1 [5s^7 + s^6 + 5s^4][p(s) + q(s)]\,ds,$$

$$B = \frac{1}{720} \int_0^1 [5s^7 + s^6 + 5s^4]r(s)\,ds.$$

By hypothesis, we have $A < 1$. Since $\phi(x,0,0) \neq 0$, there exists an interval $[a,b] \subset [0,1]$ such that $\min_{a \leq x \leq b} |\phi(x,0,0)| > 0$ and as $r(x) \geq |\phi(x,0,0)|$ a.e. $x \in [0,1]$. Hence $B > 0$.

Let $L = B(1-A)^{-1}$ and $U = \{y \in E : \|y\| < L\}$. Assume that $y \in \partial U$ and $\lambda > 1$ are such that $Ty = \lambda y$.

Then

$$\begin{aligned}
\lambda L &= \lambda \|y\| = \|Ty\| \\
&= \max_{0 \leq x \leq 1} |(Ty)(x)| \\
&\leq \frac{1}{5040} \int_0^x s^4[(x-s)^3 + 4x(x-s)^2 + 10x^2(3x-s)]|\phi(s,y(s),y''(s))|\,ds \\
&\quad + \frac{1}{5040} \int_x^1 x^4[(s-x)^3 + 4s(s-x)^2 + 10s^2(3s-x)]|\phi(s,y(s),y''(s))|\,ds \\
&\leq \frac{1}{5040} \max_{0 \leq x \leq 1} \int_0^x s^4[(x-s)^3 + 4x(x-s)^2 + 10x^2(3x-s)]|\phi(s,y(s),y''(s))|\,ds \\
&\quad + \frac{1}{5040} \max_{0 \leq x \leq 1} \int_x^1 x^4[(s-x)^3 + 4s(s-x)^2 + 10s^2(3s-x)]|\phi(s,y(s),y''(s))|\,ds \\
&= \frac{1}{5040} \int_0^1 s^4[(1-s)^3 + 4(1-s)^2 + 10(3-s)]|\phi(s,y(s),y''(s))|\,ds \\
&\quad + \frac{1}{5040} \int_0^1 s^4[s^3 + 4s(s)^2 + 10s^2(3s)]|\phi(s,y(s),y''(s))|\,ds \\
&= \frac{1}{5040} \int_0^1 [34s^7 + 7s^6 - 21s^5 + 35s^4]|\phi(s,y(s),y''(s))|\,ds \\
&\leq \frac{1}{5040} \int_0^1 [35s^7 + 7s^6 + 35s^4]|\phi(s,y(s),y''(s))|\,ds \\
&\leq \frac{1}{720} \int_0^1 [5s^7 + s^6 + 5s^4][p(s)|y(s)| + q(s)|y''(s)| + r(s)]\,ds \\
&\leq \frac{1}{720} \int_0^1 [5s^7 + s^6 + 5s^4][p(s)\max_{0 \leq s \leq 1}|y(s)| + q(s)\max_{0 \leq s \leq 1}|y''(s)| + r(s)]\,ds \\
&\leq \frac{1}{720} \int_0^1 [5s^7 + s^6 + 5s^4][p(s)|y|_\infty + q(s)|y''|_\infty + r(s)]\,ds
\end{aligned}$$

$$\leq \frac{1}{720} \int_0^1 [5s^7 + s^6 + 5s^4][p(s)\|y\| + q(s)\|y\| + r(s)]\, ds$$

$$= \frac{1}{720} \int_0^1 [5s^7 + s^6 + 5s^4][p(s) + q(s)]\|y\| + \frac{1}{720} \int_0^1 [5s^7 + s^6 + 5s^4] r(s)\, ds$$

$$= A\|y\| + B = AL + B.$$

Hence, $\lambda L \leq AL + B$

$$\lambda \leq A + \frac{B}{L} = A + \frac{B}{B(1-A)^{-1}} = A + (1-A) = 1,$$

which is a contradiction, since $\lambda > 1$, hence by Theorem 1, T has a fixed point $y^* \in \overline{U}$. Since $\phi(x,0,0) \neq 0$, the boundary value problem (Equation (1)) has a nontrivial solution $y^* \in E$. □

Theorem 3. *Let $\phi(x,0,0) \neq 0$ and there exist nonnegative functions p, q, $r \in L^1[0,1]$ such that*

$$|\phi(x,y,z)| \leq p(x)|y| + q(x)|z| + r(x) \quad \text{a.e. } (x,y,z) \in [0,1] \times \mathbb{R} \times \mathbb{R}.$$

Assume that one of the conditions given below is satisfied
(1) There exists a constant $k > -5$ such that

$$p(s) + q(s) \leq \frac{720(8+k)(7+k)(5+k)}{11k^2 + 148k + 495} s^k, \quad \text{a.e. } 0 \leq s \leq 1,$$

$$\mu\left\{ s \in [0,1] : p(s) + q(s) < \frac{720(8+k)(7+k)(5+k)}{11k^2 + 148k + 495} s^k \right\} > 0$$

where μ = measure.
(2) There exists a constant $k > -1$ such that

$$p(s) + q(s) \leq \frac{6 \prod_{i=1}^{8}(k+i)}{k^3 + 21k^2 + 152k + 594}(1-s)^k, \quad \text{a.e. } 0 \leq s \leq 1,$$

$$\mu\left\{ s \in [0,1] : p(s) + q(s) < \frac{6 \prod_{i=1}^{8}(k+i)}{k^3 + 21k^2 + 152k + 594}(1-s)^k \right\} > 0$$

where μ = measure.
(3) There exists a constant $a > 1$ such that

$$\int_0^1 [p(s) + q(s)]^a\, ds < \left[\frac{1}{\frac{1}{144}\left(\frac{1}{7b+1}\right)^{\frac{1}{b}} + \frac{1}{720}\left(\frac{1}{6b+1}\right)^{\frac{1}{b}} + \frac{1}{144}\left(\frac{1}{4b+1}\right)^{\frac{1}{b}}} \right]^a, \quad \left(\frac{1}{a} + \frac{1}{b} = 1\right).$$

Then the boundary value problem (1) has at least one nontrivial solution $y^* \in E$.

Proof. To prove this theorem it is enough to prove $A < 1$.
Let
$$A = \frac{1}{720} \int_0^1 [5s^7 + s^6 + 5s^4][p(s) + q(s)] \, ds$$

(1) Consider,

$$\begin{aligned}
A &= \frac{1}{720} \int_0^1 [5s^7 + s^6 + 5s^4][p(s) + q(s)] \, ds \\
&< \frac{720(8+k)(7+k)(5+k)}{11k^2 + 148k + 495} \left[\frac{1}{720} \int_0^1 (5s^7 + s^6 + 5s^4) s^k \, ds \right] \\
&= \frac{720(8+k)(7+k)(5+k)}{11k^2 + 148k + 495} \left[\frac{1}{720} \int_0^1 (5s^{7+k} + s^{6+k} + 5s^{4+k}) \, ds \right] \\
&= \frac{720(8+k)(7+k)(5+k)}{11k^2 + 148k + 495} \left[\frac{1}{720} \left(\frac{5}{8+k} + \frac{1}{7+k} + \frac{5}{5+k} \right) \right] \\
&= \frac{720(8+k)(7+k)(5+k)}{11k^2 + 148k + 495} \left[\frac{11k^2 + 148k + 495}{720(8+k)(7+k)(5+k)} \right]
\end{aligned}$$

Thus, $A < 1$.

(2) In this case, we have

$$\begin{aligned}
A &= \frac{1}{720} \int_0^1 [5s^7 + s^6 + 5s^4][p(s) + q(s)] \, ds \\
&< \frac{6 \prod_{i=1}^8 (k+i)}{k^3 + 21k^2 + 152k + 594} \left[\frac{1}{720} \int_0^1 (5s^7 + s^6 + 5s^4)(1-s)^k \, ds \right] \\
&< \frac{6 \prod_{i=1}^8 (k+i)}{k^3 + 21k^2 + 152k + 594} \\
&\quad \left[\frac{1}{720} \int_0^1 5s^7 (1-s)^k \, ds + \int_0^1 s^6 (1-s)^k \, ds + \int_0^1 5s^4 (1-s)^k \, ds \right] \\
&< \frac{6 \prod_{i=1}^8 (k+i)}{[k^3 + 21k^2 + 152k + 594] \, 720} \frac{1}{720} \left[\frac{120}{\prod_{i=1}^5 (k+i)} + \frac{720}{\prod_{i=1}^7 (k+i)} + \frac{720 \times 35}{\prod_{i=1}^8 (k+i)} \right] \\
&= \frac{6 \prod_{i=1}^8 (k+i)}{k^3 + 21k^2 + 152k + 594} \left[\frac{k^3 + 21k^2 + 152k + 594}{6 \prod_{i=1}^8 (k+i)} \right] = 1
\end{aligned}$$

Therefore, $A < 1$.

(3) By Hölder inequality, we have

$$A \leq \left[\int_0^1 (p(s)+q(s))^a\, ds\right]^{\frac{1}{a}} \cdot \left[\frac{1}{144}\left(\int_0^1 (s^7)^b ds\right)^{(\frac{1}{b})} + \frac{1}{720}\left(\int_0^1 (s^6)^b ds\right)^{(\frac{1}{b})} + \frac{1}{144}\left(\int_0^1 (s^4)^b ds\right)^{(\frac{1}{b})}\right]$$

$$A \leq \left[\int_0^1 (p(s)+q(s))^a\, ds\right]^{\frac{1}{a}} \cdot \left[\frac{1}{144}\left(\frac{1}{7b+1}\right)^{\frac{1}{b}} + \frac{1}{720}\left(\frac{1}{6b+1}\right)^{\frac{1}{b}} + \frac{1}{144}\left(\frac{1}{4b+1}\right)^{\frac{1}{b}}\right]$$

$$< \left(\frac{1}{\frac{1}{144}\left(\frac{1}{7b+1}\right)^{\frac{1}{b}} + \frac{1}{720}\left(\frac{1}{6b+1}\right)^{\frac{1}{b}} + \frac{1}{144}\left(\frac{1}{4b+1}\right)^{\frac{1}{b}}}\right)$$

$$\left[\frac{1}{144}\left(\frac{1}{7b+1}\right)^{\frac{1}{b}} + \frac{1}{720}\left(\frac{1}{6b+1}\right)^{\frac{1}{b}} + \frac{1}{144}\left(\frac{1}{4b+1}\right)^{\frac{1}{b}}\right]$$

$$= 1.$$

□

4. Examples

Here we have given some examples to verify the above results.

Example 1. *Consider,*

$$\begin{cases} y^{(8)}(x) = \frac{x^5}{2} y \sin \sqrt{y} + \frac{\sqrt{x}}{3} y'' \cos y'' - 5 + e^{2x}, & 0 \leq x \leq 1, \\ y(0) = y'(0) = y''(0) = y'''(0) = 0, \\ y^{(4)}(1) = y^{(5)}(1) = y^{(6)}(1) = y^{(7)}(1) = 0. \end{cases}$$

Set

$$\phi(x,y,z) = \frac{x^5}{2} y \sin \sqrt{y} + \frac{\sqrt{x}}{3} z \cos z - 5 + e^{2x},$$

$$p(x) = \frac{x^5}{2}, \quad q(x) = \frac{\sqrt{x}}{3}, \quad r(x) = 5 + e^{2x}.$$

One can easily verify that $p, q, r \in L^1[0,1]$ *are nonnegative functions, and*

$$|\phi(x,y,z)| = \left|\frac{x^5}{2} y \sin \sqrt{y} + \frac{\sqrt{x}}{3} z \cos z - 5 + e^{2x}\right|$$
$$\leq p(x)|y| + q(x)|z| + r(x), \quad a.e.\ (x,y,z) \in [0,1] \times \mathbb{R} \times \mathbb{R}.$$

Also,

$$A = \frac{1}{720} \int_0^1 [5s^7 + s^6 + 5s^4][p(s) + q(s)]\, ds$$

$$= \frac{1}{720} \int_0^1 [5s^7 + s^6 + 5s^4]\left(\frac{s^5}{2} + \frac{s^{\frac{1}{2}}}{3}\right) ds$$

$$= \frac{1}{720} \int_0^1 \left[\frac{5}{2}s^{12} + \frac{5}{3}s^{\frac{15}{2}} + \frac{s^{11}}{2} + \frac{s^{\frac{13}{2}}}{3} + \frac{5}{2}s^9 + \frac{5}{3}s^{\frac{9}{2}}\right] ds$$

$$= \frac{899251}{630115200} < 1.$$

Thus, by Theorem 2, the boundary value problem (Equation (1)) has at least one nontrivial solution $y^* \in E$.

Example 2. Consider the problem,

$$\begin{cases} y^{(8)}(x) = \frac{y^4}{(5+4y^3)\sqrt{x}}\cos y + \frac{4(y'')^3}{7\sqrt{x}} + \frac{2y''}{\sqrt{x}} - \cos\sqrt{x}, & 0 \le x \le 1, \\ y(0) = y'(0) = y''(0) = y'''(0) = 0, \\ y^{(4)}(1) = y^{(5)}(1) = y^{(6)}(1) = y^{(7)}(1) = 0. \end{cases}$$

Set

$$\phi(x,y,z) = \frac{y^4}{(5+4y^3)\sqrt{x}}\cos y + \frac{4z^3}{7\sqrt{x}} + \frac{2z}{\sqrt{x}} - \cos\sqrt{x},$$

$$p(x) = \frac{1}{5\sqrt{x}}, \quad q(x) = \frac{4}{7\sqrt{x}} + \frac{2}{\sqrt{x}}, \quad r(x) = \cos\sqrt{x}.$$

One can easily verify that $p, q, r \in L^1[0,1]$ are nonnegative functions, and

$$|\phi(x,y,z)| = \left|\frac{y^4}{(5+4y^3)\sqrt{x}}\cos y + \frac{4z^3}{7\sqrt{x}} + \frac{2z}{\sqrt{x}} - \cos\sqrt{x}\right|$$

$$\le p(x)|y| + q(x)|z| + r(x), \quad \text{a.e. } (x,y,z) \in [0,1] \times \mathbb{R} \times \mathbb{R}.$$

Let $k = -\frac{1}{2} > -5$. Then,

$$\frac{720(8+k)(7+k)(5+k)}{11k^2 + 148k + 495} = \frac{631800}{1695}$$

hence,

$$p(s) + q(s) = \frac{1}{5\sqrt{s}} + \frac{4}{7\sqrt{s}} + \frac{2}{\sqrt{s}} = \frac{97}{35}s^{-\frac{1}{2}} < \frac{631800}{1695}s^{-\frac{1}{2}}$$

$$\mu\left\{s \in [0,1] : p(s) + q(s) < \frac{720(8+k)(7+k)(5+k)}{11k^2 + 148k + 495}s^k\right\} > 0$$

where μ = measure. Thus by the Theorem 3 assumption (1), the boundary value problem (Equation (2)) has at least one nontrivial solution $y^* \in E$.

Example 3. Consider the problem,

$$\begin{cases} y^{(8)}(x) = \dfrac{y^3}{4(3+y^4)\sqrt[3]{(1-x)^2}} \sin y + \dfrac{(y'')^2}{(5+y'')\sqrt[3]{(1-x)^2}} + e^{2x} + \sin 3x, \quad 0 \leq x \leq 1, \\ y(0) = y'(0) = y''(0) = y'''(0) = 0, \\ y^{(4)}(1) = y^{(5)}(1) = y^{(6)}(1) = y^{(7)}(1) = 0. \end{cases}$$

Set

$$\phi(x,y,z) = \dfrac{y^3}{4(3+y^4)\sqrt[3]{(1-x)^2}} \sin y + \dfrac{z^2}{(5+z)\sqrt[3]{(1-x)^2}} + e^{2x} + \sin 3x$$

$$p(x) = \dfrac{1}{4\sqrt[3]{(1-x)^2}}, \quad q(x) = \dfrac{1}{5\sqrt[3]{(1-x)^2}}, \quad r(x) = e^{2x} + \sin 3x.$$

Here we can easily prove that $p, q, r \in L^1[0,1]$ are nonnegative functions, and

$$\begin{aligned} |\phi(x,y,z)| &= \left| \dfrac{y^3}{4(3+y^4)\sqrt[3]{(1-x)^2}} \sin y + \dfrac{z^2}{(5+z)\sqrt[3]{(1-x)^2}} + e^{2x} + \sin 3x \right| \\ &\leq p(x)|y| + q(x)|z| + r(x), \quad a.e. \ (x,y,z) \in [0,1] \times R \times R. \end{aligned}$$

Take $k = -\frac{2}{3} > -1$. Then

$$\dfrac{6 \prod_{i=1}^{8}(k+i)}{k^3 + 21k^2 + 152k + 594} = \dfrac{24344320}{548613}.$$

Therefore,

$$\begin{aligned} p(s) + q(s) &= \dfrac{1}{4\sqrt[3]{(1-s)^2}} + \dfrac{1}{5\sqrt[3]{(1-s)^2}} \\ &= \dfrac{9}{20}(1-s)^{-\frac{2}{3}} \\ &< \dfrac{24344320}{548613}(1-s)^{-\frac{2}{3}} \end{aligned}$$

$$\mu \left\{ s \in [0,1] : p(s) + q(s) < \dfrac{6 \prod_{i=1}^{8}(k+i)}{k^3 + 21k^2 + 152k + 594}(1-s)^{-\frac{2}{3}} \right\} > 0$$

where μ = measure. Therefore, by Theorem 3 assumption (2), the boundary value problem (Equation (3)) has at least one nontrivial solution $y^* \in E$.

Example 4. Consider the problem,

$$\begin{cases} y^{(8)}(x) = \dfrac{\sqrt[4]{2+x}}{1+y^2} y e^{\sin x} + \dfrac{3\sqrt[4]{2+x}}{(5+(y'')^2)} \cos y'' + e^{-x} \cos x - \sin 2x, \quad 0 \leq x \leq 1, \\ y(0) = y'(0) = y''(0) = y'''(0) = 0, \\ y^{(4)}(1) = y^{(5)}(1) = y^{(6)}(1) = y^{(7)}(1) = 0. \end{cases}$$

Set
$$\phi(x,y,z) = \frac{\sqrt[4]{2+x}}{1+y^2}ye^{\sin x} + \frac{3\sqrt[4]{2+x}}{(5+z^2)}\cos z + e^{-x}\cos x - \sin 2x$$
$$p(x) = \sqrt[4]{2+x}, \quad q(x) = 3\sqrt[4]{2+x}, \quad r(x) = e^{-x}\cos x + \sin 2x.$$

Here we can easily prove that $p, q, r \in L^1[0,1]$ are nonnegative functions, and

$$|\phi(x,y,z)| = \left|\frac{\sqrt[4]{2+x}}{1+y^2}ye^{\sin x} + \frac{3\sqrt[4]{2+x}}{(5+z^2)}\cos z + e^{-x}\cos x - \sin 2x\right|$$
$$\leq p(x)|y| + q(x)|z| + r(x), \quad a.e. \ (x,y,z) \in [0,1] \times \mathbb{R} \times \mathbb{R}.$$

Let $a = 4 > b = \frac{4}{3} > 1$. We have that $\frac{1}{a} + \frac{1}{b} = 1$. Then

$$\int_0^1 (p(s)+q(s))^a \, ds = \int_0^1 \left[4\sqrt[4]{2+s}\right]^4 ds = 640.$$

Also, we have

$$\left[\frac{1}{\frac{1}{144}\left(\frac{1}{7b+1}\right)^{\frac{1}{b}} + \frac{1}{720}\left(\frac{1}{6b+1}\right)^{\frac{1}{b}} + \frac{1}{144}\left(\frac{1}{4b+1}\right)^{\frac{1}{b}}}\right]^a = \left[\frac{1}{\frac{1}{144}\left(\frac{3}{31}\right)^{\frac{3}{4}} + \frac{1}{720}\left(\frac{1}{9}\right)^{\frac{3}{4}} + \frac{1}{144}\left(\frac{3}{19}\right)^{\frac{3}{4}}}\right]^4$$
$$\approx 9406732117.3529.$$

Therefore,

$$\int_0^1 (p(s)+q(s))^a \, ds < 9406732117.3529$$

Further, by Theorem 3 assumption (3), the boundary value problem (Equation (4)) has at least one nontrivial solution $y^* \in E$.

5. Conclusions

In this paper, we obtain the results to prove the existence of positive solution for the eighth-order boundary value problem with the help of the classical version of Leray–Schauder alternative fixed point theorem. By applying these results, one can easily verify that whether the given boundary value problem is solvable or not.

Author Contributions: Conceptualization, T.S.; methodology, T.S.; review and editing, M.M. and S.R.

Funding: This research received no external funding.

Acknowledgments: The author would like to express their gratitude to the editor and anonymous referees for their valuable comments and suggestions which improved the original version of the manuscript in the present form.

Conflicts of Interest: The authors declare no conflict of interest.

References

1. Chandrasekhar, S. *Hydrodynamic and Hydromagnetic Stability*; Dover: New York, NY, USA, 1961.
2. Agarwal, R.P. *Boundary Value Problems for Higher Order Differential Equations*; MATSCIENCE: 98; Institute of Mathematical Sciences: Madras, India, July 1979.
3. Reddy, S.M. Numerical Solution of Eighth Order Boundary Value Problems by Petrov-Galerkin Method with Quintic B-splines as basic functions and Septic B-Splines as weight functions. *Int. J. Eng. Comput. Sci.* **2016**, *5*, 17902–17908.
4. Porshokouhi, M.G.; Ghanbari, B.; Gholami, M.; Rashidi, M. Numerical Solution of Eighth Order Boundary Value Problems with Variational Iteration Method. *Gen. Math. Notes* **2011**, *2*, 128–133.
5. Ballem, S.; Kasi Viswanadham, K.N.S. Numerical Solution of Eighth Order Boundary Value Problems by Galerkin Method with Septic B-splines. *Procedia Eng.* **2015**, *127*, 1370–1377. [CrossRef]
6. Graef, J.R.; Henderson, J.; Yang, B. Positive Solutions for a Nonlinear Higher-Order Boundary Value Problems. *Electron. J. Diff. Equ.* **2007**, *45*, 1–10.
7. Graef, J.R.; Moussaoui, T. A classof nth-order BVPs with nonlocal conditions. *Comput. Math. Appl.* **2009**, *58*, 1662–1671. [CrossRef]
8. Hussin, C.H.C.; Mandangan, A. Solving Eighth-order Boundary Value Problems Using Differential Transformation Method. *AIP Conf. Proc.* **2014**, *1635*, 99–106.
9. Kasi Viswanadham, K.N.S.; Ballem, S. Numerical Solution of Eighth Order Boundary Value Problems by Galerkin Method with Quintic B-splines. *Int. J. Comput. Appl.* **2014**, *89*, 7–13.
10. Liu, X.; Jiang, W.; Guo, Y. Multi-Point Boundary Value Problems For Higher Order Differential Equations. *Appl. Math. E-Notes* **2004**, *4*, 106–113.
11. Napoli, A.; Abd-Elhameed, W.M. Numerical Solution of Eighth-Order Boundary Value Problems by Using Legendre Polynomials. *Int. J. Comput. Methods* **2017**, *14*, 19. [CrossRef]
12. Noor, M.A.; Mohyud-Din, S.T. Variational Iteration Decomposition Method for Solving Eighth-Order Boundary Value Problems. *Diff. Equ. Nonlinear Mech.* **2008**. [CrossRef]
13. Xu, X.; Zhou, F. Numerical Solutions for the Eighth-Order Initial and Boundary Value Problems Using the Second Kind Chebyshev Wavelets. *Adv. Math. Phys.* **2015**, *2015*, 964623. [CrossRef]
14. Agarwal, R.P.; Karapınar, E.; O'Regan, D.; Roldan-Lopez-de-Hierro, A.F. *Fixed Point Theory in Metric Type Spaces*; Springer International Publishing: Cham, Switzerland, 2015.
15. Aleksić, S.; Kadelburg, Z.; Mitrović, Z.D.; Radenović, S. A new survey: Cone metric spaces. *J. Int. Math. Virtual Inst.* **2019**, *9*, 93–121.
16. Ćirić, L. *Some Recent Results in Metrical Fixed Point Theory*; University of Belgrade: Belgrade, Serbia, 2003.
17. Kirk, W.; Shahzad, N. *Fixed Point Theory in Distance Spaces*; Springer International Publishing: Cham, Switzerland, 2014.
18. Deng, G.; Huang, H.; Cvetković, M.; Radenović, S. Cone valued measure of noncompactness and related fixed point theorems. *Bull. Int. Math. Virtual Inst.* **2018**, *8*, 233–243.
19. Birgani, O.T.; Chandok, S.; Dedović, N.; Radenović, S. A note on some recent results of the conformable derivative. *Adv. Theory Nonlinear Anal. Its Appl.* **2019**, *3*, 11–17.
20. Todorčević, V. *Harmonic Quasiconformal Mappings and Hyperbolic Type Metrics*; Springer Nature: Cham, Switzerland, 2019.
21. Bekri, Z.; Benaicha, S. Nontrivial solution of a nonlinear sixth-order boundary value problem. *Waves Wavelets Fract.* **2018**, *4*, 10–18. [CrossRef]
22. Ma, R. Existence and uniqueness theorems for some fourth-order nonlinear boundary value problems. *Int. J. Math. Math. Sci.* **2000**, *23*, 783–788. [CrossRef]
23. Zvyagin, V.G.; Baranovskii, E.S. Topological degree of condensing multi-valued perturbations of the (S) + -class maps and its applications. *J. Math. Sci.* **2010**, *170*, 405–421. [CrossRef]
24. Ahmad, B.; Ntouyas, S.K. A study of higher-order nonlinear ordinary differential equations with four-point nonlocal integral boundary conditions. *J. Appl. Math. Comput.* **2012**, *39*, 97–108. [CrossRef]

25. Cabrera, B. López, I.J.; Sadarangani, K.B. Existence of positive solutions for the nonlinear elastic beam equation via a mixed monotone operator. *J. Comput. Appl. Math.* **2018**, *327*, 306–313. [CrossRef]
26. Isac, G. *Leray–Schauder Type Alternatives, Complementarity Problems and Variational Inequalities*; Springer: New York, NY, USA, 2006.
27. Klaus, D. *Nonlinear Functional Analysis*; Springer: Berlin, Germany, 1985.

© 2019 by the authors. Licensee MDPI, Basel, Switzerland. This article is an open access article distributed under the terms and conditions of the Creative Commons Attribution (CC BY) license (http://creativecommons.org/licenses/by/4.0/).

Article

Informal Complete Metric Space and Fixed Point Theorems

Hsien-Chung Wu

Department of Mathematics, National Kaohsiung Normal University, Kaohsiung 802, Taiwan; hcwu@nknucc.nknu.edu.tw

Received: 19 October 2019; Accepted: 2 November 2019; Published: 7 November 2019

Abstract: The concept of informal vector space is introduced in this paper. In informal vector space, the additive inverse element does not necessarily exist. The reason is that an element in informal vector space which subtracts itself cannot be a zero element. An informal vector space can also be endowed with a metric to define a so-called informal metric space. The completeness of informal metric space can be defined according to the similar concept of a Cauchy sequence. A new concept of fixed point and the related results are studied in informal complete metric space.

Keywords: Cauchy sequence; near fixed point; informal metric space; informal vector space; null set

MSC: 47H10; 54H25

---

1. Introduction

The basic operations in (conventional) vector space are vector addition and scalar multiplication. Based on these two operations, the vector space should satisfy some required conditions (eight axioms in total) by referring to [1–5]. However, some spaces cannot comply with all of the axioms given in vector space. For example, the space consisting of all subsets of $\mathbb{R}$ cannot satisfy all of the axioms in vector space (Wu [6]). Also, the space consisting of all fuzzy numbers in $\mathbb{R}$ cannot satisfy all of the axioms in vector space, where the addition and scalar multiplication of fuzzy sets are considered (Wu [7]). The main reason is that the additive inverse element does not exist.

Let S and T be two subsets of $\mathbb{R}$. The addition and scalar multiplication for the subsets of $\mathbb{R}$ are defined by

$$S + T = \{s + t : s \in S \text{ and } t \in T\} \text{ and } kS = \{ks : s \in S\} \text{ for any } k \in \mathbb{R}.$$

Let $\mathcal{X}$ denote the family of all subsets of $\mathbb{R}$. Given any $S \in \mathcal{X}$, the subtraction $S - S$ by itself is given by

$$S - S = \{s_1 - s_2 : s_1, s_2 \in S\},$$

which cannot be the zero element in $\mathcal{X}$. Therefore, in this paper, we propose the concept of null set for the purpose of playing the role of a zero element in the so-called informal vector space. Since the informal metric space is a completely new concept, there are no available, relevant references for this topic. The readers may instead refer to the monographs [1–5] on topological vector spaces and the monographs [8–10] on functional analysis.

In this paper, we propose the concept of informal vector space that can include the space consisting of all bounded and closed intervals in $\mathbb{R}$ and the space consisting of all fuzzy numbers in $\mathbb{R}$. We also introduce the concept of null set that can be regarded as a kind of "zero element" of informal vector space. When the null set is degenerated as a singleton set $\{\theta\}$, an informal vector space will turn into a conventional vector space with the zero element θ. In other words, the results obtained in this paper can be reduced to the results in conventional vector space when the null set is taken to be a singleton set.

Based on the concept of null set, we can define the concept of almost identical elements in informal vector space. We can also endow a metric to the informal vector space defining the so-called informal metric space. This kind of metric is completely different from the conventional metric defined in vector space, since it involves the null set and almost identical concept. The most important triangle inequality is still included in an informal metric space. Based on this metric, the concepts of limit and class limit of a sequence in informal metric space are defined herein. Under this setting, we can similarly define the concept of a Cauchy sequence, which can be used to define the completeness of informal metric space. The main aim of this paper was to establish the so-called near-fixed point in informal, complete metric space, where the near fixed point is based on the almost identical concept. We shall also claim that if the null set is degenerated as a singleton set, then the concept of near a fixed point is identical to the concept of a (conventional) fixed point.

In Sections 2 and 3, the concept of informal vector space and informal metric space are proposed. The interesting properties are derived in order to study the new type of fixed point theorems. In Section 4, according to the informal metric, the concept of a Cauchy sequence is similarly defined. The completeness of informal metric space is also defined according to the concept of Cauchy sequences. In Section 5, we present many new types of fixed point theorems that are established using the almost identical concept in informal metric space.

2. Informal Vector Spaces

Let X be a universal set, and let $\mathbb{F}$ be a scalar field. We assume that X is endowed with the vector addition $x \oplus y$ and scalar multiplication αx for any $x, y \in X$ and $\alpha \in \mathbb{F}$. In this case, we call X a universal set over $\mathbb{F}$. In the conventional vector space over $\mathbb{F}$, the additive inverse element of x is denoted by $-x$, and it can also be shown that $-x = -1x$. In this paper, we shall not consider the concept of inverse elements. However, for convenience, we still adopt $-x = -1x$.

For $x, y \in X$, the *substraction* $x \ominus y$ is defined by $x \ominus y = x \oplus (-y)$, where $-y$ means the scalar multiplication $(-1)y$. For any $x \in X$ and $\alpha \in \mathbb{F}$, we have to mention that $(-\alpha)x \neq -\alpha x$ and $\alpha(-x) \neq -\alpha x$ in general, unless $\alpha(\beta x) = (\alpha\beta)x$ for any $\alpha, \beta \in \mathbb{F}$. In this paper, this law will not always be assumed to be true.

Example 1. Let C be a subset of complex plane $\mathbb{C}$ defined by

$$C = \{a + bi : a, b \in \mathbb{R} \text{ satisfying } a \leq b\}.$$

The usual addition and scalar multiplication in $\mathbb{C}$ are defined by

$$(a + bi) + (c + di) = (a + c) + (b + d)i \text{ and } k(a + bi) = ka + kbi \text{ for } k \in \mathbb{R}.$$

Given any $z = a + bi \in C$, its additive inverse in $\mathbb{C}$ denoted by $-z$ is

$$-z = (-1)z = -a - bi.$$

We see that $-z \notin C$. Therefore, the subset C is not closed under the above scalar multiplication. In other words, the subset C cannot form a vector space. However, if the scalar multiplication in the subset C is defined by

$$k(a + bi) = \begin{cases} ka + kbi & \text{if } k \geq 0 \\ kb + kai & \text{if } k < 0. \end{cases}$$

then the subset C is closed under the above addition and this new scalar multiplication. In this case, we shall consider the subset C as an informal vector space that will be defined below.

Example 2. *Let $\mathcal{I}$ be the set of all closed intervals in $\mathbb{R}$. The addition is given by*

$$[a,b] \oplus [c,d] = [a+c, b+d]$$

and the scalar multiplication is given by

$$k[a,b] = \begin{cases} [ka, kb] & \text{if } k \geq 0 \\ [kb, ka] & \text{if } k < 0. \end{cases}$$

We see that $\mathcal{I}$ cannot be a (conventional) vector space, since the inverse element cannot exist for any non-degenerated closed interval. On the other hand, the distributive law for scalar addition does not hold true in $\mathcal{I}$; that is, the equality $(\alpha + \beta)x = \alpha x \oplus \beta x$ cannot hold true for any $x \in \mathcal{I}$ and $\alpha, \beta \in \mathbb{R}$. This shows another reason why $\mathcal{I}$ cannot be a (conventional) vector space.

Definition 1. *Let X be a universal set over the scalar field $\mathbb{F}$. We define the null set of X as follows*

$$\Omega = \{x \ominus x : x \in X\}.$$

We say that the null set Ω satisfies the neutral condition if and only if $\omega \in \Omega$ implies $-\omega \in \Omega$.

Example 3. *Continued from Example 1, for any $z = a + bi \in \mathbb{C}$, we have*

$$z \ominus z = z + (-1)z = (a+bi) + (-b-ai) = (a-b) + (b-a)i \in \mathbb{C}.$$

Therefore, the null set Ω is given by

$$\Omega = \{-k + ki : k \in \mathbb{R} \text{ and } k \geq 0\} = \{-k + ki : k \in \mathbb{R}_+\}.$$

Now we are in a position to define the concept of informal vector space.

Definition 2. *Let X be a universal set over $\mathbb{F}$. We say that X is an informal vector space over $\mathbb{F}$ if and only if the following conditions are satisfied:*

- $1x = x$ for any $x \in X$;
- $x = y$ implies $x \oplus z = y \oplus z$ and $\alpha x = \alpha y$ for any $x, y, z \in X$ and $\alpha \in \mathbb{F}$;
- The commutative and associative laws for vector addition hold true in X; that is, $x \oplus y = y \oplus x$ and $(x \oplus y) \oplus z = x \oplus (y \oplus z)$ for any $x, y, z \in X$.

Definition 3. *Let X be an informal vector space over $\mathbb{F}$ with the null set Ω. Given any $x, y \in X$, we say that x and y are almost identical if and only if any one of the following conditions is satisfied:*

- $x = y$;
- *There exists $\omega \in \Omega$ such that $x = y \oplus \omega$ or $x \oplus \omega = y$;*
- *There exists $\omega_1, \omega_2 \in \Omega$ such that $x \oplus \omega_1 = y \oplus \omega_2$.*

In this case, we write $x \stackrel{\Omega}{=} y$.

Remark 1. *Suppose that the informal vector space X over $\mathbb{F}$ with the null set Ω contains the zero element θ; that is, $x = x \oplus \theta = \theta \oplus x$ for any $x \in X$. Then, we can simply say that $x \stackrel{\Omega}{=} y$ if and only if $\omega_1, \omega_2 \in \Omega$ exists, such that $x \oplus \omega_1 = y \oplus \omega_2$ (i.e., only the third condition is satisfied), since the first and second conditions can be rewritten as the third condition by adding the zero element θ. We also remark that if we want to discuss some properties based on $x \stackrel{\Omega}{=} y$, it suffices to consider the third condition $x \oplus \omega_1 = y \oplus \omega_2$, even though X does not contain the zero element θ. The reason is that the same arguments are still applicable for the first and second conditions.*

According to the binary relation $\stackrel{\Omega}{=}$, for any $x \in X$, we define the class

$$[x] = \left\{ y \in X : x \stackrel{\Omega}{=} y \right\}. \tag{1}$$

The family of all classes $[x]$ for $x \in X$ is denoted by $[X]$. For $y \in [x]$, it is not necessarily that the class $[y]$ is equal to the class $[x]$, unless the binary relation $\stackrel{\Omega}{=}$ is an equivalence relation.

Proposition 1. *Let X be an informal vector space over $\mathbb{F}$ with the null set Ω. If Ω is closed under the vector addition, then the binary relation $\stackrel{\Omega}{=}$ is an equivalence relation.*

Proof. For any $x \in X$, $x = x$ implies $x \stackrel{\Omega}{=} x$, which shows the reflexivity. According to the definition of the binary relation $\stackrel{\Omega}{=}$, the symmetry is obvious. Regarding the transitivity, for $x \stackrel{\Omega}{=} y$ and $y \stackrel{\Omega}{=} z$, we want to claim $x \stackrel{\Omega}{=} z$. From Remark 1, it suffices to just consider the cases of

$$x \oplus \omega_1 = y \oplus \omega_2 \text{ and } y \oplus \omega_3 = z \oplus \omega_4$$

for some $\omega_i \in \Omega$ for $i = 1, \cdots, 4$. By the associative and commutative laws for vector addition, we have

$$x \oplus \omega_1 \oplus \omega_3 = y \oplus \omega_3 \oplus \omega_2 = z \oplus \omega_4 \oplus \omega_2,$$

which shows $x \stackrel{\Omega}{=} z$, since Ω is closed under the vector addition. This completes the proof. $\square$

Let X be an informal vector space over $\mathbb{F}$ with the null set Ω such that Ω is closed under the vector addition. Proposition 1 says that the classes defined in (1) form the equivalence classes. It is clear to see that $y \in [x]$ implies $[x] = [y]$. In other words, the family of all equivalence classes form a partition of the whole set X.

We also need to remark that the space $[X]$ is still not a (conventional) vector space. The reason is that not all of the axioms taken in the vector space will be satisfied in $[X]$, since the original space X does not satisfy all of the axioms in the vector space. For example, we consider the informal vector space $\mathcal{I}$ over $\mathbb{R}$ from Example 2. The quotient set $[\mathcal{I}]$ cannot be a real vector space, since

$$(\alpha + \beta)[x] \neq \alpha[x] + \beta[x]$$

for $\alpha\beta < 0$. The reason is that $(\alpha + \beta)x \neq \alpha x + \beta x$ for $x \in \mathcal{I}$ and $\alpha\beta < 0$.

3. Informal Metric Spaces

Now, we are in a position to introduce the concept of the so-called informal metric space.

Definition 4. *Let X be an informal vector space over $\mathbb{F}$ with the null set Ω. For the non-negative, real-valued function d defined on $X \times X$, we consider the following conditions:*

(i) $d(x, y) = 0$ *if and only if* $x \stackrel{\Omega}{=} y$ *for all* $x, y \in X$;
(i') $d(x, y) = 0$ *if and only if* $x = y$ *for all* $x, y \in X$;
(ii) $d(x, y) = d(y, x)$ *for all* $x, y \in X$;
(iii) $d(x, y) \leq d(x, z) + d(z, y)$ *for all* $x, y, z \in X$.

Different kinds of metric spaces are defined below.

- *The pair (X, d) is called a **pseudo-metric space** if and only if d satisfies conditions (ii) and (iii).*
- *The pair (X, d) is called a **metric space** if and only if d satisfies conditions (i'), (ii), and (iii).*
- *The pair (X, d) is called a **informal metric space** if and only if d satisfies conditions (i), (ii), and (iii).*

(iv) We say that d satisfies the **null super-inequality** if and only if, for any $\omega_1, \omega_2 \in \Omega$ and $x, y, z \in X$, we have

$$d(x \oplus \omega_1, y \oplus \omega_2) \geq d(x,y), \quad d(x \oplus \omega_1, y) \geq d(x,y) \text{ and } d(x, y \oplus \omega_2) \geq d(x,y).$$

(iv′) We say that d satisfies the **null sub-inequality** if and only if, for any $\omega_1, \omega_2 \in \Omega$ and $x, y \in X$, we have

$$d(x \oplus \omega_1, y \oplus \omega_2) \leq d(x,y), \quad d(x \oplus \omega_1, y) \leq d(x,y) \text{ and } d(x, y \oplus \omega_2) \leq d(x,y).$$

(iv″) We say that d satisfies the **null equality** if and only if, for any $\omega_1, \omega_2 \in \Omega$ and $x, y \in X$, we have

$$d(x \oplus \omega_1, y \oplus \omega_2) = d(x,y), \quad d(x \oplus \omega_1, y) = d(x,y) \quad \text{and} \quad d(x, y \oplus \omega_2) = d(x,y).$$

Example 4. *Continued from Example 2, we take $X = \mathcal{I}$ that consists of all bounded and closed intervals, which is not a vector space. For $I_1 = [a_1^L, a_1^U]$ and $I_2 = [a_2^L, a_2^U]$ in $\mathcal{I}$, we define a nonnegative real-valued function d for $\mathcal{I} \times \mathcal{I}$ by*

$$d(I_1, I_2) = \left| a_1^L + a_1^U - a_2^L - a_2^U \right|.$$

Suppose that

$$d(I_1, I_2) = \left| a_1^L + a_1^U - a_2^L - a_2^U \right| = 0.$$

We cannot obtain $I_1 = I_2$. Therefore, condition (i′) in Definition 4 is not satisfied, which says that $(\mathcal{I}, d)$ cannot be a (standard) metric space. However, using the basic arithmetics, we can obtain $I_1 \stackrel{\Omega}{=} I_2$. For any I_1, I_2, I_3 in $\mathcal{I}$, it is not difficult to show that

$$d(I_1, I_2) = d(I_2, I_1) \text{ and } d(I_1, I_2) \leq d(I_1, I_3) + d(I_3, I_2).$$

Therefore $(\mathcal{I}, d)$ is indeed an informal metric space. Moreover, we are going to claim that d also satisfies the null equality. We first note that the null set Ω in $\mathcal{I}$ is given by

$$\Omega = \{[-k, k] : k \geq 0\}.$$

For any $k_1, k_2 \in \mathbb{R}_+$, i.e., $\omega_1 = [-k_1, k_1], \omega_2 = [-k_2, k_2] \in \Omega$, we have

$$\begin{aligned} d(I_1 \oplus \omega_1, I_2 \oplus \omega_2) &= d\left([a_1^L, a_1^U] \oplus [-k_1, k_1], [a_2^L, a_2^U] \oplus [-k_2, k_2]\right) \\ &= d\left([a_1^L - k_1, a_1^U + k_1], [a_2^L - k_2, a_2^U + k_2]\right) \\ &= \left| (a_1^L - k_1 + a_1^U + k_1) - (a_2^L - k_2 + a_2^U + k_2) \right| \\ &= \left| (a_1^L + a_1^U) - (a_2^L + a_2^U) \right| \\ &= d\left([a_1^L, a_1^U], [a_2^L, a_2^U]\right) = d(I_1, I_2), \end{aligned}$$

which shows that d indeed satisfies the null equality.

4. Cauchy Sequences

In this section, we are going to introduce the concepts of Cauchy sequences and completeness in the informal metric space. We first introduce the concept of limit in the informal metric space.

Definition 5. *Let X be an informal vector space over $\mathbb{F}$ with the null set Ω, and let (X, d) be a pseudo-metric space. The sequence $\{x_n\}_{n=1}^{\infty}$ in X is said to be **convergent** if and only if*

$$\lim_{n \to \infty} d(x_n, x) = 0 \text{ for some } x \in X.$$

The element x is called the **limit** of the sequence $\{x_n\}_{n=1}^\infty$.

The sense of uniqueness of limit will be different for the metric space and informal metric space. Let $\{x_n\}_{n=1}^\infty$ be a sequence in (X, d). If there exists $x, y \in X$ such that

$$\lim_{n \to \infty} d(x_n, x) = 0 = \lim_{n \to \infty} d(x_n, y),$$

then, by the triangle inequality (iii) in Definition 4, we have

$$0 \leq d(x, y) \leq d(x, x_n) + d(x_n, y) \to 0 + 0 = 0 \text{ as } n \to \infty, \tag{2}$$

which says that $d(x, y) = 0$.

- Suppose that (X, d) is a metric space. By condition (i') in Definition 4, we see that $x = y$. This shows the uniqueness.
- Suppose that (X, d) is an informal metric space. By condition (i) in Definition 4, we see that $x \stackrel{\Omega}{=} y$. Recall that if Ω is closed under the vector addition, then we can consider the equivalence classes. In this case, we also see that y is in the equivalence class $[x]$.

On the other hand, we further assume that d satisfies the null equality. If $\{x_n\}_{n=1}^\infty$ is a sequence in X such that $d(x_n, x) \to 0$ as $n \to \infty$, then, for any $y \in [x]$, i.e., $x \oplus \omega_1 = y \oplus \omega_2$ for some $\omega_1, \omega_2 \in \Omega$, we also have $d(x_n, y) \to 0$ as $n \to \infty$, as shown below:

$$0 \leq d(x_n, y) = d(x_n, \omega_2 \oplus y) = d(x_n, \omega_1 \oplus x) = d(x_n, x) \to 0 \text{ as } n \to \infty.$$

Therefore, we propose the following definition.

Definition 6. Let (X, d) be an informal pseudo-metric space with the null set Ω. If $\{x_n\}_{n=1}^\infty$ is a sequence in X such that

$$\lim_{n \to \infty} d(x_n, x) = 0$$

for some $x \in X$, then the class $[x]$ is called the **class limit** of $\{x_n\}_{n=1}^\infty$. We also write

$$\lim_{n \to \infty} x_n = [x] \text{ or } x_n \to [x].$$

Proposition 2. *Let (X, d) be an informal pseudo-metric space with the null set Ω such that Ω is closed under the vector addition. Then, the class limit in the informal metric space is unique.*

Proof. Let $\{x_n\}_{n=1}^\infty$ be a convergent sequence in X with the class limits $[x]$ and $[y]$. According to the definition, we have

$$\lim_{n \to \infty} d(x_n, x) = 0 \text{ and } \lim_{n \to \infty} d(x_n, y) = 0.$$

Using (2), it follows that $d(x, y) = 0$, which also implies $y \in [x]$, i.e., $[x] = [y]$. This shows the uniqueness in the sense of class limit. □

Definition 7. Let (X, d) be an informal metric space.

- A sequence $\{x_n\}_{n=1}^\infty$ in X is called a **Cauchy** sequence if and only if, given any $\epsilon > 0$, $N \in \mathbb{N}$ exists, such that $d(x_n, x_m) < \epsilon$ for all $n > N$ and $m > N$.
- A subset M of X is said to be **complete** if and only if every Cauchy sequence in M is convergent to some element in M.

Proposition 3. *Every convergent sequence in an informal metric space is a Cauchy sequence.*

Example 5. *Continued from Example 4, we see that $(\mathcal{I}, d)$ is an informal metric space such that d satisfies the null equality. We are going to claim that $(\mathcal{I}, d)$ is complete. Given a sequence $\{I_n\}_{n=1}^{\infty}$ in the informal metric space $(\mathcal{I}, d)$ by $I_n = [a_n^L, a_n^U]$ for $n = 1, 2, \cdots$, suppose that $\{I_n\}_{n=1}^{\infty}$ is a Cauchy sequence. Then, given any $\epsilon > 0$, for sufficiently large n and m, we have*

$$\epsilon > d(I_n, I_m) = d\left([a_n^L, a_n^U], [a_m^L, a_m^U]\right) = \left|\left(a_n^L + a_n^U\right) - \left(a_m^L + a_m^U\right)\right|. \tag{3}$$

Let $a_n = a_n^L + a_n^U$. Then, the expression (3) shows that $\{a_n\}_{n=1}^{\infty}$ is a Cauchy sequence in $\mathbb{R}$. The completeness of $\mathbb{R}$ says that $a \in \mathbb{R}$ exists, satisfying $|a_n - a| < \epsilon$ for sufficiently large n. Now we define a bounded closed interval $[a^L, a^U]$ satisfying $a^L + a^U = a$. Then

$$d\left([a_n^L, a_n^U], [a^L, a^U]\right) = \left|\left(a_n^L + a_n^U\right) - \left(a^L + a^U\right)\right| = |a_n - a| < \epsilon$$

for a sufficiently large n, which says that the sequence $\{I_n\}_{n=1}^{\infty}$ converges to $[a^L, a^U]$. Therefore, we conclude that the space $(\mathcal{I}, d)$ is complete.

5. Near Fixed Point Theorems

Let X be a universal set, and let $T : X \to X$ be a function from X into itself. We say that $x \in X$ is a fixed point if and only if $T(x) = x$. The well-known Banach contraction principle presents the fixed point of function T when X is taken to be a complete metric space. We shall study the Banach contraction principle when X is taken to be an informal complete metric space.

Definition 8. *Let X be an informal vector space over $\mathbb{F}$ with a null set Ω, and let $T : X \to X$ be a function defined on X into itself. A point $x \in X$ is called a near fixed point of T if and only if $T(x) \stackrel{\Omega}{=} x$.*

Example 6. *Continued from Example 5, we see that the null set Ω in $(\mathcal{I}, d)$ is given by*

$$\Omega = \{[-k, k] : k \geq 0\},$$

which is closed under the vector addition. Let $T : (\mathcal{I}, d) \to (\mathcal{I}, d)$ be a function from $\mathcal{I}$ into itself. Suppose that $[a^L, a^U]$ is a near fixed point of T, i.e., $T([a^L, a^U]) \stackrel{\Omega}{=} [a^L, a^U]$. By definition, nonnegative numbers k_1 and k_2 exist such that one of the following equalities is satisfied:

- $T([a^L, a^U]) \oplus [-k_1, k_1] = [a^L, a^U]$;
- $T([a^L, a^U]) = [a^L, a^U] \oplus [-k_1, k_1]$;
- $T([a^L, a^U]) \oplus [-k_1, k_1] = [a^L, a^U] \oplus [-k_2, k_2]$,

where $[-k_1, k_1]$ and $[-k_2, k_2]$ are in the null set Ω.

Remark 2. *We have the following observations.*

- *By definition, we see that $T(x) \stackrel{\Omega}{=} x$ if and only if $\omega_1, \omega_2 \in \Omega$ exist, such that $T(x) = x$, $T(x) \oplus \omega_1 = x$, or $T(x) = x \oplus \omega_1$ or $T(x) \oplus \omega_1 = x \oplus \omega_2$.*
- *If the informal vector space X owns a zero element θ, then the (conventional) fixed point is also a near fixed point.*
- *If the informal vector space X turns into a (conventional) vector space over $\mathbb{F}$, then the concepts of near fixed point and (conventional) fixed point are equivalent.*

Definition 9. *Let (X, d) be an informal metric space. A function $T : (X, d) \to (X, d)$ is called a contraction of X if and only if there is a real number $0 < \alpha < 1$ such that*

$$d(T(x), T(y)) \leq \alpha d(x, y)$$

for any $x, y \in X$.

Example 7. *Continued from Example 4, suppose that T is a contraction of $\mathcal{I}$. Then, a real number $0 < \alpha < 1$ exists, such that*
$$d(T([a_1^L, a_1^U]), T(a_2^L, a_2^U])) \leq \alpha \cdot d([a_1^L, a_1^U], [a_2^L, a_2^U])$$
for any $[a_1^L, a_1^U], [a_2^L, a_2^U] \in \mathcal{I}$. In particular, we take $\tilde{\mathcal{I}}$ to be a collection of all subintervals of $[0,1]$. Now, we take $T : \tilde{\mathcal{I}} \to \tilde{\mathcal{I}}$ by

$$T([a^L, a^U]) = \frac{1}{3}[a^L, a^U] \otimes [a^L, a^U]$$
$$= \frac{1}{3}\left[\min\left\{a^L a^L, a^L a^U, a^U a^U\right\}, \max\left\{a^L a^L, a^L a^U, a^U a^U\right\}\right] = \frac{1}{3}\left[a^L a^L, a^U a^U\right],$$

where $a^L, a^U \in [0,1]$. From Example 4, we have

$$d(T([a_1^L, a_1^U]), T(a_2^L, a_2^U])) = d\left(\frac{1}{3}\left[a_1^L a_1^L, a_1^U a_1^U\right], \frac{1}{3}\left[a_2^L a_2^L, a_2^U a_2^U\right]\right)$$
$$= \frac{1}{3}\left|a_1^L a_1^L + a_1^U a_1^U - a_2^L a_2^L - a_2^U a_2^U\right|$$
$$= \frac{1}{3}\left|\left(a_1^L + a_2^L\right)\left(a_1^L - a_2^L\right) + \left(a_1^U + a_2^U\right)\left(a_1^U - a_2^U\right)\right|$$
$$\leq \frac{2}{3}\left|\left(a_1^L - a_2^L\right) + \left(a_1^U - a_2^U\right)\right| \text{ (since } a_1^L, a_1^U, a_2^L, a_2^U \in [0,1])$$
$$= \frac{2}{3}d([a_1^L, a_1^U], [a_2^L, a_2^U]),$$

which says that T is a contraction of $\tilde{\mathcal{I}}$.

Given any initial element $x_0 \in X$, we define the iterative sequence $\{x_n\}_{n=1}^\infty$ using the function T as follows:
$$x_1 = T(x_0), \quad x_2 = T(x_1) = T^2(x_0), \cdots, x_n = T^n(x_0). \tag{4}$$

Under some suitable conditions, we are going to show that the sequence $\{x_n\}_{n=1}^\infty$ can converge to a near fixed point.

Theorem 1. *Let (X, d) be an informal complete metric space with the null set Ω such that d satisfies the null equality. Suppose that Ω is closed under the vector addition, and that the function $T : (X, d) \to (X, d)$ is a contraction of X. Then T has a near fixed point $x \in X$ satisfying $T(x) \stackrel{\Omega}{=} x$. More precisely, the near fixed point x can be obtained by the following limit*

$$d(x_n, x) \to 0 \text{ as } n \to \infty,$$

where the sequence $\{x_n\}_{n=1}^\infty$ is generated by the iteration (4). Moreover, we have the following properties.

- *There is a unique equivalence class $[x]$ satisfying that if $\bar{x} \notin [x]$, then $\bar{x}$ cannot be a near fixed point, which shows the sense of uniqueness.*
- *Suppose that $\bar{x} \in [x]$. Then $\bar{x}$ is also a near fixed point of T satisfying $T(\bar{x}) \stackrel{\Omega}{=} \bar{x}$ and $[\bar{x}] = [x]$.*
- *Suppose that $\bar{x}$ is a near fixed point of T. Then $\bar{x} \in [x]$; i.e., $[\bar{x}] = [x]$. In other words, if x and $\bar{x}$ are the near fixed points of T, then $x \stackrel{\Omega}{=} \bar{x}$.*

Proof. Proposition 1 says that the family of all classes $[x]$ for $x \in X$ forms the equivalence classes. Given any initial element $x_0 \in X$, we can generate the iterative sequence $\{x_n\}_{n=1}^{\infty}$ using (4). We want to claim that $\{x_n\}_{n=1}^{\infty}$ is a Cauchy sequence. Since T is a contraction of X, we have

$$\begin{aligned} d(x_{m+1}, x_m) &= d(T(x_m), T(x_{m-1})) \leq \alpha d(x_m, x_{m-1}) \\ &= \alpha d(T(x_{m-1}), T(x_{m-2})) \leq \alpha^2 d(x_{m-1}, x_{m-2}) \\ &\leq \cdots \leq \alpha^m d(x_1, x_0). \end{aligned}$$

For $n < m$, using the triangle inequality, we obtain

$$\begin{aligned} d(x_m, x_n) &\leq d(x_m, x_{m-1}) + d(x_{m-1}, x_{m-2}) + \cdots + d(x_{n+1}, x_n) \\ &\leq \left(\alpha^{m-1} + \alpha^{m-2} + \cdots + \alpha^n\right) \cdot d(x_1, x_0) \\ &= \alpha^n \cdot \frac{1 - \alpha^{m-n}}{1 - \alpha} \cdot d(x_1, x_0). \end{aligned}$$

Since $0 < \alpha < 1$, we have $1 - \alpha^{m-n} < 1$ in the numerator, which says that

$$d(x_m, x_n) \leq \frac{\alpha^n}{1 - \alpha} \cdot d(x_1, x_0) \to 0 \text{ as } n \to \infty.$$

This shows that $\{x_n\}_{n=1}^{\infty}$ is indeed a Cauchy sequence. The completeness says that $x \in X$ exists, satisfying $d(x_n, x) \to 0$, i.e., $x_n \to [x]$ from Definition 6 and Proposition 2.

Now, we want to claim that any point $\bar{x} \in [x]$ is a near fixed point. We first have $\bar{x} \oplus \omega_1 = x \oplus \omega_2$ for some $\omega_1, \omega_2 \in \Omega$. According to the triangle inequality and using the fact of contraction of X, we obtain

$$\begin{aligned} d(\bar{x}, T(\bar{x})) &= d(\bar{x} \oplus \omega_1, T(\bar{x})) \text{ (since } d \text{ satisfies the null equality)} \\ &\leq d(\bar{x} \oplus \omega_1, x_m) + d(x_m, T(\bar{x})) \\ &= d(\bar{x} \oplus \omega_1, x_m) + d(T(x_{m-1}), T(\bar{x})) \\ &\leq d(\bar{x} \oplus \omega_1, x_m) + \alpha d(x_{m-1}, \bar{x}) \\ &= d(\bar{x} \oplus \omega_1, x_m) + \alpha d(x_{m-1}, \bar{x} \oplus \omega_1) \text{ (since } d \text{ satisfies the null equality)} \\ &= d(x \oplus \omega_2, x_m) + \alpha d(x_{m-1}, x \oplus \omega_2) \\ &= d(x, x_m) + \alpha d(x_{m-1}, x) \text{ (since } d \text{ satisfies the null equality),} \end{aligned}$$

which implies $d(\bar{x}, T(\bar{x})) = 0$ as $m \to \infty$. We conclude that $T(\bar{x}) \stackrel{\Omega}{=} \bar{x}$ for any point $\bar{x} \in [x]$.

Now, we assume that there is another near fixed point $\bar{\bar{x}}$ of T satisfying $\bar{\bar{x}} \notin [x]$, i.e., $\bar{\bar{x}} \stackrel{\Omega}{=} T(\bar{\bar{x}})$. Then

$$\bar{\bar{x}} \oplus \omega_1 = T(\bar{\bar{x}}) \oplus \omega_2 \text{ and } x \oplus \omega_3 = T(x) \oplus \omega_4$$

for some $\omega_i \in \Omega, i = 1, \cdots, 4$. Since T is a contraction of X and d satisfies the null equality, we obtain

$$d(\bar{\bar{x}}, x) = d(\bar{\bar{x}} \oplus \omega_1, x \oplus \omega_3) = d(T(\bar{\bar{x}}) \oplus \omega_2, T(x) \oplus \omega_4) = d(T(\bar{\bar{x}}), T(x)) \leq \alpha d(\bar{\bar{x}}, x),$$

which implies $d(\bar{\bar{x}}, x) = 0$, since $0 < \alpha < 1$. Therefore, we obtain $\bar{\bar{x}} \stackrel{\Omega}{=} x$, which contradicts $\bar{\bar{x}} \notin [x]$. This says that any $\bar{\bar{x}} \notin [x]$ cannot be a near fixed point. Equivalently, if $\bar{x}$ is a near fixed point of T, then $\bar{x} \in [x]$. This completes the proof. □

Example 8. *Continued from Example 5, we see that $(\mathcal{I}, d)$ is a complete informal metric space, such that d satisfies the null equality. Suppose that T is a contraction of $\mathcal{I}$. Then, there exists a real number $0 < \alpha < 1$ such that*

$$d(T([a_1^L, a_1^U]), T(a_2^L, a_2^U])) \leq \alpha \cdot d([a_1^L, a_1^U], [a_2^L, a_2^U])$$

for any $[a_1^L, a_1^U], [a_2^L, a_2^U] \in \mathcal{I}$. Given any initial element $I_0 = [a_0^L, a_0^U] \in \mathcal{I}$, we can generate the iterative sequence $\{I_n\}_{n=1}^{\infty}$ using the function T, where $I_n = [a_n^L, a_n^U]$, as follows:

$$I_1 = T(I_0), \quad I_2 = T(I_1) = T^2(I_0), \cdots, I_n = T^n(I_0);$$

that is,

$$[a_1^L, a_1^U] = T([a_0^L, a_0^U]), \quad [a_2^L, a_2^U] = T([a_1^L, a_1^U]) = T^2([a_0^L, a_0^U]), \cdots, [a_n^L, a_n^U] = T^n([a_0^L, a_0^U]).$$

Using Theorem 1, the near fixed point $I = [a^L, a^U]$ is obtained by the limit

$$d(I_n, I) = d([a_n^L, a_n^U], [a^L, a^U]) \to 0 \text{ as } n \to \infty.$$

Definition 10. *Let (X, d) be an informal metric space with the null set Ω, and let Ω be closed under the vector addition. A function $T : (X, d) \to (X, d)$ is called a **weakly strict contraction** of X if and only if the following conditions are satisfied:*

- $x \stackrel{\Omega}{=} y$, *i.e., $[x] = [y]$ implies $d(T(x), T(y)) = 0$;*
- $x \stackrel{\Omega}{\neq} y$, *i.e., $[x] \neq [y]$ implies $d(T(x), T(y)) < d(x, y)$.*

We see that if T is a contraction of X, then it is also a weakly strict contraction of X.

Theorem 2. *Let (X, d) be an informal, complete metric space with the null set Ω, and let Ω be closed under the vector addition. Suppose that the function $T : (X, d) \to (X, d)$ is a weakly strict contraction of X. If $\{T^n(x_0)\}_{n=1}^{\infty}$ forms a Cauchy sequence for some $x_0 \in X$, then T has a near fixed point $x \in X$ satisfying $T(x) \stackrel{\Omega}{=} x$. More precisely, the near fixed point x can be obtained by the following limit*

$$d(T^n(x_0), x) \to 0 \text{ as } n \to \infty.$$

Assume further that d satisfies the null equality. Then, we also have the following properties.

- *There is a unique equivalence class $[x]$ satisfying that if $\bar{x} \notin [x]$ then $\bar{x}$ cannot be a near fixed point, which shows the sense of uniqueness.*
- *Suppose that $\bar{x} \in [x]$. Then $\bar{x}$ is also a near fixed point of T, satisfying $T(\bar{x}) \stackrel{\Omega}{=} \bar{x}$ and $[\bar{x}] = [x]$.*
- *Suppose that $\bar{x}$ is a near fixed point of T. Then $\bar{x} \in [x]$; i.e., $[\bar{x}] = [x]$. In other words, if x and $\bar{x}$ are the near fixed points of T, then $x \stackrel{\Omega}{=} \bar{x}$.*

Proof. The assumption says that $\{T^n(x_0)\}_{n=1}^{\infty}$ is a Cauchy sequence. Since X is complete, it follows that $x \in X$ exists, such that $d(T^n(x_0), x) \to 0$. From Definition 6 and Proposition 2, we see that $T^n(x_0) \to [x]$. Now, given any $\epsilon > 0$, there exists an integer N, such that $d(T^n(x_0), x) < \epsilon$ for $n \geq N$. Two cases will be considered.

- Suppose that $T^n(x_0) \stackrel{\Omega}{=} x$. Since T is a weakly strict contraction of X, it follows that

$$d(T^{n+1}(x_0), T(x)) = 0 < \epsilon.$$

- Suppose that $T^n(x_0) \stackrel{\Omega}{\neq} x$. Since T is a weakly strict contraction of X, we have

$$d(T^{n+1}(x_0), T(x)) < d(T^n(x_0), x) < \epsilon \text{ for } n \geq N.$$

Therefore, we conclude that $d(T^{n+1}(x_0), T(x)) \to 0$. The triangle inequality says that

$$d(T(x), x) \leq d\left(T(x), T^{n+1}(x_0)\right) + d\left(T^{n+1}(x_0), x\right) \to 0 \text{ as } n \to \infty.$$

Therefore, we obtain $d(T(x), x) = 0$, i.e., $T(x) \stackrel{\Omega}{=} x$. This shows that x is a near fixed point.

Now, we assume further that d satisfies the null equality. We want to show that each point $\tilde{x} \in [x]$ is a near fixed point of T. Since $\tilde{x} \stackrel{\Omega}{=} x$, we have $\tilde{x} \oplus \omega_1 = x \oplus \omega_2$ for some $\omega_1, \omega_2 \in \Omega$. The null equality says that

$$d(T^n(x_0), \tilde{x}) = d(T^n(x_0), \tilde{x} \oplus \omega_1) = d(T^n(x_0), x \oplus \omega_2) = d(T^n(x_0), x) \to 0 \text{ as } n \to \infty.$$

Therefore, we can also obtain $d(T^{n+1}(x_0), T(\tilde{x})) \to 0$ as $n \to \infty$ by using the above argument. On the other hand, the triangle inequality also says that

$$d(\tilde{x}, T(\tilde{x})) \leq d(\tilde{x}, T^{n+1}(x_0)) + d(T^{n+1}(x_0), T(\tilde{x})) \to 0 \text{ as } n \to \infty,$$

which implies $d(\tilde{x}, T(\tilde{x})) = 0$. Therefore, we obtain $T(\tilde{x}) \stackrel{\Omega}{=} \tilde{x}$ for any point $\tilde{x} \in [x]$.

Suppose that $\tilde{x} \notin [x]$ and $\tilde{x}$ is another near fixed point of T. Then, we have $T(\tilde{x}) \stackrel{\Omega}{=} \tilde{x}$ and $[\tilde{x}] \neq [x]$, i.e., $x \stackrel{\Omega}{\neq} \tilde{x}$. We also have $T(x) \oplus \omega_1 = x \oplus \omega_2$ and $T(\tilde{x}) \oplus \omega_3 = \tilde{x} \oplus \omega_4$, where $\omega_i \in \Omega$ for $i = 1, 2, 3, 4$. Now, we obtain

$$d(x, \tilde{x}) = d(x \oplus \omega_2, \tilde{x} \oplus \omega_4) \text{ (using the concept of null equality)}$$
$$= d(T(x) \oplus \omega_1, T(\tilde{x}) \oplus \omega_3) = d(T(x), T(\tilde{x})) \text{ (using the concept of null equality)}$$
$$< d(x, \tilde{x}) \text{ (since } T \text{ is a weakly strict contraction and } x \stackrel{\Omega}{\neq} \tilde{x}\text{)}.$$

Therefore we led to a contradiction, which says that $\tilde{x}$ cannot be a near fixed point of T. In other words, if $\tilde{x}$ is a near fixed point of T, then $\tilde{x} \in [x]$. This completes the proof. □

Meir and Keeler [11] studied the fixed point theorem for the weakly-uniformly strict contraction. Therefore, under the informal metric space (X, d), we propose the following definition by considering the fact $d(x, y) = 0$ for $x \stackrel{\Omega}{=} y$.

Definition 11. *Let (X, d) be an informal metric space with the null set Ω, and let Ω be closed under the vector addition. A function $T : (X, d) \to (X, d)$ is called a weakly uniformly strict contraction of X if and only if the following conditions are satisfied:*

- $x \stackrel{\Omega}{=} y$, *i.e., $[x] = [y]$ implies $d(T(x), T(y)) = 0$;*
- *given any $\epsilon > 0$, $\delta > 0$ exists, such that $\epsilon \leq d(x, y) < \epsilon + \delta$ implies $d(T(x), T(y)) < \epsilon$ for any $x \stackrel{\Omega}{\neq} y$, i.e., $[x] \neq [y]$.*

Remark 3. *It is clear to see that if T is a weakly uniformly strict contraction of X, then it is also a weakly strict contraction of X.*

Lemma 1. *Let (X, d) be an informal metric space with the null set Ω, and let Ω be closed under the vector addition. Let $T : (X, d) \to (X, d)$ be a weakly uniformly strict contraction of X. Then the sequence $\{d(T^n(x), T^{n+1}(x))\}_{n=1}^{\infty}$ is decreasing to zero for any $x \in X$.*

Proof. For convenience, we write $T^n(x) = x_n$ for all n. Let $c_n = d(x_n, x_{n+1})$.

- Suppose that $[x_{n-1}] \neq [x_n]$. By Remark 3, we have

$$c_n = d(x_n, x_{n+1}) = d(T^n(x), T^{n+1}(x)) < d(T^{n-1}(x), T^n(x)) = d(x_{n-1}, x_n) = c_{n-1}.$$

- Suppose that $[x_{n-1}] = [x_n]$. Then, by the first condition of Definition 11,

$$c_n = d(T^n(x), T^{n+1}(x)) = d(T(x_{n-1}), T(x_n)) = 0 < c_{n-1}.$$

Therefore, we conclude that the sequence $\{c_n\}_{n=1}^{\infty}$ is decreasing. Now, we also consider the following two cases.

- Let m be the first index in the sequence $\{x_n\}_{n=1}^{\infty}$ such that $[x_{m-1}] = [x_m]$. Then, we can show that $c_{m-1} = c_m = c_{m+1} = \cdots = 0$. Since $x_{m-1} \stackrel{\Omega}{=} x_m$, we have $c_{m-1} = d(x_{m-1}, x_m) = 0$. The first condition of Definition 11 says that

$$0 = d(T(x_{m-1}), T(x_m)) = d(T^m(x), T^{m+1}(x)) = d(x_m, x_{m+1}) = c_m,$$

which implies $x_m \stackrel{\Omega}{=} x_{m+1}$; i.e., $[x_m] = [x_{m+1}]$. We can similarly obtain $c_{m+1} = 0$ and $[x_{m+1}] = [x_{m+2}]$. Therefore, the sequence $\{c_n\}_{n=1}^{\infty}$ is decreasing to zero.

- Suppose that $[x_{m+1}] \neq [x_m]$ for all $m \geq 1$. Since the sequence $\{c_n\}_{n=1}^{\infty}$ is decreasing, we can assume that $c_n \downarrow \epsilon > 0$, i.e., $c_n \geq \epsilon > 0$ for all n, which says that $\delta > 0$ exists, such that $\epsilon \leq c_m < \epsilon + \delta$ for some m, i.e., $\epsilon \leq d(x_m, x_{m+1}) < \epsilon + \delta$. The second condition of Definition 11 says that

$$c_{m+1} = d(x_{m+1}, x_{m+2}) = d(T^{m+1}(x), T^{m+2}(x)) = d(T(x_m), T(x_{m+1})) < \epsilon,$$

which contradicts $c_{m+1} \geq \epsilon$.

This completes the proof. □

Theorem 3. *Let (X, d) be an informal complete metric space with the null set Ω, and let Ω be closed under the vector addition. Let $T : (X, d) \to (X, d)$ be a weakly uniformly strict contraction of X. Then T has a near fixed point satisfying $T(x) \stackrel{\Omega}{=} x$. More precisely, the near fixed point x is obtained by the following limit*

$$d(T^n(x_0), x) \to 0 \text{ as } n \to \infty \text{ for some } x_0.$$

Assume further that d satisfies the null equality. Then we also have the following properties.

- *There is a unique equivalence class $[x]$ satisfying that if $\bar{x} \notin [x]$, then $\bar{x}$ cannot be a near fixed point, which shows the sense of uniqueness.*
- *Suppose that $\bar{x} \in [x]$. Then $\bar{x}$ is also a near fixed point of T satisfying $T(\bar{x}) \stackrel{\Omega}{=} \bar{x}$ and $[\bar{x}] = [x]$.*
- *Suppose that $\bar{x}$ is a near fixed point of T. Then $\bar{x} \in [x]$; i.e., $[\bar{x}] = [x]$. In other words, if x and $\bar{x}$ are the near fixed points of T, then $x \stackrel{\Omega}{=} \bar{x}$.*

Proof. From Theorem 2 and Remark 3, we just need to show that if T is a weakly uniformly strict contraction, then $\{T^n(x_0)\}_{n=1}^{\infty} = \{x_n\}_{n=1}^{\infty}$ is a Cauchy sequence for $x_0 \in X$. Suppose that $\{x_n\}_{n=1}^{\infty}$ is not a Cauchy sequence. By definition, $2\epsilon > 0$ exists, such that, given any N, $m, n \geq N$ exists, satisfying $d(x_m, x_n) > 2\epsilon$. The assumption says that T is a weakly uniformly strict contraction on X. Therefore, $\delta > 0$ exists, such that

$$\epsilon \leq d(x, y) < \epsilon + \delta \text{ implies } d(T(x), T(y)) < \epsilon \text{ for any } x \stackrel{\Omega}{\neq} y.$$

Let $\delta' = \min\{\delta, \epsilon\}$. We want to show that

$$\epsilon \leq d(x, y) < \epsilon + \delta' \text{ implies } d(T(x), T(y)) < \epsilon \text{ for any } x \stackrel{\Omega}{\neq} y. \tag{5}$$

It is clear to see that if $\delta' = \epsilon$, i.e., $\epsilon < \delta$, then $\epsilon + \delta' = \epsilon + \epsilon < \epsilon + \delta$.

Let $c_n = d(x_n, x_{n+1})$. Lemma 1 says that the sequence $\{c_n\}_{n=1}^\infty$ is decreasing to zero. Therefore, we can find N such that $c_N < \delta'/3$. For $n > m \geq N$, we have

$$d(x_m, x_n) > 2\epsilon \geq \epsilon + \delta', \tag{6}$$

which implicitly says that $x_m \not\stackrel{\Omega}{=} x_n$. Since $\{c_n\}_{n=1}^\infty$ is decreasing, we obtain

$$d(x_m, x_{m+1}) = c_m \leq c_N < \frac{\delta'}{3} \leq \frac{\epsilon}{3} < \epsilon. \tag{7}$$

For j with $m < j \leq n$, we also have

$$d(x_m, x_{j+1}) \leq d(x_m, x_j) + d(x_j, x_{j+1}). \tag{8}$$

We want to show that j with $m < j \leq n$ exists, such that $x_m \not\stackrel{\Omega}{=} x_j$ and

$$\epsilon + \frac{2\delta'}{3} < d(x_m, x_j) < \epsilon + \delta'. \tag{9}$$

Let $\gamma_j = d(x_m, x_j)$ for $j = m+1, \cdots, n$. Then (6) and (7) say that

$$\gamma_{m+1} < \epsilon \text{ and } \gamma_n > \epsilon + \delta'. \tag{10}$$

Let j_0 be an index satisfying

$$j_0 = \max\left\{j \in [m+1, n] : \gamma_j \leq \epsilon + \frac{2\delta'}{3}\right\}.$$

Using (10), we have $m+1 \leq j_0 < n$. This says that j_0 is well-defined. The definition of j_0 also says that $j_0 + 1 \leq n$ and $\gamma_{j_0+1} > \epsilon + \frac{2\delta'}{3}$. Therefore, we obtain $x_m \not\stackrel{\Omega}{=} x_{j_0+1}$, which says that the expression (9) will be sound if we can show that

$$\epsilon + \frac{2\delta'}{3} < \gamma_{j_0+1} < \epsilon + \delta'.$$

Suppose that this is not true; i.e., $\gamma_{j_0+1} \geq \epsilon + \delta'$. Using (8), we obtain

$$\frac{\delta'}{3} > c_N \geq c_{j_0} = d(x_{j_0}, x_{j_0+1}) \geq \gamma_{j_0+1} - \gamma_{j_0} \geq \epsilon + \delta' - \epsilon - \frac{2\delta'}{3} = \frac{\delta'}{3},$$

which contradicts the fact that (9) is sound. Since $x_m \not\stackrel{\Omega}{=} x_j$, forms (5), we see that (9) implies

$$d(x_{m+1}, x_{j+1}) = d(T(x_m), T(x_j)) < \epsilon. \tag{11}$$

Therefore, we obtain

$$d(x_m, x_j) \leq d(x_m, x_{m+1}) + d(x_{m+1}, x_{j+1}) + d(x_{j+1}, x_j)$$
$$< c_m + \epsilon + c_j \text{ (by (11))}$$
$$< \frac{\delta'}{3} + \epsilon + \frac{\delta'}{3} = \epsilon + \frac{2\delta'}{3},$$

which contradicts (9). Therefore, every sequence $\{T^n(x)\}_{n=1}^\infty = \{x_n\}_{n=1}^\infty$ is a Cauchy sequence. This completes the proof. □

Funding: This research received no external funding.

Conflicts of Interest: The author declares no conflict of interest.

References

1. Adasch, N.; Ernst, B.; Keim, D. *Topological Vector Spaces: The Theory without Convexity Conditions*; Springer-Verlag: Berlin/Heidelberg, Germany, 1978.
2. Khaleelulla, S.M. *Counterexamples in Topological Vector Spaces*; Springer-Verlag: Berlin/Heidelberg, Germany, 1982.
3. Schaefer, H.H. *Topological Vector Spaces*; Springer-Verlag: Berlin/Heidelberg, Germany, 1966.
4. Peressini, A.L. *Ordered Topological Vector Spaces*; Harper and Row: New York, NY, USA, 1967.
5. Wong, Y.-C.; Ng, K.-F. *Partially Ordered Topological Vector Spaces*; Oxford University Press: Oxford, UK, 1973.
6. Wu, H.-C. Near Fixed Point Theorems in Hyperspaces. *Mathematics* **2018**, *6*, 90. [CrossRef]
7. Wu, H.-C. Near Fixed Point Theorems in the Space of Fuzzy Numbers. *Mathematics* **2018**, *6*, 108. [CrossRef]
8. Aubin, J.-P. *Applied Functional Analysis*, 2nd ed.; John Wiley & Sons: Hoboken, NJ, USA, 2000.
9. Conway, J.B. *A Course in Functional Analysis*, 2nd ed.; Springer-Verlag: Berlin/Heidelberg, Germany, 1990.
10. Riesz, F.; Sz.-Nagy, B. *Functional Analysis*; Dover Publications, Inc.: New York, NY, USA, 1955.
11. Meir, A.; Keeler, E. A Theorem on Contraction Mappings. *J. Math. Anal. Appl.* **1969**, *28*, 326–329. [CrossRef]

© 2019 by the authors. Licensee MDPI, Basel, Switzerland. This article is an open access article distributed under the terms and conditions of the Creative Commons Attribution (CC BY) license (http://creativecommons.org/licenses/by/4.0/).

Article
A New Approach to the Interpolative Contractions

Yaé Ulrich Gaba [1,2,*,†] and Erdal Karapınar [3,*,†]

1. Institut de Mathématiques et de Sciences Physiques (IMSP)/UAC, BP 613, Porto-Novo, Benin
2. African Center for Advanced Studies, PO Box 4477, Yaounde, Cameroon
3. Department of Medical Research, China Medical University Hospital, China Medical University, Taichung 40402, Taiwan
* Correspondence: yaeulrich.gaba@gmail.com (Y.U.G.); karapinar@mail.cmuh.org.tw (E.K.)
† These authors contributed equally to this work.

Received: 17 September 2019; Accepted: 2 October 2019; Published: 10 October 2019

Abstract: We propose a refinement in the interpolative approach in fixed-point theory. In particular, using this method, we prove the existence of fixed points and common fixed points for Kannan-type contractions and provide examples to support our results.

Keywords: interpolative contraction; contraction; fixed point

1. Preliminaries

Kannan fixed-point theorem is the first significant variant of the outstanding result of Banach on the metric fixed-point theory [1,2]. Kannan's theorem has been generalized in different ways. In the present note, we zoom in on one of the recent generalizations that was proposed by Karapınar [3] as *interpolative Kannan-type contraction*. It was indicated in [3] that each interpolative Kannan-type contraction in a complete metric space admits a fixed point (see also e.g., [4–7]). More precisely, we have:

Theorem 1 ([3], Theorem 2.2). *Let (X,d) be a complete metric space and $T : X \to X$ an interpolative Kannan-type contraction, i.e., T is a self-map such that there exist $\lambda \in [0,1)$, $\alpha \in (0,1)$ with*

$$d(Tx, Ty) \leq \lambda d(x, Tx)^\alpha d(y, Ty)^{1-\alpha} \qquad (1)$$

for all $x, y \in X \backslash Fix(T)$, where $Fix(T) := \{x \in X : Tx = x\}$.
Then T has a fixed point in X.

Our contribution in the present manuscript aims at sharpening the inequality (1) by increasing the degree of freedom of the powers appearing in the right-hand side in the framework of standard metric spaces. We also indicate the novelty of our results by expressing some examples.

2. Main Results

We start with the following definition.

Definition 1. *Let (X, d) a metric space and $T : X \to X$ a self-map. We shall call T a (λ, α, β)-interpolative Kannan contraction, if there exist $\lambda \in [0,1), \alpha, \beta \in (0,1)$ with $\alpha + \beta < 1$ such that*

$$d(Tx, Ty) \leq \lambda d(x, Tx)^\alpha d(y, Ty)^\beta \qquad (2)$$

for all $x, y \in X$ with $x \neq Tx, y \neq Ty$.

We are now ready to state the main result of this paper.

Theorem 2. *Let (X, d) a complete metric space and $T : X \to X$ be a (λ, α, β)-interpolative Kannan contraction with $\lambda \in [0, 1), \alpha, \beta \in (0, 1)$ so that $\alpha + \beta < 1$. Then T has a fixed point in X.*

Proof. Following the steps of the proof of ([3], Theorem 2.2), we construct the sequence $(x_n)_{n \geq 1}$ by iterating $x_n = T^n x_0$ where $x_0 \in X$ is an arbitrary starting point. Then, we observe that

$$d(x_n, x_{n+1}) = d(Tx_{n-1}, Tx_n) \leq \lambda d(x_{n-1}, x_n)^\alpha d(x_n, x_{n+1})^\beta,$$

i.e.,

$$d(x_n, x_{n+1})^{1-\beta} \leq \lambda d(x_{n-1}, x_n)^\alpha \leq \lambda d(x_{n-1}, x_n)^{1-\beta}$$

since $\alpha < 1 - \beta$.

As already elaborated in the proof of ([3], Theorem 2.2), the classical procedure leads to the existence of a unique fixed point $x^* \in X$. □

We conclude this section by presenting an example explaining why our approach is more general.

Example 1 (Compare ([3], Example 2.3)). *Take $X = \{x, y, z, w\}$ and endow it with the following metric:*

	x	y	z	w
x	0	5/2	4	5/2
y	5/2	0	3/2	1
z	4	3/2	0	3/2
w	5/2	1	3/2	0

We also define the self-map T on X as

$$Tx = x; \ Ty = w; \ Tz = x; \ Tw = y.$$

We observe that the inequality:

$$d(Tx, Ty) \leq \lambda d(x, Tx)^\alpha d(y, Ty)^\beta$$

is satisfied for:

$$\alpha = \frac{1}{8}, \ \beta = \frac{3}{4}, \ \lambda = \frac{8}{9} \leq \frac{9}{10};$$

$$\alpha = \frac{1}{9}, \ \beta = \frac{3}{4}, \ \lambda = \frac{8}{9} \leq \frac{9}{10};$$

$$\alpha = \frac{1}{8}, \ \beta = \frac{4}{5}, \ \lambda = \frac{8}{9} \leq \frac{9}{10}.$$

In all these cases, $\alpha + \beta < 1$ i.e., $\beta < 1 - \alpha$ and the map obviously has a unique fixed point. In other words, the inequality

$$d(Tx, Ty) \leq \lambda d(x, Tx)^\alpha d(y, Ty)^{1-\alpha}$$

could just be replaced by the existence of two reals α, β such that $\alpha + \beta < 1$,

$$d(Tx, Ty) \leq \lambda d(x, Tx)^\alpha d(y, Ty)^\beta.$$

Inspired by the above question, we introduce the idea of "optimal interpolative triplet (α, β, λ)" for a (λ, α, β)-interpolative Kannan contraction.

Definition 2. Let (X, d) be a metric space and $T : X \to X$ be a self-map. We shall call T a relaxed (λ, α, β)-interpolative Kannan contraction, if there exist $0 \leq \lambda, \alpha, \beta$ such that

$$d(Tx, Ty) \leq \lambda d(x, Tx)^\alpha d(y, Ty)^\beta. \tag{3}$$

Definition 3. Let (X, d) be a metric space and $T : X \to X$ be a relaxed (λ, α, β)-interpolative Kannan contraction. The triplet (λ, α, β) will be called "optimal interpolative triplet" if for any $\varepsilon > 0$, the inequality (3) fails for at least one of the triplet

$$(\lambda - \varepsilon, \alpha, \beta), \ (\lambda, \alpha - \varepsilon, \beta), \ (\lambda, \alpha, \beta - \varepsilon).$$

Therefore, we formulate the following conjecture for which we currently do not have any proof.

Theorem 3. Let (X, d) be a complete metric space. Let $T : X \to X$ be a map such that for any $n \geq 0$, T^n admits an optimal interpolative triplet $(\lambda_n, \alpha_n, \beta_n)$. If $\sum \lambda_n < \infty$ and $\sum \alpha_n + \beta_n < \infty$, then T has a unique fixed point. Moreover, this fixed point can be obtained via the Picard iteration.

Theorem 2 can easily be generalized to the case of two maps. More precisely:

Definition 4. Let (X, d) be a metric space and $R, T : X \to X$ be two self-maps. We shall call (R, T) a (λ, α, β)-interpolative Kannan contraction pair, if there exist $\lambda \in [0, 1)$, $\alpha, \beta \in (0, 1)$ with $\alpha + \beta < 1$ such that

$$d(Rx, Ty) \leq \lambda d(x, Rx)^\alpha d(y, Ty)^\beta \tag{4}$$

for all $x, y \in X$ with $x \neq Rx, y \neq Ty$.

Our result then goes as follows:

Theorem 4. Let (X, d) be a complete metric space and (R, T) be a (λ, α, β)-interpolative Kannan contraction pair. Then R and T have a common fixed point in X, i.e., there exists $x^* \in X$ such that $Rx^* = x^* = Tx^*$.

Proof. We construct the sequence $(x_n)_{n \geq 1}$ by iterating

$$x_{2n+1} = Rx_{2n}, \ x_{2n+2} = Tx_{2n+1}$$

where $x_0 \in X$ is an arbitrary starting point.

$$d(x_{2n+1}, x_{2n+2}) \leq \lambda d(x_{2n}, x_{2n+1})^\alpha d(x_{2n+1}, x_{2n+2})^\beta \leq \lambda d(x_{2n}, x_{2n+1})^\alpha d(x_{2n+1}, x_{2n+2})^{1-\alpha}.$$

The proof then follows the same steps as ([8], Theorem 2.1). As already elaborated in the proof of ([8], Theorem 2.1), the classical procedure leads to the existence of a unique fixed point $x^* \in X$. □

Example 2. We use the metric defined in Example 1. We also define on X the self-maps T as

$$Tx = x; \ Ty = y; \ Tz = w; \ Tw = w$$

and R as

$$Rx = x; \ Ry = w; \ Rz = z; \ Rw = w.$$

We observe that the inequality:

$$d(Rx, Ty) \leq \lambda d(x, Rx)^\alpha d(y, Ty)^\beta$$

is satisfied for:

$$\alpha = \frac{1}{8}, \beta = \frac{3}{4}, \lambda = \frac{8}{9};$$

$$\alpha = \frac{1}{9}, \beta = \frac{5}{6}, \lambda = \frac{9}{10};$$

$$\alpha = \frac{10}{11}, \beta = \frac{1}{2}, \lambda = \frac{5}{7}.$$

R and T have two common fixed points x and w.

The above conjecture (Theorem 3) motives us in the investigation of interpolative Kannan contraction for a family of maps. Indeed Noorwali [8] used interpolation to obtain a common fixed-point result for a Kannan-type contraction mapping. We aim at generalizing ([8], Theorem 2.1) and Theorem 4 with the use of a (λ, α, β)-interpolative Kannan contraction for a family of maps. More precisely:

Problem 1. *Let (X, d) be a complete metric space. Let $T_n : X \to X, n \geq 1$ be a family of self-maps such for any $x, y \in X$*

$$d(T_i x, T_j y) \leq \lambda_{i,j}\, d(x, T_i x)^{\alpha_i} d(y, T_j y)^{\beta_j}.$$

What are the conditions on $\lambda_{i,j}, \alpha_i \beta_j$ for T_n to have a (unique)common fixed point.

Author Contributions: Y.U.G. writing–original draft preparation; E.K. writing–review and editing.

Funding: This research received no external funding.

Acknowledgments: The authors thanks anonymous referees for their remarkable comments, suggestion, and ideas that help to improve this paper. Y.U.G. wishes to thank the African Institute for Mathematical Sciences (AIMS), in South Africa, which accepted him as a visitor in May 2019 and provided full funding for his stay.

Conflicts of Interest: The authors declare no conflict of interest.

References

1. Ćirić, L. *Some Recent Results in Metrical Fixed Point Theory*; University of Belgrade: Beograd, Serbia, 2003.
2. Todorcević, V. *Harmonic Quasiconformal Mappings and Hyperbolic Type Metrics*; Springer Nature Switzerland AG: Cham, Switzerland, 2019.
3. Karapınar, E. Revisiting the Kannan Type Contractions via Interpolation. *Adv. Theory Nonlinear Anal. Appl.* **2018**, *2*, 85–87. [CrossRef]
4. Aydi, H.; Karapınar, E.; Roldán López de Hierro, A.F. ω-Interpolative Ćirić-Reich-Rus-Type Contractions. *Mathematics* **2019**, *7*, 57. [CrossRef]
5. Karapınar, E.; Fulga, A. New Hybrid Contractions on b-Metric Spaces. *Mathematics* **2019**, *7*, 578. [CrossRef]
6. Karapınar, E.; Alqahtani, O.; Aydi, H. On Interpolative Hardy-Rogers Type Contractions. *Symmetry* **2019**, *11*, 8, doi:10.3390/sym11010008. [CrossRef]
7. Debnath, P.; Radenović, S.; Mitrović, Z.D. Interpolative Hardy-Rogers and Reich-Rus-Ćirić Type Contractions in Rectangular B-Metric Space and B-Metric Spaces. *Mat. Vesnik.* **2019**. in press.
8. Noorwali, M. Common Fixed Point for Kannan type contraction via Interpolation. *J. Math. Anal.* **2018**, *9*, 92–94.

© 2019 by the authors. Licensee MDPI, Basel, Switzerland. This article is an open access article distributed under the terms and conditions of the Creative Commons Attribution (CC BY) license (http://creativecommons.org/licenses/by/4.0/).

Article

C^*-Algebra Valued Fuzzy Soft Metric Spaces and Results for Hybrid Pair of Mappings

Daripally Ram Prasad [1], Gajula Naveen Venkata Kishore [2], Hüseyin Işık [3,4,*], Bagathi Srinuvasa Rao [5] and Gorantla Adi Lakshmi [1]

[1] Research Scholar, Department of Mathematics, Koneru Lakshmaiah Education Foundation, Vaddeswaram, Guntur District, Aandhra Pradesh 522 502, India
[2] Department of Mathematics, Sagi Rama Krishnam Raju Engineering College, China Amiram, Bhimavaram, Andhra Pradesh 534 204, India
[3] Nonlinear Analysis Research Group, Ton Duc Thang University, Ho Chi Minh City 700000, Vietnam
[4] Faculty of Mathematics and Statistics, Ton Duc Thang University, Ho Chi Minh City 700000, Vietnam
[5] Department of Mathematics, Dr. B. R. Ambedkar University, Srikakulam, Etcherla, Andhra Pradesh 532410, India
* Correspondence: huseyin.isik@tdtu.edu.vn

Received: 16 June 2019; Accepted: 24 July 2019; Published: 16 August 2019

Abstract: In this paper, we establish some results on coincidence point and common fixed point theorems for a hybrid pair of single valued and multivalued mappings in complete C^*-algebra valued fuzzy soft metric spaces. In addition, we provided some coupled fixed point theorems. Finally, we have given examples which support our main results.

Keywords: fuzzy soft points; C^*-algebra-valued fuzzy soft metric; ω-compatible; coincidence point; common fixed point; multi-valued map.

MSC: 22E46; 53C35; 57S20

---

1. Introduction and Preliminaries

We know that the fixed points that can be discussed are divided into two types. The first type deals with contraction and is referred to as Banach fixed point theorems, the second type deals with compact mappings and more involved. Metric fixed point theorems plays very important role, many authors proved fixed point theorems in various spaces (see e.g., [1–36]).

The study of fixed points for multivalued mappings using the Hausdorff metric was initiated by Nadler ([14]). The theory of multivalued mappings has a wide range of applications, it has been applied in control theory, convex optimization, differential inclusions, economics, etc. The existence of fixed points for various multivalued contractive mappings has been studied by many authors under different conditions (see [15–30]).

In the year 2014, Ma et al. [7] introduced the concept of C^*-algebra valued metric space and established some fixed point results. Later, Alsulami et al. [32] suggested some remarks on C^*-algebras and proved Banach type contraction result, this line of research was continued in (see [8,10–12,31,34,35]).

Fuzzy set theory was introduced by Zadeh [36] and the theory of soft sets initiated by Molodstov [37] which helps to solve problems in all areas. Maji et al. [38,39] introduced several operations in soft sets and as also coined fuzzy soft sets. In [1] Thangaraj Beaula et al. defined fuzzy soft metric space in terms of fuzzy soft points and proved some results. On the other hand several authors proved smany results in fuzzy soft sets and fuzzy soft metric spaces (see [1,2,5,6,40–44]).

Recently, R.P.Agarwal et al. [25] introduced the concept of C^*-algebra valued fuzzy soft metric space based on C^*-algebras and fuzzy soft elements and described the convergence and completeness properties in this space also they provided some fixed point theorems (see [25,26]).

The main aim of this paper is to introduce the concept of multi-valued mappings in C^*-algebra valued fuzzy soft metric spaces and proved some coincidence and common fixed point theorems for a two-pair of multi-valued and single-valued maps satisfying new type of contractive conditions. Also we provided some coupled fixed point theorems and finally we are initiate some examples which supports our main results.

Throughout this paper, we use the following notations as in C^*-algebras:

U refers to an initial universe, E the set of all parameters for U and $P(\tilde{U})$ the set of all fuzzy set of U. (U, E) means the universal set U and parameter set E, $\tilde{C}$ refer to C^*-algebras. Details on C^*-algebras are available in [27]. An algebra '$\tilde{C}$' together with a conjugate linear involution map $*: \tilde{C} \to \tilde{C}$, defined by $\tilde{a} \to \tilde{a}^*$ such that for all $\tilde{a}, \tilde{b} \in \tilde{C}$, we have $(\tilde{a}\tilde{b})^* = \tilde{b}^*\tilde{a}^*$ and $(\tilde{a}^*)^* = \tilde{a}$, is called a $*$-algebra. Moreover, if $\tilde{C}$ an identity element $\tilde{I}_{\tilde{C}}$, then the pair $(\tilde{C}, *)$ is called a unital $*$-algebra. A unital $*$-algebra $(\tilde{C}, *)$ together with a complete sub multiplicative norm satisfying $\tilde{a} = \tilde{a}^*$ for all $\tilde{a} \in \tilde{C}$ is called a Banach $*$-algebra. A C^*-algebra is a Banach $*$-algebra $(\tilde{C}, *)$ such that $\tilde{a}^*\tilde{a} = \tilde{a}^2$ for all $\tilde{a} \in \tilde{C}$, An element $\tilde{a} \in \tilde{C}$ is called a positive element if $\tilde{a} = \tilde{a}^*$ and $\sigma(\tilde{a}) \subset R(C)^*$ is set of non-negative fuzzy soft real numbers, where $\sigma(\tilde{a}) = \{\lambda \in R(C)^* : \lambda\tilde{I} - \tilde{a},$ is non-invertible$\}$. If $\tilde{a} \in \tilde{C}$ is positive, we write it as $\tilde{a} \geq \tilde{0}_{\tilde{C}}$. Using positive elements, one can define partial ordering on $\tilde{C}$ as follows; $\tilde{a} \leq \tilde{b}$ if and only if $\tilde{0}_{\tilde{C}} \leq \tilde{b} - \tilde{a}$. Each positive element '$\tilde{a}$' of a C^*-algebra $\tilde{C}$ has a unique positive square root. Subsequently, $\tilde{C}$ will denote a unital C^*-algebra with the identity element $\tilde{I}_{\tilde{C}}$. Furthermore, $\tilde{C}_+$ and $\tilde{C}'$ will denote the set $\{\tilde{a} \in \tilde{C} : \tilde{0}_{\tilde{C}} \leq \tilde{a}\}$ and set $\{\tilde{a} \in \tilde{C} : \tilde{a}\tilde{b} = \tilde{b}\tilde{a}\}$, respectively.

Definition 1 ([37]). *A Fuzzy set A in U is characterized by a function with domain as U and values in $[0, 1]$. The collection of all fuzzy set U is $P(\tilde{U})$.*

Definition 2 ([38]). *A pair (F, E) is called a soft set over U if and only if $F: E \to P(U)$ is mapping from E into $P(U)$ the set of all sub set of U.*

Definition 3 ([43]). *Let $C \subseteq E$ then the mapping $F_E: C \to P(\tilde{U})$, defined by $F_E(e) = \mu^e F_E$ (a fuzzy sub set of U), is called fuzzy soft set over (U, E) where, $\mu^e F_E = \tilde{0}$ if $e \in E - C$ and $\mu^e F_E \neq \tilde{0}$ if $e \in C$. The set of all fuzzy soft set over (U, E) is denoted by $FS(U, E)$.*

Definition 4 ([43]). *Let $F_E \in FS(U, E)$ and $F_E(e) = \tilde{1}$ for all $e \in E$. Then F_E is called absolute fuzzy soft set. It is denoted by $\tilde{E}$.*

Now we recall some basic definitions and properties of C^*-algebra-valued Fuzzy soft metric spaces.

Definition 5 ([25]). *Let $C \subseteq E$ and $\tilde{E}$ be the absolute fuzzy soft set that is $F_E(e) = \tilde{1}$ for all $e \in E$. Let $\tilde{C}$ denote the C^*-algebra. The C^*-algebra valued fuzzy soft metric using fuzzy soft points is defined as a mapping $\tilde{d}_{c^*}: \tilde{E} \times \tilde{E} \to \tilde{C}$ satisfying the following conditions.*

(M_0) $\tilde{0}_{\tilde{C}} \leq \tilde{d}(F_{e_1}, F_{e_2})$ for all $F_{e_1}, F_{e_2} \in \tilde{E}$.
(M_1) $\tilde{d}_{c^*}(F_{e_1}, F_{e_2}) = \tilde{0}_{\tilde{C}} \Leftrightarrow F_{e_1} = F_{e_2}$
(M_2) $\tilde{d}_{c^*}(F_{e_1}, F_{e_2}) = \tilde{d}_{c^*}(F_{e_2}, F_{e_1})$
(M_3) $\tilde{d}_{c^*}(F_{e_1}, F_{e_3}) \leq \tilde{d}_{c^*}(F_{e_1}, F_{e_2}) + \tilde{d}_{c^*}(F_{e_2}, F_{e_3}) \; \forall \; F_{e_1}, F_{e_2}, F_{e_3} \in \tilde{E}$.

The fuzzy soft set $\tilde{E}$ with the C^*-algebra valued fuzzy soft metric $\tilde{d}_{c^*}$ is called the C^*-algebra valued fuzzy soft metric space. It is denoted by $(\tilde{E}, \tilde{C}, \tilde{d}_{c^*})$.

Definition 6 ([25]). *A sequence $\{F_{e_n}\}$ in a C^*-algebra valued fuzzy soft metric space $(\tilde{E}, \tilde{C}, \tilde{d}_{c^*})$ is said to converges to $F_{e'}$ in $\tilde{E}$ with respect to $\tilde{C}$. If $\|\tilde{d}_{c^*}(F_{e_n}, F_{e'})\|_{\tilde{C}} \to \tilde{0}_{\tilde{C}}$ is said to converges to as $n \to \infty$ that is for every*

$\tilde{0}_{\tilde{C}} \prec \tilde{e}$ there exists $\tilde{0}_{\tilde{C}} \prec \tilde{\delta}$ and a positive integer $N = N(\tilde{e})$, such that $\|\tilde{d}_{c*}(F_{e_n}, F_{e'})\| < \tilde{\delta}$ implies that $\|\mu^a_{F_{e_n}}(s) - \mu^a_{F_{e'}}(s)\| < \tilde{e}$, whenever $n \geq N$. It is usually denoted as $\lim_{n\to\infty} F_{e_n} = F_{e'}$.

Definition 7 ([25]). *A sequence $\{F_{e_n}\}$ in a C^*-algebra valued fuzzy soft metric space $(\tilde{E}, \tilde{C}, \tilde{d}_{c*})$ is said to be Cauchy sequence. If to every $\tilde{0}_{\tilde{C}} \prec \tilde{e}$ there exist $\tilde{0}_{\tilde{C}} \prec \tilde{\delta}$ and a positive integer $N = N(\tilde{e})$ such that $\|\tilde{d}_{c*}(F_{e_n}, F_{e_m})\| < \tilde{\delta}$ implies that $\|\mu^a_{F_{e_n}}(s) - \mu^a_{F_{e_m}}(s)\| < \tilde{e}$ whenever $n, m \geq N$. That is $\|\tilde{d}_{c*}(F_{e_n}, F_{e_m})\|_{\tilde{C}} \to \tilde{0}_{\tilde{C}}$ as $n, m \to \infty$.*

Definition 8 ([25]). *A C^*-algebra valued fuzzy soft metric space $(\tilde{E}, \tilde{C}, \tilde{d}_{c*})$ is said to be complete. If every Cauchy sequence in $\tilde{E}$ converges to some fuzzy soft point of $\tilde{E}$.*

Example 1 ([25]). *Let $C \subseteq R$ and $E \subseteq R$, let $\tilde{E}$ be an absolute fuzzy soft set that is $\tilde{E}(e) = \tilde{1}$ for all $e \in E$, and $\tilde{C} = M_2(R(C)^*)$, define $\tilde{d}_{c*}: \tilde{E} \times \tilde{E} \to \tilde{C}$ by*

$$\tilde{d}_{c*}(F_{e_1}, F_{e_2}) = \begin{bmatrix} i & 0 \\ 0 & i \end{bmatrix},$$

where $i = \inf\{|\mu^a_{F_{e_1}}(s) - \mu^a_{F_{e_2}}(s)|/s \in C\}$ and $F_{e_1}, F_{e_2} \in \tilde{E}$. Then $\tilde{d}_{c*}$ is a C^*-algebra valued fuzzy soft metric and $(\tilde{E}, \tilde{C}, \tilde{d}_{c*})$ is a complete C^*-algebra valued fuzzy soft metric space by the completeness of $R(C)^*$.

Lemma 1 ([25]). *Let $\tilde{C}$ be a C^*-algebra with the identity element $\tilde{I}_{\tilde{C}}$ and $\tilde{x}$ be a positive element of $\tilde{C}$. If $\tilde{a} \in \tilde{C}$ is such that $\|\tilde{a}\| < 1$ then for $m < n$, we have*

$$\lim_{n\to\infty} \sum_{k=m}^{n} (\tilde{a}^*)^k \tilde{x}(\tilde{a})^k = \tilde{I}_{\tilde{C}} \|(\tilde{x})^{\frac{1}{2}}\|^2 \left(\frac{\|\tilde{a}\|^m}{1-\|\tilde{a}\|}\right) \quad (1)$$

and

$$\sum_{k=m}^{n} (\tilde{a}^*)^k \tilde{x}(\tilde{a})^k \to \tilde{0}_{\tilde{C}} \text{ as } m \to \infty. \quad (2)$$

Lemma 2 ([25]). *Suppose that $\tilde{C}$ is a unital C^*-algebra with unit $\tilde{1}$.*

(i) *If $\tilde{a} \in \tilde{C}_+$ with $\|\tilde{a}\| < \frac{1}{2}$ then $\tilde{I} - \tilde{a}$ is invertible and $\|\tilde{a}(\tilde{I}-\tilde{a})^{-1}\| < 1$*
(ii) *suppose that $\tilde{a}, \tilde{b} \in \tilde{C}$ with $\tilde{a}, \tilde{b} \geq \tilde{0}_{\tilde{C}}$ and $\tilde{a}\tilde{b} = \tilde{b}\tilde{a}$ then $\tilde{a}\tilde{b} \geq \tilde{0}_{\tilde{C}}$*
(iii) *$\tilde{C}'$ we denote the set $\{\tilde{a} \in \tilde{C}/\tilde{a}\tilde{b} = \tilde{b}\tilde{a} \ \forall \ \tilde{b} \in \tilde{C}\}$. Let $\tilde{a} \in \tilde{C}'$, if $\tilde{b}, \tilde{c} \in \tilde{C}$ with $\tilde{b} \geq \tilde{c} \geq \tilde{0}$ and $\tilde{I} - \tilde{a} \in \tilde{C}'_+$ is an invertible operator, then $(\tilde{I} - \tilde{a})^{-1}\tilde{b} \geq (\tilde{I} - \tilde{a})^{-1}\tilde{c}$, where $\tilde{C}'_+ = \tilde{C}_+ \cap \tilde{C}'$.*

Notice that in c^*-algebra, if $\tilde{0} \leq \tilde{a}, \tilde{b}$, one cannot conclude that $\tilde{0} \leq \tilde{a}\tilde{b}$. Indeed, consider the c^*-algebra $M_2(R(C)^*)$ and set

$$\tilde{a} = \begin{bmatrix} F_{e_1}(a) & F_{e_2}(a) \\ F_{e_2}(a) & F_{e_1}(b) \end{bmatrix} = \begin{bmatrix} 0.3 & 0.1 \\ 0.1 & 0.2 \end{bmatrix}$$

$$\text{and } \tilde{b} = \begin{bmatrix} F_{e_1}(c) & F_{e_2}(c) \\ F_{e_2}(c) & F_{e_1}(d) \end{bmatrix} = \begin{bmatrix} 0.4 & 0.5 \\ 0.5 & 0.6 \end{bmatrix}$$

then clearly $\tilde{a} \geq \tilde{0}$ and $\tilde{b} \geq \tilde{0}$ but $\tilde{a}, \tilde{b} \in M_2(R(C)^*)_+$ while $\tilde{a}\tilde{b} \notin M_2(R(C)^*)_+$.

2. Main Results

In this section, first we give the notion of Hausdorff metric in C^*-algebra valued fuzzy soft metric spaces.

Let $(\tilde{E}, \tilde{C}, \tilde{d}_{c*})$ be a C^*-algebra valued fuzzy soft metric space. We denote by $CB(\tilde{E})$ be a class of all nonempty closed and bounded subsets of $\tilde{E}$. For a points $F_{e_1}, F_{e_2} \in \tilde{E}$ and $\tilde{X}, \tilde{Y} \in CB(\tilde{E})$, define $\tilde{D}_{c^*}(F_{e_1}, \tilde{Y}) = \inf_{G_{e_1} \in \tilde{Y}} \tilde{d}_{c^*}(F_{e_1}, G_{e_1})$. Let $\tilde{H}_{c^*}$ be the Hausdorff C^*-algebra valued fuzzy soft metric induced by the C^*-algebra valued fuzzy soft metric $\tilde{d}_{c^*}$ on $\tilde{E}$ that is

$$\tilde{H}_{c^*}(\tilde{X}, \tilde{Y}) = \max\left\{\sup_{F_{e_1} \in \tilde{X}} \tilde{D}_{c^*}(F_{e_1}, \tilde{Y}), \sup_{G_{e_1} \in \tilde{Y}} \tilde{D}_{c^*}(\tilde{X}, G_{e_1})\right\}$$

for every $\tilde{X}, \tilde{Y} \in CB(\tilde{E})$. It is well known that $(CB(\tilde{E}), \tilde{C}, \tilde{H}_{c^*})$ is a complete C^*-algebra valued fuzzy soft metric space, whenever $(\tilde{E}, \tilde{C}, \tilde{d}_{c^*})$ is a complete C^*-algebra valued fuzzy soft metric space.

Definition 9. *Let $T: \tilde{E} \to CB(\tilde{E})$ be a multivalued map. An element $F_{e_1} \in \tilde{E}$ is fixed point of F if $F_{e_1} \in TF_{e_1}$.*

Definition 10. *Let $T: \tilde{E} \to CB(\tilde{E})$ and $f: \tilde{E} \to \tilde{E}$ be a multivalued map and single valued maps. An element $F_{e_1} \in \tilde{E}$ is coincidence point of F and f if $fF_{e_1} \in TF_{e_1}$. We denote*

$$C\{f, T\} = \{F_{e_1} \in \tilde{E}/fF_{e_1} \in TF_{e_1}\}$$

Definition 11. *The mappings $T: \tilde{E} \to CB(\tilde{E})$ and $f: \tilde{E} \to \tilde{E}$ are weakly compatible if they commute at their coincidence points, i.e., if $fTF_{e_1} = TfF_{e_1}$, whenever $fF_{e_1} \in TF_{e_1}$.*

Definition 12. *Let $T: \tilde{E} \to CB(\tilde{E})$ and $f: \tilde{E} \to \tilde{E}$ be a multivalued map and single valued maps. The map f is said to be T-weakly commuting at $F_{e_1} \in \tilde{E}$ if $ffF_{e_1} \in TfF_{e_1}$.*

Definition 13. *An element $F_{e_1} \in \tilde{E}$ is a common fixed point of $T, S: \tilde{E} \to CB(\tilde{E})$ and $f: \tilde{E} \to \tilde{E}$ if $F_{e_1} = fF_{e_1} \in TF_{e_1} \cap SF_{e_1}$.*

Example 2. *Let $U = R^+$ and $E = C = [0, 4]$, let $\tilde{E}$ be an absolute fuzzy soft set that is $\tilde{E}(e) = \tilde{1}$ for all $e \in E$, and $\tilde{C} = M_2(R(C)^*)$, define $\tilde{d}_{c^*}: \tilde{E} \times \tilde{E} \to \tilde{C}$ by $\tilde{d}_{c^*}(F_{e_1}(a), F_{e_2}(a))(s) = \begin{bmatrix} i & 0 \\ 0 & i \end{bmatrix}$ where $i = \inf\{|\mu^a_{F_{e_1}}(s) - \mu^a_{F_{e_2}}(s)|/s \in C\}$ then $(\tilde{E}, \tilde{C}, \tilde{d}_{c^*})$ is a C^*-algebra valued fuzzy soft metric space and define $f: \tilde{E} \to \tilde{E}$ and $T: \tilde{E} \to CB(\tilde{E})$*

$$fF_e(a) = \begin{cases} \tilde{0} & \text{if } F_e(a) \in [0, \frac{1}{2}] \\ \frac{F_e(a)}{2} & \text{if } F_e(a) \in (\frac{1}{2}, 1] \end{cases}, TF_e(a) = \begin{cases} \{F_e(a)\} & \text{if } F_e(a) \in [0, \frac{1}{2}] \\ [0, \tilde{1} - \frac{F_e(a)}{4}] & \text{if } F_e(a) \in (\frac{1}{2}, 1] \end{cases}$$

We have

- $f\tilde{1} = \frac{1}{2} \in [0, \frac{3}{4}] = T\tilde{1}$ that is, $F_e(a) = \tilde{1}$ is a coincidence point of f and T;
- $fT\tilde{1} = [0, \frac{1}{2}] \neq [0, \frac{7}{8}] = Tf\tilde{1}$ that is, f and T are not weakly compatible mappings;
- $ff\tilde{1} = \frac{1}{4} \in [0, \frac{7}{8}] = Tf\tilde{1}$ that is, f is T-weakly commuting at $\tilde{1}$.

Theorem 1. *Let $(\tilde{E}, \tilde{C}, \tilde{d}_{c^*})$ be a complete C^*-algebra valued fuzzy soft metric space, and $T: \tilde{E} \to CB(\tilde{E})$ be a multivalued map satisfying*

$$\tilde{H}_{c^*}(TF_{e_1}, TF_{e_2}) \leq \tilde{a}^* \tilde{d}_{c^*}(F_{e_1}, F_{e_2}) \tilde{a} \tag{3}$$

for all $F_{e_1}, F_{e_2} \in \tilde{E}$, where $\tilde{a} \in \tilde{C}$ with $\|\tilde{a}\| < 1$. Then T has a unique fixed point in $\tilde{E}$.

Lemma 3. *If $\tilde{X}, \tilde{Y} \in CB(\tilde{E})$ and $F_{e_1} \in \tilde{X}$, then for any fixed $\tilde{b} \in \tilde{C}'_+$ with $\|\tilde{b}\| < 1$, there exists $F_{e_2} = F_{e_2}(F_{e_1}) \in \tilde{Y}$ such that*

$$\tilde{d}_{c^*}(F_{e_1}, F_{e_2}) \leq \tilde{b} \tilde{H}_{c^*}(\tilde{X}, \tilde{Y}). \tag{4}$$

Theorem 2. Let $(\tilde{E}, \tilde{C}, \tilde{d}_{c^*})$ be a complete C^*-algebra valued fuzzy soft metric space. Let $S, T: \tilde{E} \to CB(\tilde{E})$ be a pair of multivalued maps and $f, g : \tilde{E} \to \tilde{E}$ be a single-valued maps. Suppose that

$$\tilde{H}_{c^*}(SF_{e_1}, TF_{e_2}) \leq \tilde{a}\tilde{d}_{c^*}(fF_{e_1}, gF_{e_2}) + \tilde{a}\left(\tilde{D}_{c^*}(fF_{e_1}, SF_{e_1}) + \tilde{D}_{c^*}(gF_{e_2}, TF_{e_2})\right) \\ + \tilde{a}\left(\tilde{D}_{c^*}(fF_{e_1}, TF_{e_2}) + \tilde{D}_{c^*}(gF_{e_2}, SF_{e_1})\right), \quad (5)$$

for all $F_{e_1}, F_{e_2} \in \tilde{E}$, where $\tilde{a} \in \tilde{C}_+'$ with $\|\tilde{a}\| < 1$. Suppose that

(A_1) $S\tilde{E} \subseteq g\tilde{E}$, $T\tilde{E} \subseteq f\tilde{E}$;
(A_2) $f(\tilde{E})$ and $g(\tilde{E})$ are closed.

Then, there exist points $F_{e'}, G_{e'} \in \tilde{E}$, such that $fF_{e'} \in SF_{e'}$, $gG_{e'} \in TG_{e'}$ and $fF_{e'} = gG_{e'}$, $SF_{e'} = TG_{e'}$.

Proof. Let $F_{e_0} \in \tilde{E}$ be an arbitrary. From (A_1) and Lemma 3, there exist $F_{e_1}, F_{e_2} \in \tilde{E}$, such that $gF_{e_1} \in SF_{e_0}$, $fF_{e_2} \in TF_{e_1}$ and

$$\tilde{d}_{c^*}(gF_{e_1}, fF_{e_2}) \leq \tilde{b}\tilde{H}_{c^*}(SF_{e_0}, TF_{e_1}). \quad (6)$$

From (5) and (6), we have

$$\tilde{d}_{c^*}(gF_{e_1}, fF_{e_2}) \leq \tilde{b}\tilde{H}_{c^*}(SF_{e_0}, TF_{e_1}) \\ \leq \tilde{b}\tilde{a}\tilde{d}_{c^*}(fF_{e_0}, gF_{e_1}) + \tilde{b}\tilde{a}\left(\tilde{D}_{c^*}(fF_{e_0}, SF_{e_0}) + \tilde{D}_{c^*}(gF_{e_1}, TF_{e_1})\right) \\ + \tilde{b}\tilde{a}\left(\tilde{D}_{c^*}(fF_{e_0}, TF_{e_1}) + \tilde{D}_{c^*}(gF_{e_1}, SF_{e_0})\right). \quad (7)$$

In contrast, we have

$$\tilde{D}_{c^*}(fF_{e_0}, SF_{e_0}) \leq \tilde{d}_{c^*}(fF_{e_0}, gF_{e_1}) \\ \tilde{D}_{c^*}(gF_{e_1}, TF_{e_1}) \leq \tilde{d}_{c^*}(gF_{e_1}, fF_{e_2}) \\ \tilde{D}_{c^*}(gF_{e_1}, SF_{e_0}) \leq \tilde{d}_{c^*}(gF_{e_1}, gF_{e_1}) = 0 \\ \tilde{D}_{c^*}(fF_{e_0}, TF_{e_1}) \leq \tilde{d}_{c^*}(fF_{e_0}, fF_{e_2}) \leq \tilde{d}_{c^*}(fF_{e_0}, gF_{e_1}) + \tilde{d}_{c^*}(gF_{e_1}, fF_{e_2}). \quad (8)$$

From (7) and (8), we have

$$\tilde{d}_{c^*}(gF_{e_1}, fF_{e_2}) \leq \tilde{b}\tilde{a}\tilde{d}_{c^*}(fF_{e_0}, gF_{e_1}) + \tilde{b}\tilde{a}\left(\tilde{d}_{c^*}(fF_{e_0}, gF_{e_1}) + \tilde{d}_{c^*}(gF_{e_1}, fF_{e_2})\right) \\ + \tilde{b}\tilde{a}\left(\tilde{d}_{c^*}(fF_{e_0}, gF_{e_1}) + \tilde{d}_{c^*}(gF_{e_1}, fF_{e_2})\right) \\ = 3\tilde{b}\tilde{a}\tilde{d}_{c^*}(fF_{e_0}, gF_{e_1}) + 2\tilde{b}\tilde{a}\tilde{d}_{c^*}(gF_{e_1}, fF_{e_2}). \quad (9)$$

Therefore,

$$(1 - 2\tilde{b}\tilde{a})\tilde{d}_{c^*}(gF_{e_1}, fF_{e_2}) \leq 3\tilde{b}\tilde{a}\tilde{d}_{c^*}(fF_{e_0}, gF_{e_1}).$$

Since $\|\tilde{b}\|\|\tilde{a}\| < \frac{1}{2}$ Then $1 - 2\tilde{b}\tilde{a}$ is invertible, and can expressed as $(1 - 2\tilde{b}\tilde{a})^{-1} = \sum_{m=0}^{\infty}(2\tilde{b}\tilde{a})^m$, which together with $2\tilde{b}\tilde{a} \in \tilde{C}_+'$ can yields $(1 - 2\tilde{b}\tilde{a})^{-1} \in \tilde{C}_+'$. By Lemma 2 (iii), we know

$$\tilde{d}_{c^*}(gF_{e_1}, fF_{e_2}) \leq \tilde{\kappa}\tilde{d}_{c^*}(fF_{e_0}, gF_{e_1}),$$

where $\tilde{\kappa} = 3\tilde{b}\tilde{a}(1 - 2\tilde{b}\tilde{a})^{-1} \in \tilde{C}_+'$ with $\|3\tilde{b}\tilde{a}(1 - 2\tilde{b}\tilde{a})^{-1}\| < 1$. Again from (A_1) and Lemma 3 with $\|\tilde{b}\| < 1$, as $fF_{e_2} \in TF_{e_1}$, there exists $F_{e_3} \in \tilde{E}$ such that $gF_{e_3} \in SF_{e_2}$ and

$$\tilde{d}_{c^*}(fF_{e_2}, gF_{e_3}) \leq \tilde{b}\tilde{H}_{c^*}(SF_{e_2}, TF_{e_1}). \quad (10)$$

From (5) and (10), we get

$$\tilde{d}_{c^*}(fF_{e_2}, gF_{e_3}) \leq \tilde{b}\tilde{H}_{c^*}(SF_{e_2}, TF_{e_1})$$
$$\leq \tilde{b}\tilde{a}\tilde{d}_{c^*}(fF_{e_2}, gF_{e_1}) + \tilde{b}\tilde{a}\left(\tilde{D}_{c^*}(fF_{e_2}, SF_{e_2}) + \tilde{D}_{c^*}(gF_{e_1}, TF_{e_1})\right)$$
$$+ \tilde{b}\tilde{a}\left(\tilde{D}_{c^*}(fF_{e_2}, TF_{e_1}) + \tilde{D}_{c^*}(gF_{e_1}, SF_{e_2})\right). \tag{11}$$

In contrast, we have

$$\tilde{D}_{c^*}(fF_{e_2}, SF_{e_2}) \leq \tilde{d}_{c^*}(fF_{e_2}, gF_{e_3})$$
$$\tilde{D}_{c^*}(gF_{e_1}, TF_{e_1}) \leq \tilde{d}_{c^*}(gF_{e_1}, fF_{e_2})$$
$$\tilde{D}_{c^*}(fF_{e_2}, TF_{e_1}) \leq \tilde{d}_{c^*}(fF_{e_2}, fF_{e_2}) = 0$$
$$\tilde{D}_{c^*}(gF_{e_1}, SF_{e_2}) \leq \tilde{d}_{c^*}(gF_{e_1}, gF_{e_3}) \leq \tilde{d}_{c^*}(gF_{e_1}, fF_{e_2}) + \tilde{d}_{c^*}(fF_{e_2}, gF_{e_3}). \tag{12}$$

Similarly as above, from (11) and (12), we get

$$\tilde{d}_{c^*}(fF_{e_2}, gF_{e_3}) \leq \tilde{\kappa}\tilde{d}_{c^*}(gF_{e_1}, fF_{e_2}).$$

Continuing this process, we can construct a sequence $\{G_{e_n}\}$ in $\tilde{E}$, such that $G_{e_0} = gF_{e_1}$ and, for each $n \in N$,

$$G_{e_{2n}} = gF_{e_{2n+1}} \in SF_{e_{2n}} \quad G_{e_{2n+1}} = fF_{e_{2n+2}} \in TF_{e_{2n+1}} \tag{13}$$

and

$$\tilde{d}_{c^*}(G_{e_{2n}}, G_{e_{2n+1}}) = \tilde{d}_{c^*}(gF_{e_{2n+1}}, fF_{e_{2n+2}}) \leq \tilde{\kappa}\tilde{d}_{c^*}(gF_{e_{2n+1}}, fF_{e_{2n}})$$
$$\tilde{d}_{c^*}(G_{e_{2n-1}}, G_{e_{2n}}) = \tilde{d}_{c^*}(fF_{e_{2n}}, gF_{e_{2n+1}}) \leq \tilde{\kappa}\tilde{d}_{c^*}(gF_{e_{2n-1}}, fF_{e_{2n}}).$$

Therefore, we have

$$\tilde{d}_{c^*}(G_{e_n}, G_{e_{n+1}}) \leq \tilde{\kappa}\tilde{d}_{c^*}(G_{e_{n-1}}, G_{e_n}) \text{ for all } n \geq 1. \tag{14}$$

From (14), by induction and Lemma 2 (iii), we get

$$\tilde{d}_{c^*}(G_{e_n}, G_{e_{n+1}}) \leq \tilde{\kappa}^n \tilde{d}_{c^*}(G_{e_0}, G_{e_1}) \text{ for all } n \in N. \tag{15}$$

Now, we shall show that $\{G_{e_n}\}$ is a Cauchy sequence in $\tilde{E}$.
For $m > n$, by using triangle inequality and (15), we have

$$\tilde{d}_{c^*}(G_{e_n}, G_{e_m}) \leq \tilde{d}_{c^*}(G_{e_n}, G_{e_{n+1}}) + \tilde{d}_{c^*}(G_{e_{n+1}}, G_{e_{n+2}}) + \cdots + \tilde{d}_{c^*}(G_{e_{m-1}}, G_{e_m})$$
$$\leq (\tilde{\kappa}^n + \tilde{\kappa}^{n+1} + \tilde{\kappa}^{n+2} + \cdots + \tilde{\kappa}^{m-1})\tilde{d}_{c^*}(G_{e_0}, G_{e_1})$$
$$\leq \|\tilde{\kappa}^n + \tilde{\kappa}^{n+1} + \tilde{\kappa}^{n+2} + \cdots + \tilde{\kappa}^{m-1}\|\|\tilde{d}_{c^*}(G_{e_0}, G_{e_1})\|\tilde{I}_{\tilde{C}}$$
$$\leq \|\|\tilde{\kappa}^n\| + \|\tilde{\kappa}^{n+1}\| + \cdots + \|\tilde{\kappa}^{m-1}\|\|\|\tilde{d}_{c^*}(G_{e_0}, G_{e_1})\|\tilde{I}_{\tilde{C}}$$
$$= \frac{\|\tilde{\kappa}\|^n}{1-\|\tilde{\kappa}\|}\|\tilde{d}_{c^*}(G_{e_0}, G_{e_1})\|\tilde{I}_{\tilde{C}} \to 0 \text{ as } n \to \infty.$$

Hence $\{G_{e_n}\}$ is a Cauchy sequence. Now as, $(\tilde{E}, \tilde{C}, \tilde{d}_{c^*})$ be a complete C^*-algebra valued fuzzy soft metric space, $\{G_{e_n}\}$ converges to some $G_{e'} \in \tilde{E}$. Therefore,

$$\lim_{n \to \infty} G_{e_n} = \lim_{n \to \infty} gF_{e_{2n+1}} = \lim_{n \to \infty} fF_{e_{2n+2}} = G_{e'}. \tag{16}$$

As $G_{e_{2n}} = gF_{e_{2n+1}}$, $G_{e_{2n+1}} = fF_{e_{2n+2}}$ and $f(\tilde{E})$, $g(\tilde{E})$ are closed, then $G_{e'} \in f(\tilde{E})$ and $G_{e'} \in g(\tilde{E})$. Therefore, there exist $F_{e'}, F_{e''} \in \tilde{E}$, such that $fF_{e'} = G_{e'}$ and $gF_{e''} = G_{e'}$. Thus, we have proved that

$$fF_{e'} = gF_{e''}. \tag{17}$$

From the contraction type condition (5) and (13), we obtain

$$\begin{aligned}
\tilde{D}_{c*}(fF_{e'}, SF_{e'}) &\leq \tilde{d}_{c*}(fF_{e'}, fF_{e_{2n+2}}) + \tilde{D}_{c*}(fF_{e_{2n+2}}, SF_{e'}) \\
&\leq \tilde{d}_{c*}(fF_{e'}, fF_{e_{2n+2}}) + \tilde{H}_{c*}(SF_{e'}, TF_{e_{2n+1}}) \\
&\leq \tilde{d}_{c*}(fF_{e'}, fF_{e_{2n+2}}) + \tilde{a}\tilde{d}_{c*}(fF_{e'}, gF_{e_{2n+1}}) \\
&\quad + \tilde{a}\left(\tilde{D}_{c*}(fF_{e'}, SF_{e'}) + \tilde{D}_{c*}(gF_{e_{2n+1}}, TF_{e_{2n+1}})\right) \\
&\quad + \tilde{a}\left(\tilde{D}_{c*}(fF_{e'}, TF_{e_{2n+1}}) + \tilde{D}_{c*}(gF_{e_{2n+1}}, SF_{e'})\right) \\
&\leq \tilde{d}_{c*}(fF_{e'}, fF_{e_{2n+2}}) + \tilde{a}\tilde{d}_{c*}(fF_{e'}, gF_{e_{2n+1}}) \\
&\quad + \tilde{a}\left(\tilde{D}_{c*}(fF_{e'}, SF_{e'}) + \tilde{D}_{c*}(gF_{e_{2n+1}}, fF_{e_{2n+2}})\right) \\
&\quad + \tilde{a}\left(\tilde{D}_{c*}(fF_{e'}, fF_{e_{2n+2}}) + \tilde{D}_{c*}(gF_{e_{2n+1}}, SF_{e'})\right).
\end{aligned}$$

which implies

$$\begin{aligned}
\tilde{D}_{c*}(fF_{e'}, SF_{e'}) &\leq (1-\tilde{a})^{-1}\tilde{d}_{c*}(fF_{e'}, fF_{e_{2n+2}}) + (1-\tilde{a})^{-1}\tilde{a}\tilde{d}_{c*}(fF_{e'}, gF_{e_{2n+1}}) \\
&\quad + (1-\tilde{a})^{-1}\tilde{a}\left(\tilde{D}_{c*}(gF_{e_{2n+1}}, fF_{e_{2n+2}})\right) \\
&\quad + (1-\tilde{a})^{-1}\tilde{a}\left(\tilde{D}_{c*}(fF_{e'}, fF_{e_{2n+2}}) + \tilde{D}_{c*}(gF_{e_{2n+1}}, SF_{e'})\right).
\end{aligned}$$

Letting $n \to \infty$ in the above inequality and using (16) and (17), we obtain

$$\|\tilde{D}_{c*}(fF_{e'}, SF_{e'})\| \leq \|(1-\tilde{a})^{-1}\tilde{a}\|\|\tilde{D}_{c*}(fF_{e'}, SF_{e'})\|.$$

Then $\tilde{D}_{c*}(fF_{e'}, SF_{e'}) = 0$. Hence, as $SF_{e'}$ is closed,

$$fF_{e'} \in SF_{e'}. \tag{18}$$

Similarly, we can prove that

$$gF_{e''} \in TF_{e''}. \tag{19}$$

Now, we have to prove that

$$SF_{e'} = TF_{e''}. \tag{20}$$

Using (5), (17)–(19), we get

$$\begin{aligned}
\tilde{H}_{c*}(SF_{e'}, TF_{e''}) &\leq \tilde{a}\tilde{d}_{c*}(fF_{e'}, gF_{e''}) + \tilde{a}\left(\tilde{D}_{c*}(fF_{e'}, SF_{e'}) + \tilde{D}_{c*}(gF_{e''}, TF_{e''})\right) \\
&\quad + \tilde{a}\left(\tilde{D}_{c*}(fF_{e'}, TF_{e''}) + \tilde{D}_{c*}(gF_{e''}, SF_{e'})\right) \\
&\leq \tilde{a}\left(\tilde{D}_{c*}(gF_{e''}, TF_{e''}) + \tilde{D}_{c*}(fF_{e'}, SF_{e'})\right) = \tilde{0}_{\tilde{c}}.
\end{aligned}$$

Hence, $SF_{e'} = TF_{e''}$. Thus, by (17)–(20), we have proved that

$$fF_{e'} \in SF_{e'} \quad gF_{e''} \in TF_{e''} \quad fF_{e'} = gF_{e''} \quad SF_{e'} = TF_{e''}.$$

□

Example 3. *Let $E = \{e_1, e_2, e_3\}, U = \{a, b, c, d\}$ and C and D are two subset of E where $C = \{e_1, e_2, e_3\}$, $D = \{e_1, e_2,\}$. Define fuzzy soft set as,*

$$(F_E, C) = \left\{ \begin{array}{l} e_1 = \{a_{0.1}, b_{0.3}, c_{0.4}, d_{0.5}\}, e_2 = \{a_{0.3}, b_{0.4}, c_{0.6}, d_{0.7}\}, \\ e_3 = \{a_{0.6}, b_{0.7}, c_{0.8}, d_{0.9}\} \end{array} \right\}$$

$$(G_E, D) = \{e_1 = \{a_{0.4}, b_{0.5}, c_{0.2}, d_{0.6}\}, e_2 = \{a_{0.5}, b_{0.6}, c_{0.3}, d_{0.7}\}\}$$

$$F_{e_1} = \mu_{F_{e_1}} = \{a_{0.1}, b_{0.3}, c_{0.4}, d_{0.5}\}, F_{e_2} = \mu_{F_{e_2}} = \{a_{0.3}, b_{0.4}, c_{0.6}, d_{0.7}\}$$

$$F_{e_3} = \mu_{F_{e_3}} = \{a_{0.6}, b_{0.7}, c_{0.8}, d_{0.9}\}$$

$$G_{e_1} = \mu_{G_{e_1}} = \{a_{0.4}, b_{0.5}, c_{0.2}, d_{0.6}\}, G_{e_2} = \mu_{G_{e_2}} = \{a_{0.5}, b_{0.6}, c_{0.3}, d_{0.7}\}$$

and $FSC(F_E) = \{F_{e_1}, F_{e_2}, F_{e_3}, G_{e_1}, G_{e_2}\}$, let $\tilde{E}$ be absolute fuzzy soft set that is $\tilde{E}(e) = \tilde{1}$, for all $e \in E$, and $\tilde{C} = M_2(R(C)^*)$, be the C^*-algebra. Define $\tilde{d}_{c^*}: \tilde{E} \times \tilde{E} \to \tilde{C}$ by $\tilde{d}_{c^*}(F_{e_1}, F_{e_2}) = (\inf\{|F_{e_1}(a) - F_{e_2}(a)|/a \in C\}, 0)$, then obviously $(\tilde{E}, \tilde{C}, \tilde{d}_{c^*})$ is a complete C^*-algebra valued fuzzy soft metric space.

We define $S: \tilde{E} \to CB(\tilde{E})$ by $SF_{e_1}(a) = F_{e_1}^2 + \frac{1}{4}$, $T: \tilde{E} \to CB(\tilde{E})$ by $TF_{e_1}(a) = F_{e_1}^3 + \frac{1}{4}$, $f: \tilde{E} \to \tilde{E}$ by $fF_{e_1} = 2F_{e_1}^2$ and $g: \tilde{E} \to \tilde{E}$ by $gF_{e_1} = 2F_{e_1}^3$ for all $a \in U$ and $F_{e_1} \in \tilde{E}$. Notice that $fF_{e_1} = 2F_{e_1}^2 = \{0.02, 0.18, 0.32, 0.50\}$ and $gF_{e_2} = 2F_{e_1}^3 = \{0.054, 0.128, 0.432, 0.686\}$. Thus, $\inf\{|\mu_{fF_{e_1}}^a(s) - \mu_{gF_{e_2}}^a(s)|/s \in C\}$

$= \inf\{0.034, 0.052, 0.112, 0.186\} = 0.034$. Hence $\tilde{d}_{c^*}(fF_{e_1}, gF_{e_2}) = \begin{bmatrix} 0.034 & 0 \\ 0 & 0.034 \end{bmatrix}$.

Also, we have

$$\begin{aligned}
\tilde{d}_{c^*}(SF_{e_1}, TF_{e_2})(a) &= (\inf\{|SF_{e_1}(a) - TF_{e_2}(a)|/a \in C\}, 0) \\
&= (\inf\{0.017, 0.026, 0.056, 0.093\}, 0) = \begin{bmatrix} 0.017 & 0 \\ 0 & 0.017 \end{bmatrix} \\
&\leq \begin{bmatrix} 0.027 & 0 \\ 0 & 0.027 \end{bmatrix} \\
&\leq \begin{bmatrix} 0.8 & 0 \\ 0 & 0.8 \end{bmatrix} \begin{bmatrix} 0.034 & 0 \\ 0 & 0.034 \end{bmatrix} \\
&\leq \tilde{c} \tilde{d}_{c^*}(fF_{e_1}, gF_{e_2}).
\end{aligned}$$

Here $\tilde{c} = \begin{bmatrix} 0.8 & 0 \\ 0 & 0.8 \end{bmatrix}$ with $\|\tilde{c}\| = 0.8 < 1$.

Therefore, (5) holds for all $F_{e_1}, F_{e_2} \in \tilde{E}$. Also, the other Hypotheses (A_1) and (A_2) are satisfied. It is seen that $S(0.5) = f(0.5) = 0.5$ and $T(0.63) = g(0.63) = 0.5$. Therefore, S and f have the coincidence at the point $F_{e'} = 0.5$, T and g at the point $F_{e''} = 0.63$, and $S(0.5) = T(0.63)$.

Theorem 3. *Let $(\tilde{E}, \tilde{C}, \tilde{d}_{c^*})$ be a complete C^*-algebra valued fuzzy soft metric space. Let $S, T: \tilde{E} \to CB(\tilde{E})$ be a pair of multivalued maps and $f: \tilde{E} \to \tilde{E}$ be a single-valued map. Suppose that*

$$\tilde{H}_{c^*}(SF_{e_1}, TF_{e_2}) \leq \tilde{a}\tilde{d}_{c^*}(fF_{e_1}, fF_{e_2}) + \tilde{a}\left(\tilde{D}_{c^*}(fF_{e_1}, SF_{e_1}) + \tilde{D}_{c^*}(fF_{e_2}, TF_{e_2})\right) \\ + \tilde{a}\left(\tilde{D}_{c^*}(fF_{e_1}, TF_{e_2}) + \tilde{D}_{c^*}(fF_{e_2}, SF_{e_1})\right) \tag{21}$$

for all $F_{e_1}, F_{e_2} \in \tilde{E}$, where $\tilde{a} \in \tilde{C}_+'$ with $\|\tilde{a}\| < 1$. Suppose that

(B_1) $S\tilde{E} \cup T\tilde{E} \subseteq f\tilde{E}$;
(B_2) $f(\tilde{E})$ is closed.

Then, f, T and S have a coincidence in $\tilde{E}$. Moreover, if f is both T-weakly commuting and S-weakly commuting at each $F_{e'} \in C(f, T)$, and $ffF_{e'} = fF_{e'}$, then, f, T and S have a common fixed point in $\tilde{E}$.

Proof. If $f = g$ in Theorem (2), we obtain that there exist points $F_{e'}, G_{e'} \in \tilde{E}$, such that $fF_{e'} \in SF_{e'}$, $fG_{e'} \in TG_{e'}$ and $fF_{e'} = fG_{e'}$, $SF_{e'} = TG_{e'}$. As $F_{e'} \in C(f, T)$, f is T-weakly commuting at $F_{e'}$ and

$ffF_{e'} = fF_{e'}$. Set $G_{e'} = fF_{e'}$. Then, we have $fG_{e'} = G_{e'}$ and $G_{e'} = ffF_{e'} \in T(fF_{e'}) = TG_{e'}$. Now, since also $F_{e'} \in C(f,S)$, then f is S-weakly commuting at $F_{e'}$, and so we obtain $G_{e'} = fG_{e'} = ffF_{e'} \in S(fF_{e'}) = SG_{e'}$. Thus, we have proved that $G_{e'} = fG_{e'} \in TG_{e'} \cap SG_{e'}$, that is, $G_{e'}$ is a common fixed point of f, T and S. □

Corollary 1. *Let $(\tilde{E}, \tilde{C}, \tilde{d}_{c*})$ be a complete C^*-algebra valued fuzzy soft metric space. Let $S, T: \tilde{E} \to CB(\tilde{E})$ be a pair of multivalued maps. Suppose that*

$$\tilde{H}_{c*}\left(SF_{e_1}, TF_{e_2}\right) \preceq \tilde{a}\tilde{d}_{c*}(F_{e_1}, F_{e_2}) + \tilde{a}\left(\tilde{D}_{c*}(F_{e_1}, SF_{e_1}) + \tilde{D}_{c*}(F_{e_2}, TF_{e_2})\right) \\ + \tilde{a}\left(\tilde{D}_{c*}(F_{e_1}, TF_{e_2}) + \tilde{D}_{c*}(F_{e_2}, SF_{e_1})\right) \quad (22)$$

for all $F_{e_1}, F_{e_2} \in \tilde{E}$, where $\tilde{a} \in \tilde{C}_+'$ with $\|\tilde{a}\| < 1$. Then there exist a point $F_{e'} \in \tilde{E}$ such that $F_{e'} \in SF_{e'} \cap TF_{e'}$ and $SF_{e'} = TF_{e'}$.

Proof. If $f = g = \tilde{I}_{\tilde{C}}$ ($\tilde{I}_{\tilde{C}}$ being the identity map on $\tilde{E}$) in Theorem 2, then, we obtain the common fixed-point result. □

Corollary 2. *Let $(\tilde{E}, \tilde{C}, \tilde{d}_{c*})$ be a complete C^*-algebra valued fuzzy soft metric space. Let $S: \tilde{E} \to CB(\tilde{E})$ be a pair of multivalued map. Suppose that*

$$\tilde{H}_{c*}\left(SF_{e_1}, SF_{e_2}\right) \preceq \tilde{a}\tilde{d}_{c*}(F_{e_1}, F_{e_2}) + \tilde{a}\left(\tilde{D}_{c*}(F_{e_1}, SF_{e_1}) + \tilde{D}_{c*}(F_{e_2}, SF_{e_2})\right) \\ + \tilde{a}\left(\tilde{D}_{c*}(F_{e_1}, SF_{e_2}) + \tilde{D}_{c*}(F_{e_2}, SF_{e_1})\right) \quad (23)$$

for all $F_{e_1}, F_{e_2} \in \tilde{E}$, where $\tilde{a} \in \tilde{C}_+'$ with $\|\tilde{a}\| < 1$. Then there exist a point $F_{e'} \in \tilde{E}$ such that $F_{e'} \in SF_{e'}$.

3. Coupled Fixed Point Results

In this section, we shall prove some coupled fixed point theorems in C^*-algebra valued fuzzy soft metric spaces by using different contractive conditions.

Definition 14. *$(\tilde{E}, \tilde{C}, \tilde{d}_{c*})$ be a C^*-algebra valued fuzzy soft metric space. Let $S: \tilde{E} \times \tilde{E} \to \tilde{E}$ be a mapping, an element $(F_{e_1}, G_{e_1}) \in \tilde{E} \times \tilde{E}$ is called coupled fixed point of S if $S(F_{e_1}, G_{e_1}) = F_{e_1}$ and $S(G_{e_1}, F_{e_1}) = G_{e_1}$.*

Definition 15. *$\tilde{E}$ be an absolute fuzzy soft set. An element $(F_{e_1}, G_{e_1}) \in \tilde{E} \times \tilde{E}$ is called*

(i) *a coupled coincidence point of mappings $S: \tilde{E} \times \tilde{E} \to \tilde{E}$ and $f: \tilde{E} \to \tilde{E}$ if $fF_{e_1} = S(F_{e_1}, G_{e_1})$ and $fG_{e_1} = S(G_{e_1}, F_{e_1})$*
(ii) *a common coupled fixed point of mappings $S: \tilde{E} \times \tilde{E} \to \tilde{E}$ and $f: \tilde{E} \to \tilde{E}$ if $F_{e_1} = fF_{e_1} = S(F_{e_1}, G_{e_1})$ and $G_{e_1} = fG_{e_1} = S(G_{e_1}, F_{e_1})$.*

Definition 16. *Let $\tilde{E}$ be an absolute fuzzy soft set and $S: \tilde{E} \times \tilde{E} \to \tilde{E}$ and $f: \tilde{E} \to \tilde{E}$. Then $\{S, f\}$ is said to be ω-compatible pairs if $f\left(S(F_{e_1}, G_{e_1})\right) = S(fF_{e_1}, fG_{e_1})$ and $f\left(S(G_{e_1}, F_{e_1})\right) = S(fG_{e_1}, fF_{e_1})$.*

Theorem 4. *Let $(\tilde{E}, \tilde{C}, \tilde{d}_{c*})$ be a C^*-algebra valued fuzzy soft metric space. Suppose $S, T: \tilde{E} \times \tilde{E} \to \tilde{E}$ and $f, g: \tilde{E} \to \tilde{E}$ be satisfying*

(1) *$S(\tilde{E} \times \tilde{E}) \subseteq g(\tilde{E})$ and $T(\tilde{E} \times \tilde{E}) \subseteq f(\tilde{E})$*
(2) *$\{S, f\}$ and $\{T, g\}$ are ω-compatible pairs.*
(3) *one of $f(\tilde{E})$ or $g(\tilde{E})$ is complete C^*-algebra valued fuzzy soft metric of $\tilde{E}$*
(4) *$\tilde{d}_{c*}\left(S(F_{e_1}, G_{e_1}), T(F_{e_2}, G_{e_2})\right) \preceq \tilde{a}^* \tilde{d}_{c*}(fF_{e_1}, gF_{e_2})\tilde{a} + \tilde{a}^* \tilde{d}_{c*}(fG_{e_1}, gG_{e_2})\tilde{a}$*
  for all $F_{e_1}, F_{e_2}, G_{e_1}, G_{e_2} \in \tilde{E}$,

where $\tilde{a} \in \tilde{C}$ with $\|\sqrt{2}\tilde{a}\| < 1$. Then S, T, f and g have a unique common coupled fixed point in $\tilde{E} \times \tilde{E}$.

Proof. Let $F_{e_0}, G_{e_0} \in \tilde{E}$. From (Theorem 4 (1)), we can construct the sequences $\{F_{e_{2n}}\}_{2n=1}^{\infty}$, $\{G_{e_{2n}}\}_{2n=1}^{\infty}$, $\{I_{e_{2n}}\}_{2n=1}^{\infty}$, $\{J_{e_{2n}}\}_{2n=1}^{\infty}$ such that

$$S(F_{e_{2n}}, G_{e_{2n}}) = gF_{e_{2n+1}} = I_{e_{2n}} \quad T(F_{e_{2n+1}}, G_{e_{2n+1}}) = fF_{e_{2n+2}} = I_{e_{2n+1}}$$
$$S(G_{e_{2n}}, F_{e_{2n}}) = gG_{e_{2n+1}} = J_{e_{2n}} \quad T(G_{e_{2n+1}}, F_{e_{2n+1}}) = fG_{e_{2n+2}} = J_{e_{2n+1}},$$

for $n = 0, 1, 2, \cdots$

Notices that in C^*-algebra, if $\tilde{a}, \tilde{b} \in \tilde{C}_+$ and $\tilde{a} \leq \tilde{b}$, then for any $\tilde{x} \in \tilde{C}_+$ both $\tilde{x}^* \tilde{a} \tilde{x}$ and $\tilde{x}^* \tilde{b} \tilde{x}$ are positive elements and $\tilde{x}^* \tilde{a} \tilde{x} \leq \tilde{x}^* \tilde{b} \tilde{x}$.

From (Theorem 4 (4)), we get

$$\begin{aligned}
\tilde{d}_{c^*}(I_{e_{2n+1}}, I_{e_{2n+2}}) &= \tilde{d}_{c^*}\left(S(F_{e_{2n+1}}, G_{e_{2n+1}}), T(F_{e_{2n+2}}, G_{e_{2n+2}})\right) \\
&\leq \tilde{a}^* \tilde{d}_{c^*}(fF_{e_{2n+1}}, gF_{e_{2n+2}})\tilde{a} + \tilde{a}^* \tilde{d}_{c^*}(fG_{e_{2n+1}}, gG_{e_{2n+2}})\tilde{a} \\
&\leq \tilde{a}^* \left(\tilde{d}_{c^*}(I_{e_{2n}}, I_{e_{2n+1}}) + \tilde{d}_{c^*}(J_{e_{2n}}, J_{e_{2n+1}})\right) \tilde{a}.
\end{aligned} \quad (24)$$

Similarly,

$$\tilde{d}_{c^*}(J_{e_{2n+1}}, J_{e_{2n+2}}) \leq \tilde{a}^* \left(\tilde{d}_{c^*}(J_{e_{2n}}, J_{e_{2n+1}}) + \tilde{d}_{c^*}(I_{e_{2n}}, I_{e_{2n+1}})\right) \tilde{a}. \quad (25)$$

Let $\alpha_{2n+1} = \tilde{d}_{c^*}(I_{e_{2n+1}}, I_{e_{2n+2}}) + \tilde{d}_{c^*}(J_{e_{2n+1}}, J_{e_{2n+2}})$.
Now from (24) and (25), we have

$$\begin{aligned}
\alpha_{2n+1} &= \tilde{d}_{c^*}(I_{e_{2n+1}}, I_{e_{2n+2}}) + \tilde{d}_{c^*}(J_{e_{2n+1}}, J_{e_{2n+2}}) \\
&\leq \tilde{a}^* \left(\tilde{d}_{c^*}(I_{e_{2n}}, I_{e_{2n+1}}) + \tilde{d}_{c^*}(J_{e_{2n}}, J_{e_{2n+1}})\right) \tilde{a} \\
&\quad + \tilde{a}^* \left(\tilde{d}_{c^*}(J_{e_{2n}}, J_{e_{2n+1}}) + \tilde{d}_{c^*}(I_{e_{2n}}, I_{e_{2n+1}})\right) \tilde{a} \\
&\leq (\sqrt{2}\tilde{a})^* \alpha_{2n} (\sqrt{2}\tilde{a}) \\
&\vdots \\
&\leq \left[(\sqrt{2}\tilde{a})^*\right]^{2n+1} \alpha_0 (\sqrt{2}\tilde{a})^{2n+1}.
\end{aligned}$$

Now, we can obtain for any $n \in \mathbb{N}$

$$\begin{aligned}
\alpha_n &= \tilde{d}_{c^*}(I_{e_n}, I_{e_{n+1}}) + \tilde{d}_{c^*}(J_{e_n}, J_{e_{n+1}}) \\
&\leq (\sqrt{2}\tilde{a})^* \alpha_{n-1} (\sqrt{2}\tilde{a}) \\
&\vdots \\
&\leq \left[(\sqrt{2}\tilde{a})^*\right]^n \alpha_0 (\sqrt{2}\tilde{a})^n.
\end{aligned}$$

If $\alpha_0 = \tilde{0}_{\tilde{C}}$, then from Definition-1 of S_2 we know $(I_{\alpha_0}, J_{\alpha_0})$ is a coupled fixed point of S, T, f and g. Now letting $\tilde{0}_{\tilde{C}} \leq \alpha_0$, we get for any $n \in \mathbb{N}$, for any $p \in \mathbb{N}$ and using triangle inequality

$$\begin{aligned}
\tilde{d}_{c^*}(I_{e_{2n+p}}, I_{e_{2n}}) &\leq \tilde{d}_{c^*}(I_{e_{2n+p}}, I_{e_{2n+p-1}}) \\
&\quad + \tilde{d}_{c^*}(I_{e_{2n+p-1}}, I_{e_{2n+p-2}}) + \cdots + \tilde{d}_{c^*}(I_{e_{2n+1}}, I_{e_{2n}}).
\end{aligned}$$

$$\begin{aligned}
\tilde{d}_{c^*}(J_{e_{2n+p}}, J_{e_{2n}}) &\leq \tilde{d}_{c^*}(J_{e_{2n+p}}, J_{e_{2n+p-1}}) \\
&\quad + \tilde{d}_{c^*}(J_{e_{2n+p-1}}, J_{e_{2n+p-2}}) + \cdots + \tilde{d}_{c^*}(J_{e_{2n+1}}, J_{e_{2n}}).
\end{aligned}$$

Consequently,

$$\begin{aligned}
\tilde{d}_{c^*}(I_{e_{2n+p}}, I_{e_{2n}}) + \tilde{d}_{c^*}(J_{e_{2n+p}}, J_{e_{2n}}) &\leq \alpha_{2n+p-1} + \alpha_{2n+p-2} + \cdots + \alpha_{2n} \\
&\leq \sum_{m=2n}^{2n+p-1} \left[(\sqrt{2}\tilde{a})^*\right]^m \alpha_0 (\sqrt{2}\tilde{a})^m
\end{aligned}$$

and then

$$\|\tilde{d}_{c^*}(I_{e_{2n+p}}, I_{e_{2n}}) + \tilde{d}_{c^*}(J_{e_{2n+p}}, J_{e_{2n}})\| \leq \alpha_{2n+p-1} + \alpha_{2n+p-2} + \cdots + \alpha_{2n}$$

$$\leq \sum_{m=2n}^{2n+p-1} \|\sqrt{2}\tilde{a}\|^{2m} \alpha_0$$

$$\leq \sum_{m=n}^{\infty} \|\sqrt{2}\tilde{a}\|^{2m} \alpha_0$$

$$= \frac{\|\sqrt{2}\tilde{a}\|^{2n}}{1-\|\sqrt{2}\tilde{a}\|^2} \alpha_0 \to 0 \text{ as } n \to \infty,$$

which together with $\tilde{d}_{c^*}(I_{e_{2n+p}}, I_{e_{2n}}) \leq \tilde{d}_{c^*}(I_{e_{2n+p}}, I_{e_{2n}}) + \tilde{d}_{c^*}(J_{e_{2n+p}}, J_{e_{2n}})$ and $\tilde{d}_{c^*}(J_{e_{2n+p}}, J_{e_{2n}}) \leq \tilde{d}_{c^*}(I_{e_{2n+p}}, I_{e_{2n}}) + \tilde{d}_{c^*}(J_{e_{2n+p}}, J_{e_{2n}})$ implies $\{I_{e_{2n}}\}$ and $\{J_{e_{2n}}\}$ are Cauchy sequences in $\tilde{E}$ with respect to $\tilde{C}$. It follows that $\{I_{e_{2n+1}}\}$ and $\{J_{e_{2n+1}}\}$ are also Cauchy sequences in $\tilde{E}$ with respect to $\tilde{C}$. Thus, $\{I_{e_n}\}$ and $\{J_{e_n}\}$ are Cauchy sequences in $(\tilde{E}, \tilde{C}, \tilde{d}_{c^*})$.

Suppose $f(\tilde{E})$ is complete subspace of $(\tilde{E}, \tilde{C}, \tilde{d}_{c^*})$. Then the sequences $\{I_{e_n}\}$ and $\{J_{e_n}\}$ are converge to $I_{e'}, J_{e'}$ respectively in $f(\tilde{E})$. Thus, there exist $F_{e'}, G_{e'}$ in $f(\tilde{E})$ Such that

$$\lim_{n\to\infty} I_{e_n} = I_{e'} = fF_{e'} \text{ and } \lim_{n\to\infty} J_{e_n} = J_{e'} = fG_{e'}. \tag{26}$$

We now claim that $S(F_{e'}, G_{e'}) = I_{e'}$ and $S(G_{e'}, F_{e'}) = J_{e'}$.
From (Theorem 4 (4)) and using the triangular inequality

$$\tilde{0}_{\tilde{C}} \leq \tilde{d}_{c^*}(S(F_{e'}, G_{e'}), I_{e'})$$
$$\leq \tilde{d}_{c^*}(S(F_{e'}, G_{e'}), I_{e_{2n+1}}) + \tilde{d}_{c^*}(I_{e_{2n+1}}, I_{e'})$$
$$\leq \tilde{d}_{c^*}(S(F_{e'}, G_{e'}), T(F_{e_{2n+1}}, G_{e_{2n+1}})) + \tilde{d}_{c^*}(I_{e_{2n+1}}, I_{e'})$$
$$\leq \tilde{a}^* \tilde{d}_{c^*}(fF_{e'}, gF_{e_{2n+1}})\tilde{a} + \tilde{a}^* \tilde{d}_{c^*}(fG_{e'}, gG_{e_{2n+1}})\tilde{a} + \tilde{d}_{c^*}(I_{e_{2n+1}}, I_{e'})$$
$$\leq \tilde{a}^* \tilde{d}_{c^*}(I_{e'}, I_{e_{2n}})\tilde{a} + \tilde{a}^* \tilde{d}_{c^*}(J_{e'}, J_{e_{2n}})\tilde{a} + \tilde{d}_{c^*}(I_{e_{2n+1}}, I_{e'}).$$

Taking the limit as $n \to \infty$ in the above relation, we obtain $\tilde{d}_{c^*}(S(F_{e'}, G_{e'}), I_{e'}) = \tilde{0}_{\tilde{C}}$ and hence $S(F_{e'}, G_{e'}) = I_{e'}$. Similarly, we prove $S(G_{e'}, F_{e'}) = J_{e'}$. Therefore, it follows $S(F_{e'}, G_{e'}) = I_{e'} = fI_{e'}$ and $S(G_{e'}, F_{e'}) = J_{e'} = fJ_{e'}$. Since $\{S, f\}$ is ω-compatible pair, we have $S(I_{e'}, J_{e'}) = fI_{e'}$ and $S(J_{e'}, I_{e'}) = fJ_{e'}$. Now to prove that $fI_{e'} = I_{e'}$ and $fJ_{e'} = J_{e'}$.

$$\tilde{0}_{\tilde{C}} \leq \tilde{d}_{c^*}(fI_{e'}, I_{e_{2n+1}}) \leq \tilde{d}_{c^*}(S(I_{e'}, J_{e'}), T(F_{e_{2n+1}}, G_{e_{2n+1}}))$$
$$\leq \tilde{a}^* \tilde{d}_{c^*}(fI_{e'}, gF_{e_{2n+1}})\tilde{a} + \tilde{a}^* \tilde{d}_{c^*}(fJ_{e'}, gG_{e_{2n+1}})\tilde{a}$$
$$\leq \tilde{a}^* \tilde{d}_{c^*}(fI_{e'}, I_{e_{2n}})\tilde{a} + \tilde{a}^* \tilde{d}_{c^*}(fJ_{e'}, J_{e_{2n}})\tilde{a}.$$

Taking the limit as $n \to \infty$ in the above relation, we obtain $\tilde{d}_{c^*}(fI_{e'}, I_{e'}) = 0_{\tilde{C}}$ which implies $fI_{e'} = I_{e'}$. Similarly we can prove $fJ_{e'} = J_{e'}$. Therefore, $S(I_{e'}, J_{e'}) = fI_{e'} = I_{e'}$ and $S(J_{e'}, I_{e'}) = fJ_{e'} = J_{e'}$. Thus, $(I_{e'}, J_{e'})$ is common coupled fixed point of S and f. Since $S(\tilde{E} \times \tilde{E}) \subseteq g(\tilde{E})$. So there exist $K_{e'}, L_{e'} \in \tilde{E}$ such that $S(I_{e'}, J_{e'}) = I_{e'} = gK_{e'}$ and $S(J_{e'}, I_{e'}) = J_{e'} = gL_{e'}$. Now from (Theorem 4 (4)) and using the triangular inequality

$$\tilde{0}_{\tilde{C}} \leq \tilde{d}_{c^*}(I_{e'}, T(K_{e'}, L_{e'})) \leq \tilde{d}_{c^*}(S((I_{e'}, J_{e'})), T(K_{e'}, L_{e'}))$$
$$\leq \tilde{a}^* \tilde{d}_{c^*}(fI_{e'}, gK_{e'})\tilde{a} + \tilde{a}^* \tilde{d}_{c^*}(fJ_{e'}, gL_{e'})\tilde{a}$$
$$\leq \tilde{a}^* \tilde{d}_{c^*}(I_{e'}, I_{e'})\tilde{a} + \tilde{a}^* \tilde{d}_{c^*}(J_{e'}, J_{e'})\tilde{a}.$$

We have $\tilde{d}_{c^*}(I_{e'}, T(K_{e'}, L_{e'})) = 0$, which means $I_{e'} = T(K_{e'}, L_{e'})$. Similarly, we can prove $T(L_{e'}, K_{e'}) = J_{e'}$. Since $\{T, g\}$ is ω-compatible pair, we have $T(I_{e'}, J_{e'}) = gI_{e'}$ and $T(J_{e'}, I_{e'}) = gJ_{e'}$. Now we prove that $gI_{e'} = I_{e'}$ and $gJ_{e'} = J_{e'}$.

$$\tilde{0}_{\tilde{C}} \leq \tilde{d}_{c^*}(I_{e'}, gI_{e'}) \leq \tilde{d}_{c^*}(S((I_{e'}, J_{e'})), T(I_{e'}, J_{e'}))$$
$$\leq \tilde{a}^* \tilde{d}_{c^*}(fI_{e'}, gI_{e'})\tilde{a} + \tilde{a}^* \tilde{d}_{c^*}(fJ_{e'}, gJ_{e'})\tilde{a}$$
$$\leq \tilde{a}^* \tilde{d}_{c^*}(I_{e'}, gI_{e'})\tilde{a} + \tilde{a}^* \tilde{d}_{c^*}(J_{e'}, gJ_{e'})\tilde{a} \tag{27}$$

and

$$\tilde{0}_{\tilde{C}} \leq \tilde{d}_{c^*}(J_{e'}, gJ_{e'}) \leq \tilde{d}_{c^*}(S((J_{e'}, I_{e'})), T(J_{e'}, I_{e'}))$$
$$\leq \tilde{a}^* \tilde{d}_{c^*}(fJ_{e'}, gJ_{e'})\tilde{a} + \tilde{a}^* \tilde{d}_{c^*}(fI_{e'}, gI_{e'})\tilde{a}$$
$$\leq \tilde{a}^* \tilde{d}_{c^*}(J_{e'}, gJ_{e'})\tilde{a} + \tilde{a}^* \tilde{d}_{c^*}(I_{e'}, gI_{e'})\tilde{a}. \tag{28}$$

From (27) and (28)

$$\tilde{0}_{\tilde{C}} \leq \tilde{d}_{c^*}(I_{e'}, gI_{e'}) + \tilde{d}_{c^*}(J_{e'}, gJ_{e'}) \leq (\sqrt{2}\tilde{a}^*)\left(\tilde{d}_{c^*}(I_{e'}, gI_{e'}) + \tilde{d}_{c^*}(J_{e'}, gJ_{e'})\right)(\sqrt{2}\tilde{a}).$$

Therefore,

$$\tilde{0} \leq \|\tilde{d}_{c^*}(I_{e'}, gI_{e'}) + \tilde{d}_{c^*}(J_{e'}, gJ_{e'})\|$$
$$\leq \|(\sqrt{2}\tilde{a}^*)\left(\tilde{d}_{c^*}(I_{e'}, gI_{e'}) + \tilde{d}_{c^*}(J_{e'}, gJ_{e'})\right)(\sqrt{2}\tilde{a})\|$$
$$\leq \|(\sqrt{2}\tilde{a})\|^2 \|\tilde{d}_{c^*}(I_{e'}, gI_{e'}) + \tilde{d}_{c^*}(J_{e'}, gJ_{e'})\|.$$

Since $\|(\sqrt{2}\tilde{a})\| < 1$, then $\|\tilde{d}_{c^*}(I_{e'}, gI_{e'}) + \tilde{d}_{c^*}(J_{e'}, gJ_{e'})\| = 0$. Hence $gI_{e'} = I_{e'}$ and $gJ_{e'} = J_{e'}$.

Therefore, we have $T(I_{e'}, J_{e'}) = gI_{e'} = I_{e'}$ and $T(J_{e'}, I_{e'}) = gJ_{e'} = J_{e'}$. Thus, $(I_{e'}, J_{e'})$ is common coupled fixed point of S, T, f and g. In the following we will show the uniqueness of common coupled fixed point in $\tilde{E}$. For this purpose, assume that there is another coupled fixed point $(I_{e''}, J_{e''})$ of S, T, f and g. Then

$$\tilde{d}_{c^*}(I_{e'}, I_{e''}) \leq \tilde{d}_{c^*}(S(I_{e'}, J_{e'}), T(I_{e''}, J_{e''}))$$
$$\leq \tilde{a}^* \tilde{d}_{c^*}(fI_{e'}, gI_{e''})\tilde{a} + \tilde{a}^* \tilde{d}_{c^*}(gJ_{e'}, gJ_{e''})\tilde{a}$$
$$\leq \tilde{a}^* \tilde{d}_{c^*}(I_{e'}, I_{e''})\tilde{a} + \tilde{a}^* \tilde{d}_{c^*}(J_{e'}, J_{e''})\tilde{a} \tag{29}$$

and

$$\tilde{d}_{c^*}(J_{e'}, J_{e''}) \leq \tilde{d}_{c^*}(S(J_{e'}, I_{e'}), T(J_{e''}, I_{e''}))$$
$$\leq \tilde{a}^* \tilde{d}_{c^*}(fJ_{e'}, gJ_{e''})\tilde{a} + \tilde{a}^* \tilde{d}_{c^*}(gI_{e'}, gI_{e''})\tilde{a}$$
$$\leq \tilde{a}^* \tilde{d}_{c^*}(J_{e'}, J_{e''})\tilde{a} + \tilde{a}^* \tilde{d}_{c^*}(I_{e'}, I_{e''})\tilde{a}. \tag{30}$$

From (29) and (30), we have that

$$\tilde{d}_{c^*}(I_{e'}, I_{e''}) + \tilde{d}_{c^*}(J_{e'}, J_{e''}) \leq (\sqrt{2}\tilde{a})^*\left(\tilde{d}_{c^*}(I_{e'}, I_{e''}) + \tilde{d}_{c^*}(J_{e'}, J_{e''})\right)(\sqrt{2}\tilde{a}),$$

which further induces that

$$\|\tilde{d}_{c^*}(I_{e'}, I_{e''}) + \tilde{d}_{c^*}(J_{e'}, J_{e''})\| \leq \|\sqrt{2}\tilde{a}\|^2 \|\tilde{d}_{c^*}(I_{e'}, I_{e''}) + \tilde{d}_{c^*}(J_{e'}, J_{e''})\|.$$

Since $\|\sqrt{2}\tilde{a}\| < 1$ then $\|\tilde{d}_{c^*}(I_{e'}, I_{e''}) + \tilde{d}_{c^*}(J_{e'}, J_{e''})\| = 0$. Hence we get $(I_{e'}, J_{e'}) = (I_{e''}, J_{e''})$ which means the coupled fixed point is unique.

To prove that S, T, f and g have a unique fixed point, we only have to prove $I_{e'} = J_{e'}$.

Now

$$\tilde{d}_{c^*}(I_{e'}, J_{e'}) = \tilde{d}_{c^*}(S(I_{e'}, J_{e'}), T(J_{e'}, I_{e'}))$$
$$\leq \tilde{a}^* \tilde{d}_{c^*}(fI_{e'}, gJ_{e'})\tilde{a} + \tilde{a}^* \tilde{d}_{c^*}(fJ_{e'}, gI_{e'})\tilde{a}$$
$$\leq \tilde{a}^* \tilde{d}_{c^*}(I_{e'}, J_{e'})\tilde{a} + \tilde{a}^* \tilde{d}_{c^*}(J_{e'}, I_{e'})\tilde{a},$$

then
$$\|\tilde{d}_{c^*}(I_{e'}, J_{e'})\| \leq \|\tilde{a}\|^2 \|\tilde{d}_{c^*}(I_{e'}, J_{e'})\| + \|\tilde{a}\|^2 \|\tilde{d}_{c^*}(J_{e'}, I_{e'})\|$$
$$\leq 2\|\tilde{a}\|^2 \|\tilde{d}_{c^*}(I_{e'}, J_{e'})\|.$$

It follows from the fact $\|a\| < \frac{1}{\sqrt{2}}$ that $\|\tilde{d}_{c^*}(I_{e'}, J_{e'})\| = 0$, thus $I_{e'} = J_{e'}$. Which means that S, T, f and g have a unique common fixed point. □

Corollary 3. *Let $(\tilde{E}, \tilde{C}, \tilde{d}_{c^*})$ be a C^*-algebra valued fuzzy soft metric space. Suppose $S: \tilde{E} \times \tilde{E} \to \tilde{E}$ and $f, g: \tilde{E} \to \tilde{E}$ be satisfying*

(1) $S(\tilde{E} \times \tilde{E}) \subseteq f(\tilde{E})$ and $S(\tilde{E} \times \tilde{E}) \subseteq g(\tilde{E})$
(2) $\{S, f\}$ and $\{S, g\}$ are ω-compatible pairs.
(3) one of $f(\tilde{E})$ or $g(\tilde{E})$ is complete C^*-algebra valued fuzzy soft metric of $\tilde{E}$
(4) $\tilde{d}_{c^*}\left(S(F_{e_1}, G_{e_1}), S(F_{e_2}, G_{e_2})\right) \leq \tilde{a}^* \tilde{d}_{c^*}(fF_{e_1}, gF_{e_2})\tilde{a} + \tilde{a}^* \tilde{d}_{c^*}(fG_{e_1}, gG_{e_2})\tilde{a}$
for all $F_{e_1}, F_{e_2}, G_{e_1}, G_{e_2} \in \tilde{E}$,

where $\tilde{a} \in \tilde{C}$ with $\|\sqrt{2}\tilde{a}\| < 1$. Then S and f, g have a unique common fixed point in $\tilde{E}$.

Corollary 4. *Let $(\tilde{E}, \tilde{C}, \tilde{d}_{c^*})$ be a C^*-algebra valued fuzzy soft metric space. Suppose $S: \tilde{E} \times \tilde{E} \to \tilde{E}$ and $f: \tilde{E} \to \tilde{E}$ be satisfying*

(1) $S(\tilde{E} \times \tilde{E}) \subseteq f(\tilde{E})$
(2) $\{S, f\}$ is ω-compatible pairs.
(3) $f(\tilde{E})$ is complete C^*-algebra valued fuzzy soft metric of $\tilde{E}$
(4) $\tilde{d}_{c^*}\left(S(F_{e_1}, G_{e_1}), S(F_{e_2}, G_{e_2})\right) \leq \tilde{a}^* \tilde{d}_{c^*}(fF_{e_1}, fF_{e_2})\tilde{a} + \tilde{a}^* \tilde{d}_{c^*}(fG_{e_1}, fG_{e_2})\tilde{a}$
for all $F_{e_1}, F_{e_2}, G_{e_1}, G_{e_2} \in \tilde{E}$,

where $\tilde{a} \in \tilde{C}$ with $\|\sqrt{2}\tilde{a}\| < 1$. Then S and f have a unique common fixed point in $\tilde{E}$.

Corollary 5. *Let $(\tilde{E}, \tilde{C}, \tilde{d}_{c^*})$ be a complete C^*-algebra valued fuzzy soft metric space. Suppose $S, T: \tilde{E} \times \tilde{E} \to \tilde{E}$ satisfies*

(1) $\tilde{d}_{c^*}\left(S(F_{e_1}, G_{e_1}), T(F_{e_2}, G_{e_2})\right) \leq \tilde{a}^* \tilde{d}_{c^*}(F_{e_1}, F_{e_2})\tilde{a} + \tilde{a}^* \tilde{d}_{c^*}(G_{e_1}, G_{e_2})\tilde{a}$

for all $F_{e_1}, F_{e_2}, G_{e_1}, G_{e_2} \in \tilde{E}$, where $\tilde{a} \in \tilde{C}$ with $\|\sqrt{2}\tilde{a}\| < 1$. Then S and T have a unique fixed point in $\tilde{E}$.

Corollary 6. *Let $(\tilde{E}, \tilde{C}, \tilde{d}_{c^*})$ be a complete C^*-algebra valued fuzzy soft metric space. Suppose $S: \tilde{E} \times \tilde{E} \to \tilde{E}$ satisfies*

(1) $\tilde{d}_{c^*}\left(S(F_{e_1}, G_{e_1}), S(F_{e_2}, G_{e_2})\right) \leq \tilde{a}^* \tilde{d}_{c^*}(F_{e_1}, F_{e_2})\tilde{a} + \tilde{a}^* \tilde{d}_{c^*}(G_{e_1}, G_{e_2})\tilde{a}$

for all $F_{e_1}, F_{e_2}, G_{e_1}, G_{e_2} \in \tilde{E}$, where $\tilde{a} \in \tilde{C}$ with $\|\sqrt{2}\tilde{a}\| < 1$. Then S has a unique fixed point in $\tilde{E}$.

Example 4. Let $E = \{e_1, e_2, e_3\}$, $U = \{p, q, r, s\}$ and C and D are two subset of E where $C = \{e_1, e_2, e_3\}$, $D = \{e_1, e_2\}$. Define fuzzy soft set as,

$$(F_E, C) = \left\{ \begin{array}{c} e_1 = \{p_{0.1}, q_{0.3}, r_{0.4}, s_{0.5}\}, e_2 = \{p_{0.3}, q_{0.4}, r_{0.6}, s_{0.8}\}, \\ e_3 = \{p_{0.6}, q_{0.7}, r_{0.8}, s_{0.9}\} \end{array} \right\}$$

$$(G_E, D) = \{e_1 = \{p_{0.4}, q_{0.5}, r_{0.2}, s_{0.6}\}, e_2 = \{p_{0.5}, q_{0.6}, r_{0.3}, s_{0.7}\}\}$$

$$F_{e_1} = \mu_{F_{e_1}} = \{p_{0.1}, q_{0.3}, r_{0.4}, s_{0.5}\}, F_{e_2} = \mu_{F_{e_2}} = \{p_{0.3}, q_{0.4}, r_{0.6}, s_{0.8}\}$$

$$F_{e_3} = \mu_{F_{e_3}} = \{p_{0.6}, q_{0.7}, r_{0.8}, s_{0.9}\}$$

$$G_{e_1} = \mu_{G_{e_1}} = \{p_{0.4}, q_{0.5}, r_{0.2}, s_{0.6}\}, G_{e_2} = \mu_{G_{e_2}} = \{p_{0.5}, q_{0.6}, r_{0.3}, s_{0.7}\}$$

and $FSC(F_E) = \{F_{e_1}, F_{e_2}, F_{e_3}, G_{e_1}, G_{e_2}\}$, let for all $e \in E$, $\tilde{E}(e) = \tilde{1}$ be absolute fuzzy soft set, and $\tilde{C} = M_2(R(C)^*)$, be the C^*-algebra. Define $\tilde{d}_{c^*}: \tilde{E} \times \tilde{E} \to \tilde{C}$ by $\tilde{d}_{c^*}(G_{e_1}, G_{e_2}) = (\inf\{|G_{e_1}(p) - G_{e_2}(p)|/p \in C\}, 0)$, then obviously $(\tilde{E}, \tilde{C}, \tilde{d}_{c^*})$ is a complete C^*-algebra valued fuzzy soft metric space.

We define $S: \tilde{E} \times \tilde{E} \to \tilde{E}$ by $S(F_{e_1}, G_{e_1})(p) = \frac{F_{e_1}^2 + G_{e_1}^2}{5}$, $T: \tilde{E} \times \tilde{E} \to \tilde{E}$ by $T(F_{e_1}, G_{e_1})(p) = \frac{F_{e_1}^2 + G_{e_1}^2}{3}$, $f: \tilde{E} \to \tilde{E}$ by $fF_{e_1} = \frac{F_{e_1}}{2}$ and $g: \tilde{E} \to \tilde{E}$ by $gF_{e_1} = F_{e_1}$ for all $p \in U$ and $F_{e_1}, G_{e_1} \in \tilde{E}$. Notice that $fF_{e_1} = \frac{F_{e_1}}{2} = \{0.05, 0.15, 0.20, 0.25\}$ and $gF_{e_2} = F_{e_2} = \{0.3, 0.4, 0.6, 0.8\}$. Thus, $\inf\{|\mu^p_{fF_{e_1}}(t) - \mu^p_{gF_{e_2}}(t)|/t \in C\} = \inf\{0.25, 0.25, 0.4, 0.55\} = 0.25$.

Hence $\tilde{d}_{c^*}(fF_{e_1}, gF_{e_2}) = \begin{bmatrix} 0.25 & 0 \\ 0 & 0.25 \end{bmatrix}$.

Also, $fG_{e_1} = \frac{G_{e_1}}{2} = \{0.2, 0.25, 0.10, 0.30\}$ and $gG_{e_2} = G_{e_2} = \{0.5, 0.6, 0.3, 0.7\}$. Thus, $\inf\{|\mu^p_{fG_{e_1}}(t) - \mu^p_{gG_{e_2}}(t)|/t \in C\} = \inf\{0.3, 0.35, 0.2, 0.4\} = 0.20$ and $\tilde{d}_{c^*}(fG_{e_1}, gG_{e_2}) = \begin{bmatrix} 0.20 & 0 \\ 0 & 0.20 \end{bmatrix}$.

Moreover, $S(F_{e_1}, G_{e_1})(p) = \frac{F_{e_1}^2 + G_{e_1}^2}{5} = \{0.034, 0.068, 0.040, 0.122\}$ and $T(F_{e_2}, G_{e_2})(p) = \frac{F_{e_2}^2 + G_{e_2}^2}{3} = \{0.11, 0.17, 0.15, 0.37\}$. Then

$$\tilde{d}_{c^*}(S(F_{e_1}, G_{e_1}), T(F_{e_2}, G_{e_2})) = \begin{bmatrix} 0.08 & 0 \\ 0 & 0.08 \end{bmatrix}$$

$$\leq \begin{bmatrix} \frac{\sqrt{3}}{3} & 0 \\ 0 & \frac{\sqrt{3}}{3} \end{bmatrix} \begin{bmatrix} 0.45 & 0 \\ 0 & 0.45 \end{bmatrix} \begin{bmatrix} \frac{\sqrt{3}}{3} & 0 \\ 0 & \frac{\sqrt{3}}{3} \end{bmatrix}$$

$$\leq \begin{bmatrix} \frac{\sqrt{3}}{3} & 0 \\ 0 & \frac{\sqrt{3}}{3} \end{bmatrix} \left(\begin{bmatrix} 0.25 & 0 \\ 0 & 0.25 \end{bmatrix} + \begin{bmatrix} 0.20 & 0 \\ 0 & 0.20 \end{bmatrix} \right) \begin{bmatrix} \frac{\sqrt{3}}{3} & 0 \\ 0 & \frac{\sqrt{3}}{3} \end{bmatrix}$$

$$\leq \tilde{c}^* \left(\tilde{d}_{c^*}(fF_{e_1}, gF_{e_2}) + \tilde{d}_{c^*}(fG_{e_1}, gG_{e_2}) \right) \tilde{c}.$$

Here $\tilde{c} = \begin{bmatrix} \frac{\sqrt{3}}{3} & 0 \\ 0 & \frac{\sqrt{3}}{3} \end{bmatrix}$ with $\|\tilde{c}\| = \frac{1}{\sqrt{3}} < \frac{1}{\sqrt{2}}$. Therefore, all the conditions of Theorem 4 satisfied. Hence S, T, f and g have a unique coupled fixed point.

Theorem 5. *Let $(\tilde{E}, \tilde{C}, \tilde{d}_{c^*})$ be a C^*-algebra valued fuzzy soft metric space. Suppose $S, T: \tilde{E} \times \tilde{E} \to \tilde{E}$ be satisfying*

(1) $S(\tilde{E} \times \tilde{E}) \subseteq T(\tilde{E} \times \tilde{E})$
(2) $\{S, T\}$ *is ω-compatible pairs.*
(3) *one of $S(\tilde{E} \times \tilde{E})$ or $T(\tilde{E} \times \tilde{E})$ is complete.*
(4) $\tilde{d}_{c^*}\left(S(F_{e_1}, G_{e_1}), S(F_{e_2}, G_{e_2})\right) \leq \tilde{a}^* \tilde{d}_{c^*}(T(F_{e_1}, G_{e_1}), T(F_{e_2}, G_{e_2})) \tilde{a}$
for all $F_{e_1}, F_{e_2}, G_{e_1}, G_{e_2} \in \tilde{E}$,

where $\tilde{a} \in \tilde{C}$ with $\|\tilde{a}\| < 1$. Then S and T have a unique common coupled fixed point in $\tilde{E} \times \tilde{E}$. Moreover, S and T have a unique common fixed point in $\tilde{E}$.

Proof. Similar to Theorem 4. □

Theorem 6. *Let $(\tilde{E}, \tilde{C}, \tilde{d}_{c^*})$ be a C^*-algebra valued fuzzy soft metric space. Suppose $S, T: \tilde{E} \times \tilde{E} \to \tilde{E}$ and $f, g: \tilde{E} \to \tilde{E}$ be satisfying*

(1) $S(\tilde{E} \times \tilde{E}) \subseteq g(\tilde{E})$ *and* $T(\tilde{E} \times \tilde{E}) \subseteq f(\tilde{E})$
(2) $\{S, f\}$ *and* $\{T, g\}$ *are ω-compatible pairs.*
(3) *one of $f(\tilde{E})$ or $g(\tilde{E})$ is complete C^*-algebra valued fuzzy soft metric of $\tilde{E}$*
(4) $\tilde{d}_{c^*}\left(S(F_{e_1}, G_{e_1}), T(F_{e_2}, G_{e_2})\right) \leq \tilde{a} \tilde{d}_{c^*}(S(F_{e_1}, G_{e_1}), fF_{e_1}) + \tilde{a} \tilde{d}_{c^*}(T(F_{e_2}, G_{e_2}), gF_{e_2})$
for all $F_{e_1}, F_{e_2}, G_{e_1}, G_{e_2} \in \tilde{E}$,

where $\tilde{a} \in \tilde{C}$ with $\|\tilde{a}\| < \frac{1}{2}$. Then S, T, f and g have a unique common coupled fixed point in $\tilde{E} \times \tilde{E}$. Moreover, S, T, f and g have a unique common fixed point in $\tilde{E}$.

Proof. Similar to Theorem 4. □

Corollary 7. Let $(\tilde{E}, \tilde{C}, \tilde{d_{c^*}})$ be a C^*-algebra valued fuzzy soft metric space. Suppose $S: \tilde{E} \times \tilde{E} \to \tilde{E}$ and $f, g: \tilde{E} \to \tilde{E}$ be satisfying

(1) $S(\tilde{E} \times \tilde{E}) \subseteq f(\tilde{E})$ and $S(\tilde{E} \times \tilde{E}) \subseteq g(\tilde{E})$
(2) $\{S, f\}$ and $\{S, g\}$ are ω-compatible pairs.
(3) one of $f(\tilde{E})$ or $g(\tilde{E})$ is complete C^*-algebra valued fuzzy soft metric of $\tilde{E}$
(4) $\tilde{d_{c^*}}\left(S(F_{e_1}, G_{e_1}), S(F_{e_2}, G_{e_2})\right) \leq \tilde{a}\tilde{d_{c^*}}(S(F_{e_1}, G_{e_1}), fF_{e_1}) + \tilde{a}\tilde{d_{c^*}}(S(F_{e_2}, G_{e_2}), gF_{e_2})$
for all $F_{e_1}, F_{e_2}, G_{e_1}, G_{e_2} \in \tilde{E}$,

where $\tilde{a} \in \tilde{C}$ with $\|\tilde{a}\| < \frac{1}{2}$. Then S and f, g have a unique common fixed point in $\tilde{E}$.

Corollary 8. Let $(\tilde{E}, \tilde{C}, \tilde{d_{c^*}})$ be a C^*-algebra valued fuzzy soft metric space. Suppose $S: \tilde{E} \times \tilde{E} \to \tilde{E}$ and $f: \tilde{E} \to \tilde{E}$ be satisfying

(1) $S(\tilde{E} \times \tilde{E}) \subseteq f(\tilde{E})$
(2) $\{S, f\}$ is ω-compatible pairs.
(3) $f(\tilde{E})$ is complete C^*-algebra valued fuzzy soft metric of $\tilde{E}$
(4) $\tilde{d_{c^*}}\left(S(F_{e_1}, G_{e_1}), S(F_{e_2}, G_{e_2})\right) \leq \tilde{a}\tilde{d_{c^*}}(S(F_{e_1}, G_{e_1}), fF_{e_1}) + \tilde{a}\tilde{d_{c^*}}(S(F_{e_2}, G_{e_2}), fF_{e_2})$
for all $F_{e_1}, F_{e_2}, G_{e_1}, G_{e_2} \in \tilde{E}$,

where $\tilde{a} \in \tilde{C}$ with $\|\tilde{a}\| < \frac{1}{2}$. Then S and f have a unique common fixed point in $\tilde{E}$.

Corollary 9. Let $(\tilde{E}, \tilde{C}, \tilde{d_{c^*}})$ be a complete C^*-algebra valued fuzzy soft metric space. Suppose $S: \tilde{E} \times \tilde{E} \to \tilde{E}$ satisfies

(1) $\tilde{d_{c^*}}\left(S(F_{e_1}, G_{e_1}), S(F_{e_2}, G_{e_2})\right) \leq \tilde{a}\tilde{d_{c^*}}(S(F_{e_1}, G_{e_1}), F_{e_1}) + \tilde{a}\tilde{d_{c^*}}(GS(F_{e_2}, G_{e_2}), F_{e_2})$

for all $F_{e_1}, F_{e_2}, G_{e_1}, G_{e_2} \in \tilde{E}$, where $\tilde{a} \in \tilde{C}$ with $\|\tilde{a}\| < \frac{1}{2}$. Then S has a unique fixed point in $\tilde{E}$.

4. Applications to Integral Equations

Theorem 7. Let us Consider the integral equation

$$F_{e_1}(x) = \int_C \left(T_1(x, y, F_{e_1}(y)) + T_1(x, y, F_{e_1}(y))\right) dy, x \in C$$

$$F_{e_1}(x) = \int_C \left(I_1(x, y, F_{e_1}(y)) + I_2(x, y, F_{e_1}(y))\right) dy, x \in C.$$

where C is a Lebesgue measurable set. Suppose that

(i) $T_1, T_2 : C \times C \times R(C)^* \to R(C)^*$ and $I_1, I_2 : C \times C \times R(C)^* \to R(C)^*$.
(ii) there exist two continuous function $\phi, \varphi : C \times C \to R(C)^*$ and $r \in (0, 1)$ such that for $u, v \in C$ and $F_{e_1}(v), F_{e_2}(v) \in R(C)^*$

$$\inf\{|T_1(u, v, F_{e_1}(v)) - I_1(u, v, F_{e_2}(v))|\} \leq r \inf\{|\phi(u, v)|\} \cdot \inf\{|(F_{e_1}(v) - F_{e_2}(v))|\},$$
$$\inf\{|T_2(u, v, F_{e_1}(v)) - I_2(u, v, F_{e_2}(v))|\} \leq r \inf\{|\varphi(u, v)|\} \cdot \inf\{|(F_{e_1}(v) - F_{e_2}(v))|\}$$

(iii) $\sup_{x \in C} \int_C \inf\{|\phi(u, v)|\} dv \leq 1$ and $\sup_{x \in C} \int_C \inf\{|\varphi(u, v)|\} dv \leq 1$

then the integral equation has a unique solutions in $L^\infty(C)$.

Proof. Let $E = C = [0,1]$ and $\tilde{E} = L^\infty(C)$ be the set of essential bounded measurable function on C and $H = L^2(C)$. The set of bounded linear operators on Hilbert space H denoted by $L(H)$. Consider $\tilde{d}_{c^*} : \tilde{E} \times \tilde{E} \to L(H)$ by $\tilde{d}_{c^*}(F_{e_1}, F_{e_2}) = M_{\inf\{|\mu^p_{F_{e_1}}(y) - \mu^p_{F_{e_2}}(y)|/y \in C\}}$ for all $F_{e_1}, F_{e_2} \in \tilde{E}$, where $M_h : H \to H$ is the multiplication operator defined by $M_h(\phi) = h \cdot \phi$ for $\phi \in H$. Then $\tilde{d}_{c^*}$ is a C^*-algebra valued fuzzy soft metric and $(\tilde{E}, L(H), \tilde{d}_{c^*})$ is a complete C^*-algebra valued fuzzy soft metric space. Define two self mappings $S, T : \tilde{E} \times \tilde{E} \to \tilde{E}$ by

$$S(F_{e_1}, G_{e_1})(x) = \int_C \left(T_1(x, y, F_{e_1}(y)) + T_2(x, y, G_{e_1}(y))\right) dy, \quad x \in C,$$

$$T(F_{e_2}, G_{e_2})(x) = \int_C \left(I_1(x, y, F_{e_2}(y)) + I_2(x, y, G_{e_2}(y))\right) dy, \quad x \in C.$$

Notice that
$$\tilde{d}_{c^*}(S(F_{e_1}, G_{e_1}), T(F_{e_2}, G_{e_2})) = M_{\inf\{|\mu^p_{S(F_{e_1}, G_{e_1})}(y) - \mu^p_{T(F_{e_2}, G_{e_2})}(y)|/y \in C\}}$$

$$\|\tilde{d}_{c^*}(S(F_{e_1}, G_{e_1}), T(F_{e_2}, G_{e_2}))\|$$

$$= \sup_{\|h\|=1} \left(M_{\inf\{|\mu^p_{S(F_{e_1}, G_{e_1})}(y) - \mu^p_{T(F_{e_2}, G_{e_2})}(y)|/y \in C\}} h, h\right)$$

$$= \sup_{\|h\|=1} \int_C \left[\inf\{|\mu^p_{S(F_{e_1}, G_{e_1})}(y) - \mu^p_{T(F_{e_2}, G_{e_2})}(y)|/y \in C\}\right] h(x) \overline{h(x)} dx$$

$$\leq \sup_{\|h\|=1} \int_C \left[\int_C \inf\{|T_1(x, y, F_{e_1}(y)) - I_1(x, y, F_{e_2}(y))|\} dy\right] |h(x)|^2 dx$$

$$+ \sup_{\|h\|=1} \int_C \left[\int_C \inf\{|T_2(x, y, G_{e_1}(y)) - I_2(x, y, G_{e_2}(y))|\} dy\right] |h(x)|^2 dx$$

$$\leq \sup_{\|h\|=1} \int_C \left[\int_C r \inf\{|\phi(x, y)(F_{e_1}(y) - F_{e_2}(y))|\} dy\right] |h(x)|^2 dx$$

$$+ \sup_{\|h\|=1} \int_C \left[\int_C r \inf\{|\varphi(x, y)(G_{e_1}(y) - G_{e_2}(y))|\} dy\right] |h(x)|^2 dx$$

$$\leq r \sup_{\|h\|=1} \int_C \left[\int_C \inf\{|\phi(x, y)|\} \inf\{|F_{e_1}(y) - F_{e_2}(y)|\} dy\right] |h(x)|^2 dx$$

$$+ r \sup_{\|h\|=1} \int_C \left[\int_C \inf\{|\varphi(x, y)|\} \inf\{|G_{e_1}(y) - G_{e_2}(y)|\} dy\right] |h(x)|^2 dx$$

$$\leq r \sup_{\|h\|=1} \int_C \left[\int_C \inf\{|\phi(x, y)|\} dy\right] |h(x)|^2 dx. \|\inf\{|F_{e_1}(y) - F_{e_2}(y)|\}\|_\infty$$

$$+ r \sup_{\|h\|=1} \int_C \left[\int_C \inf\{|\varphi(x, y)|\} dy\right] |h(x)|^2 dx. \|\inf\{|G_{e_1}(y) - G_{e_2}(y)|\}\|_\infty$$

$$\leq r \sup_{\|h\|=1} \int_C \inf\{|\phi(x, y)|\} dy. \sup_{\|h\|=1} \int_C |h(x)|^2 dx. \|\inf\{|F_{e_1}(y) - F_{e_2}(y)|\}\|_\infty$$

$$+ r \sup_{\|h\|=1} \int_C \inf\{|\varphi(x,y)|\} dy \cdot \sup_{\|h\|=1} \int_C |h(x)|^2 dx \cdot \|\inf\{|G_{e_1}(y) - G_{e_2}(y)|\}\|_\infty$$

$$\leq \quad r.\|\inf\{|F_{e_1}(y) - F_{e_2}(y)|\}\|_\infty + r.\|\inf\{|G_{e_1}(y) - G_{e_2}(y)|\}\|_\infty.$$

Set $\tilde{a} = \sqrt{r} 1_{L(H)}$, then $\tilde{a} \in L(H)$ and $\|\tilde{a}\| = \sqrt{r} < \frac{1}{\sqrt{2}}$. Hence, applying our Corollary 5, we get the desired result. □

5. Conclusions

In the present work, we proved some existing and uniqueness fixed point results for these new type of contractive mappings in complete C^*-algebra valued fuzzy soft metric spaces. Furthermore, the examples illustrate the validity of the obtained results. We hope that the results of this paper will support researchers and promote future study on C^*-algebra valued fuzzy soft metric spaces.

Author Contributions: R.P.D. analyzed and prepared/edited the manuscript, N.V.K.G. analyzed and prepared/edited the manuscript, H.I. analyzed and prepared/edited the manuscript, S.R.B. analyzed and prepared/edited the manuscript, A.L.G. analyzed and prepared/edited the manuscript.

Acknowledgments: The authors are very thanks to the reviewers and editors for valuable comments, remarks and suggestions for improving the content of the paper.

Conflicts of Interest: The authors declare no conflict of interest.

References

1. Beaula, T.; Gunaseeli, C. On fuzzy soft metric spaces. *Malaya J. Mat.* **2015**, *2*, 438–442.
2. Chang, S.S.; Cho, Y.J.; Lee, B.S.; Jung, J.S.; Kang, S.M. Coincidence point and minimization theorems in fuzzy metric spaces. *Fuzzy Sets Syst.* **1997**, *88*, 119–128. [CrossRef]
3. Abbas, M.; Murtaza, G.; Romaguera, S. On the fixed point theory of soft metric spaces. *Fixed Point Theory Appl.* **2016**, *1*, 17. [CrossRef]
4. Işık, H.; Turkoglu, D. Generalized weakly α-contractive mappings and applications to ordinary differential equations. *Miskolc Math. Notes* **2016**, *17*, 365–379. [CrossRef]
5. Beaulaa, T.; Rajab, R. Completeness in fuzzy soft metric space. *Malaya J. Mat.* **2015**, *S*, 438–442.
6. Neog, T.J.; Sut, D.K.; Hazarika, G.C. Fuzzy soft topological space. *Int. J. Latest Tend. Math.* **2012**, *2*, 54–67.
7. Ma, Z.; Jiang, L.; Sun, H. C^*-algebra valued metric space and related fixed point theoerms. *Fixed Point Theory Appl.* **2014**, *2014*, 206. [CrossRef]
8. Zada, A.; Saifullah, S.; Ma, Z. Common fixed point theorems for G-contraction in C^*-algebra valued metric spaces. *Int. J. Anal. Appl.* **2016**, *11*, 23–27.
9. Işık, H.; Turkoglu, D. Some fixed point theorems in ordered partial metric spaces. *J. Inequal. Spec. Funct.* **2013**, *4*, 13–18.
10. Cao, T. Some coupled fixed point theorems in C^*-algebra valued metric spaces. *arXiv* **2016**, arXiv:1601.07168v1.
11. Batul, S.; Kamran, T. C^*-valued contractive type mappings. *Fixed Point Theory Appl.* **2015**, *2015*. [CrossRef]
12. Kadelburg, Z.; Radenović, S. Fixed point results in C^*-algebra valued metric spaces are direct consequence of their standard counterparts. *Fixed Point Theory Appl.* **2016**, *2016*. [CrossRef]
13. Işık, H.; Hussain, N.; Kutbi, M.A. Generalized rational contractions endowed with a graph and an application to a system of integral equations. *J. Comput. Anal. Appl.* **2017**, *22*, 1158–1175.
14. Nadler, S.B. Multivalued contraction mappings. *Pac. J. Math.* **1969**, *30*, 475–488. [CrossRef]
15. Mizoguchi, N.; Takahashi, W. Fixed point theorems for multivalued mappings on complete metric spaces. *J. Math. Anal. Appl.* **1989**, *141*, 177–188. [CrossRef]
16. Ćirić, L. Fixed point theorems for multi-valued contractions in complete metric spaces. *J. Math. Anal. Appl.* **2008**, *348*, 499–507. [CrossRef]
17. Damjanović, B.; Samet, B.; Vetro, C. Common fixed point theorems for multi-valued maps. *Acta Math. Sci.* **2012**, *32*, 818–824. [CrossRef]

18. Aydi, H.; Abbas, M.; Vetro, C. Common fixed points for multi-valued generalized contractions on partial metric spaces. *Rev. Real Acad. Cienc. Exactas Fis. Nat. Ser. Mat.* **2014**, *108*, 483–501. [CrossRef]
19. Kaewcharoen, A.; Yuying, T. Coincidence points and fixed point theorems for multi-valued mappings. *Intern. J. Pure Appl. Math.* **2013**, *89*, 531–546. [CrossRef]
20. Işık, H.; Ionescu, C. New type of multivalued contractions with related results and applications. *U.P.B. Sci. Bull. Ser. A* **2018**, *80*, 13–22.
21. Popa, V. Coincidence and fixed point theorems for noncontinuous hybrid contractions. *Nonlinear Anal. Forum* **2002**, *7*, 153–158.
22. Rao, K.P.R.; Rao, K.R.K. Unique common fixed point theorems for pairs of hybrid maps under new conditions in partial metric spaces. *Demonstr. Math.* **2014**, *47*, 715–725. [CrossRef]
23. Cirić, L.B.; Ume, J.S. Common fixed point theorems for multi-valued non-self mappings. *Publ. Math. Debr.* **2002**, *60*, 359–371.
24. Pathak, H.K.; Agarwal, R.P.; Cho, Y.J.E. Coincidence and fixed point for multivalued mappings and its applications to non convex integral inclusions. *J. Comput. Appl. Math.* **2015**, *283*, 201–221. [CrossRef]
25. Agarwal, R.P.; Kishore, G.N.V.; Rao, B.S. Convergence properties on C^*-algebras valued fuzzy soft metric spaces and reated fixed point theorems. *Malaya J. Mat.* **2018**, *6*, 310–320. [CrossRef]
26. Rao, B.S.; Kishore, G.N.V.; Prasad, T.V. Fixed point theorems under Caristi's type map on C^*-algebra valued fuzzy soft metric space. *Int. J. Eng. Technol.* **2017**, *7*, 111–114.
27. Murphy, G.J. *C^*-Algebras and Operator Theory*; Academic Press: London, UK, 1990.
28. Saleem, N.; Abbas, M.; De la Sen, M. Optimal approximate solution of coincidence point equations in fuzzy metric spaces. *Mathematics* **2019**, *7*, 327. [CrossRef]
29. De la Sen, M.; Abbas, M.; Saleem, N. On optimal fuzzy best proximity coincidence points of proximal contractions involving cyclic mappings in non-Archimedean fuzzy metric spaces. *Mathematics* **2017**, *5*, 22. [CrossRef]
30. Basha, S.S. Best proximity points: Optimal solutions. *J. Optim. Theory Appl.* **2011**, *151*, 210.
31. Shehwar, D.; Kamran, T. C^*-valued G-contraction and fixed points. *J. Inequal. Appl.* **2015**, *2015*, 304. [CrossRef]
32. Alsulami, H.H.; Agarwal, R.P.; Karapınar, E.; Khojasteh, F. A short note on C*-valued contraction mappings. *J. Inequal. Appl.* **2016**, *2016*, 50. [CrossRef]
33. Dur-e-Shehwar Batul, S.; Kamran, T.; Ghiura, A. Ceristi's fixed point theorem on C^*-algebra valued metric spaces. *J. Nonlinear Sci. Appl.* **2016**, *9*, 584–588.
34. Kamran, T.; Postolache, M.; Ghiura, A.; Batul, S.; Ali, R. The Banach contraction principle in C^*-algebra-valued b-metric spaces with application. *Fixed Point Theory Appl.* **2016**, *2016*, 10. [CrossRef]
35. Lal Shateri, T. C^*-algebra-valued modular spaces and fixed point theorems. *J. Fixed Point Theory Appl.* **2017**, *19*, 1551–1560. [CrossRef]
36. Zadeh, L.A. Fuzzy soft. *Inf. Control* **1965**, *8*, 338–353. [CrossRef]
37. Molodstov, D.A. Fuzzy soft sets–First results. *Comput. Math. Appl.* **1999**, *37*, 19–31.
38. Maji, P.K.; Biswas, R.; Roy, A.R. Fuzzy soft sets. *J. Fuzzy Math.* **2001**, *9*, 589–602.
39. Maji, P.K.; Roy, A.R.; Biswas, R. An application of soft sets in a decision making problem. *Comput. Math. Appl.* **2002**, *44*, 1077–1083. [CrossRef]
40. Das, S.; Samanta, S.K. Soft metric. *Ann. Fuzzy Math. Inform.* **2013**, *6*, 77–94.
41. Das, S.; Samanta, S.K. Soft real sets, soft real numbers and their properties. *J. Fuzzy Math.* **2012**, *20*, 551–576.
42. Ghosh, J.; Dinda, B.; Samant, T.K. Fuzzy soft rings and fuzzy soft ideals. *Int. J. Pure Appl. Sci. Technol.* **2011**, *2*, 66–74.
43. Roy, S.; Samanta, T.K. A note on fuzzy soft topological spaces. *Ann. Fuzzy Math. Inform.* **2012**, *3*, 305–311.
44. Tanay, B.; Kandemir, M.B. Topological structure of fuzzy soft sets. *Comput. Math. Appl.* **2011**, *61*, 2952–2957. [CrossRef]

© 2019 by the authors. Licensee MDPI, Basel, Switzerland. This article is an open access article distributed under the terms and conditions of the Creative Commons Attribution (CC BY) license (http://creativecommons.org/licenses/by/4.0/).

Article

Relatively Cyclic and Noncyclic *P*-Contractions in Locally $\mathbb{K}$-Convex Space

Edraoui Mohamed [1], Aamri Mohamed [1] and Lazaiz Samih [2,*]

[1] Laboratory of Algebra, Analysis and Applications, Department of Mathematics, Ben M'sik Faculty of Sciences, University Hassan II, Casablanca 20000, Morocco
[2] Laboratory of Mathematical Analysis and Applications, Department of Mathematics, Dhar El Mahraz Faculty of Sciences, University Sidi Mohamed Ben Abdellah, Fes 30050, Morocco
* Correspondence: samih.lazaiz@gmail.com; Tel.: +212-062-2438-649

Received: 24 May 2019; Accepted: 20 July 2019; Published: 6 August 2019

Abstract: Our main goal of this research is to present the theory of points for relatively cyclic and relatively relatively noncyclic *p*-contractions in complete locally $\mathbb{K}$-convex spaces by providing basic conditions to ensure the existence and uniqueness of fixed points and best proximity points of the relatively cyclic and relatively noncyclic *p*-contractions map in locally $\mathbb{K}$-convex spaces. The result of this paper is the extension and generalization of the main results of Kirk and A. Abkar.

Keywords: fixed point; locally $\mathbb{K}$-convex spaces; relatively cyclic and relatively noncyclic *p*-contractions; best proximity point

1. Introduction

Let $\mathbb{K}$ be a non-archimedean valued field, i.e., $\mathbb{K}$ is neither $\mathbb{R}$ nor $\mathbb{C}$, endowed with an absolute valued function $|.|$ such that

$$|x+y| \leq \max\{|x|,|y|\} \quad (x,y \in \mathbb{K})$$

Let X be a topological vector space over $\mathbb{K}$. A seminorm on the $\mathbb{K}$-vector space X is a map $p: X \to [0.\infty)$ satisfies

(i) $p(\lambda x) = |\lambda| p(x)$, $x \in X$ and $\lambda \in \mathbb{K}$.
(ii) $p(x+y) \leq \max\{p(x), p(y)\}$, $x, y \in X$

For a seminorm p we have $p(0) = 0$ but $p(x)$ is allowed to be 0 for non-zero x. Note that each norm is a seminorm that vanishes only at 0.

Recall that a topological vector space (X, τ) over $\mathbb{K}$ is called a (non-archimedean) locally $\mathbb{K}$-convex space if τ has a basis of absolutely convex neighborhoods (a subset $A \subset X$ is called absolutely $\mathbb{K}$-convex if $0 \in A$ and $ax + by \in A$ for all $x, y \in A$ and $a, b \in B_{\mathbb{K}}$ where $B_{\mathbb{K}} = \{a \in \mathbb{K} : |a| \leq 1\}$). Every locally $\mathbb{K}$-convex topology can be generated in a natural way by some system of non-archimedean seminorms $\Gamma = \{p_\alpha\}$. A locally $\mathbb{K}$-convex space X is Hausdorff if and only if for each non-zero $x \in X$ there is a continuous seminorm p on X such that $p(x) \neq 0$. A sequence $\{a_1, a_2, \ldots\}$ in X is called Cauchy net if and only if $\lim_n p(a_{n+1} - a_n) = 0$ for any seminorm p. This follows from

$$p(a_m - a_n) \leq \max\{p(a_m - a_{m-1}), \ldots, p(a_{n+1} - a_n)\}, \quad m > n.$$

A subset S of a Hausdorff locally $\mathbb{K}$-convex space is called complete if each Cauchy net in S converges to a limit that lies in S. For details, see [1–4].

On the other hand, the most fundamental fixed point theorem is the so-called Banach contraction principle (BCP for short), this result played an important role in various fields in mathematics. Due to its importance and simplicity, several authors have obtained many interesting extensions and generalizations of the Banach contraction principle. Ciric [5] introduced quasi-contraction map, which allowed him to generalize the Banach contraction principle.

In the absence of a fixed point, i.e., the equation $Tx = x$ has no solution, it is interesting to ask whether it is possible to find $(a,b) \in A \times B$ such that

$$p(a - Ta) = p(b - Tb) = D_p(A,B). \tag{1}$$

A point $(\bar{a}, \bar{b}) \in A \times B$ is said to be a best proximity pair for the mapping $T : A \cup B \to A \cup B$ if it is solution to the problem (1). Another interesting subject of the fixed point theory is the concept of cyclic contractions maps and the best points of proximity provided by Kirk et al. [6,7].

$(A; B)$ a nonempty pair of subsets of a locally $\mathbb{K}$-convex space (X, Γ), we say that a mapping $T : A \cup B \to A \cup B$ is cyclic (resp. noncyclic) provided that $T(A) \subset B$ and $T(B) \subset A$ (resp. $T(A) \subset A$ and $T(B) \subset B$).

There are many results in this area see [8–12].

2. Fixed Point Results for Relatively Cyclic P-Contractions

In this section, we derive some fixed point theorems of certain relatively cyclic-type p-contractions in a complete locally $\mathbb{K}$-convex space.

Definition 1. *Let A and B be non empty subsets of locally $\mathbb{K}$-convex space (X, Γ). A relatively cyclic map $T : A \cup B \to A \cup B$ is said to be relatively cyclic p-contraction if there exists $0 \le \gamma_p < 1$ such that for all $p \in \Gamma$ and $a \in A$ and $b \in B$ we have*

$$p(Ta - Tb) \le \gamma_p p(a - b). \tag{2}$$

Theorem 1. *Let (X, Γ) be a complete Hausdorff locally $\mathbb{K}$-convex space, A and B be non empty closed subsets of X and $T : A \cup B \to A \cup B$ a relatively cyclic p-contraction map. Then T has a unique fixed point in $A \cap B$.*

Proof. Taking a point $a \in A$ since T is p-contraction, we have

$$p\left(T^2 a - Ta\right) = p(T(Ta) - Ta) \le \gamma_p p(Ta - a)$$

and

$$\begin{aligned} p(T^3 a - T^2 a) &= p\left(T(T^2 a) - T(Ta)\right) \\ &\le \gamma_p (T^2 a - Ta) \\ &\le \gamma_p^2 p(Ta - a) \end{aligned}$$

Inductively, using this process for all $n \in \mathbb{N}$ we have

$$p\left(T^{n+1} a - T^n a\right) \le \gamma_p^n p(Ta - a)$$

Let $n \le m$

$$p(T^m a - T^n a) \leq \max\{p(T^m a - T^{m-1} a), p(T^{m-1} a - T^{m-2} a), ..., p(T^{n+1} a - T^n a)\}$$
$$\leq \max\{\gamma_p^{m-1} p(Ta - a), \gamma_p^{m-2} p(Ta - a), ..., \gamma_p^n p(Ta - a)\}$$
$$\leq \gamma_p^n p(Ta - a)$$

Since $0 \leq \gamma_p < 1$, $\gamma_p^n \to 0$ as $n \to \infty$, we get $p(T^m a - T^n a) \to 0$, thus $\{T^n a\}$ is a p-Cauchy sequence. Since (X, Γ) is complete, we have $\{T^n a\} \to \bar{a} \in X$. We note, that $\{T^{2n} a\}$ is a sequence in A and $\{T^{2n-1} a\}$ is a sequence in B in a way that both sequences tend to same limit $\bar{a}$. Since A and B are closed, we have that $\bar{a} \in A \cap B$. Hence $A \cap B \neq \emptyset$.

We claim that $T\bar{a} = \bar{a}$. Considering the condition relatively cyclic p-contraction we have

$$p(T^{2n} a - T\bar{a}) = p(TT^{2n-1} a - T\bar{a})$$
$$\leq \gamma_p p(T^{2n-1} a - \bar{a})$$

Taking limit as $n \to \infty$ in above inequality, we have

$$p(\bar{a} - T\bar{a}) \leq \gamma_p p(\bar{a} - T\bar{a}) < p(\bar{a} - T\bar{a})$$

This implies that $p(\bar{a} - T\bar{a}) = 0$. Since X is Hausdorff, $T\bar{a} = \bar{a}$.

We shall prove that $\bar{a}$ is the existence of a unique fixed point of T. Clearly from (2) if $\bar{a}$ and $\bar{b}$ be two fixed points of T we have

$$p(\bar{a} - \bar{b}) = p(T\bar{a} - T\bar{b}) \leq \gamma_p p(\bar{a} - \bar{b})$$

Since $0 \leq \gamma_p < 1$ this implies $\bar{a} = \bar{b}$. Hence the proof is completed. □

Corollary 1. *Let A and B be two non-empty closed subsets of a complete Hausdorff locally $\mathbb{K}$-convex space X. Let $T_1 : A \to B$ and $T_2 : B \to A$ be two functions such that*

$$p(T_1(a) - T_2(b)) \leq \gamma_p p(a - b) \tag{3}$$

for all $p \in \Gamma$, $a \in A$ and $b \in B$ where $0 \leq \gamma_p < 1$. Then there exists a unique $\bar{a} \in A \cap B$ such that

$$T_1(\bar{a}) = T_2(\bar{a}) = \bar{a}$$

Proof. Apply Theorem 1 to the mapping $T : A \cup B \to A \cup B$ defined by:

$$T(a) = \begin{cases} T_1(a) & \text{if } a \in A \\ T_2(a) & \text{if } a \in B. \end{cases}$$

Observe that condition (3) is reduced to condition (2). Then T has a unique fixed $\bar{a} \in A \cap B$ such that

$$T_1(\bar{a}) = T_2(\bar{a}) = \bar{a}.$$

□

Theorem 2. *Let (X, Γ) be a complete Hausdorff locally $\mathbb{K}$-convex space, A and B two non empty closed subsets of X and $T : A \cup B \to A \cup B$ be a relatively cyclic mapping that satisfies the condition*

$$p(Ta - Tb) \leq \gamma_p \max\{p(a - b), p(a - Ta), p(b - Tb)\} \tag{4}$$

for all $p \in \Gamma$, $a \in A$ and $b \in B$ and $0 \leq \gamma_p < 1$. Then, T has a unique fixed point in $A \cap B$.

Proof. Let $a \in A$. By condition (4), we have

$$\begin{aligned} p\left(T^2a - Ta\right) &= p\left(T\left(Ta\right) - Ta\right) \\ &\leq \gamma_p \max\left\{p\left(Ta - a\right), p\left(Ta - T^2a\right)\right\} \\ &\leq \gamma_p p\left(Ta - a\right). \end{aligned}$$

Similarly, we get $p\left(T^3a - T^2a\right) \leq \gamma_p^2 p\left(Ta - a\right)$.

Inductively, using this process for all $n \in \mathbb{N}$ we have

$$p\left(T^{n+1}a - T^n a\right) \leq \gamma_p \max\left\{p\left(Ta - a\right), p\left(Ta - T^2a\right)\right\}$$

thus

$$\begin{aligned} p\left(T^m a - T^n a\right) &\leq \max\left\{p\left(T^m a - T^{m-1}a\right), p\left(T^{m-1}a - T^{m-2}a\right), ..., p\left(T^{n+1}a - T^n a\right)\right\} \\ &\leq \max\left\{\gamma_p^{m-1}p\left(Ta - a\right), \gamma_p^{m-2}p\left(Ta - a\right), ..., \gamma_p^n p\left(Ta - a\right)\right\} \\ &\leq \gamma_p^n p\left(Ta - a\right) \end{aligned}$$

Since $0 \leq \gamma_p < 1$, $\gamma_p^n \mapsto 0$ as $n \mapsto \infty$, we get $p\left(T^m a - T^n a\right) \to 0$. Hence $\{T^n a\}$ is a p-Cauchy sequence. As (X, Γ) is complete, we have $\{T^n a\} \to \bar{a} \in X$. We note, that $\{T^{2n}a\}$ is a sequence in A and $\{T^{2n-1}a\}$ is a sequence in B so that the two sequences tend to the same limit $\bar{a}$. Since A and B are closed, we have that $\bar{a} \in A \cap B$ that is $A \cap B \neq \emptyset$.

Considering the condition (4) we have:

$$\begin{aligned} p\left(T^{2n}a - T\bar{a}\right) &= p\left(TT^{2n-1}a - T\bar{a}\right) \\ &\leq \gamma_p \max\left\{p\left(T^{2n-1}a - \bar{a}\right), p\left(T^{2n-1}a - T^{2n}a\right), p\left(\bar{a} - T\bar{a}\right)\right\} \end{aligned}$$

Taking limit as $n \to \infty$ in above inequality, we have

$$p\left(z - Tz\right) \leq \gamma_p p\left(z - Tz\right) < p\left(z - Tz\right)$$

which implies that $p\left(\bar{a} - T\bar{a}\right) = 0$, since X is Hausdorff, $T\bar{a} = \bar{a}$.

Clearly from (4) if u and v be fixed points of T we have

$$\begin{aligned} p\left(u - v\right) &= p\left(Tu - Tv\right) \\ &\leq \gamma_p \max\left\{p\left(u - v\right), p\left(u - Tu\right), p\left(v - Tv\right)\right\} \\ &\leq \gamma_p p\left(u - v\right) \end{aligned}$$

Since $0 \leq \gamma_p < 1$ this implies $u = v$. □

Corollary 2. *Let A and B be two non-empty closed subsets of a complete Hausdorff locally $\mathbb{K}$-convex space X. let $T_1 : A \to B$ and $T_2 : B \to A$ be two functions such that*

$$p\left(T_1\left(a\right) - T_2\left(b\right)\right) \leq \gamma_p \max\left\{p\left(a - b\right), p\left(a - T_1\left(a\right)\right), p\left(b - T_2\left(b\right)\right)\right\} \quad (5)$$

for all $p \in \Gamma$ and $a \in A$ and $b \in B$ where $0 < \gamma_p < 1$. Then there exists a unique $\bar{a} \in A \cap B$ such that

$$T_1\left(\bar{a}\right) = T_2\left(\bar{a}\right) = \bar{a}$$

Proof. Let $T: A \cup B \to A \cup B$ defined by

$$T(a) = \begin{cases} T_1(a) & \text{if } a \in A \\ T_2(a) & \text{if } a \in B \end{cases}$$

Then T satisfies condition (4), we can now apply Theorem 2 to deduce that T has a unique fixed point $\bar{a} \in A \cap B$ such that

$$T_1(\bar{a}) = T_2(\bar{a}) = \bar{a}$$

□

3. Fixed Points of Relatively Noncyclic Mappings

In this section motivated by Theorem 3.1 [13], we prove the existence of a best proximity point of relatively noncyclic mappings and studied the existence of solution of problem (1) for relatively p-nonexpansive mappings in locally $\mathbb{K}$-convex.

Definition 2. *Let (X, Γ) be a complete Hausdorff locally $\mathbb{K}$-convex space, $A, B \subset X$, we set*

$$\begin{aligned} A_0^p &= \{a \in A : p(a-b) = D_p(A,B), \text{ for some } b \in B\} \\ B_0^p &= \{a \in B : p(a-b) = D_p(A,B), \text{ for some } a \in A\} \end{aligned}$$

We extend the well known notion of p-property introduced in [5] for metric spaces to the case of locally $\mathbb{K}$-convex spaces.

Definition 3. *Let (A, B) be a pair of nonempty subsets of a locally convex space (X, Γ) with $A_0^p \neq \emptyset$. The pair (A, B) is said to have p-property iff*

$$\begin{cases} p(a_1 - b_1) = D_p(A,B) \\ p(a_2 - b_2) = D_p(A,B) \end{cases} \implies p(a_1 - a_2) = p(b_1 - b_2) \quad (\forall p \in \Gamma).$$

where $a_1, a_2 \in A_0^p$ and $b_1, b_2 \in B_0^p$

Definition 4. *Let (A, B) be a pair of nonempty subsets of a locally convex space (X, Γ). A mapping $T: A \cup B \to A \cup B$ is called relatively p-nonexpansive iff $p(Ta - Tb) \le p(a - b)$ for all $p \in \Gamma$ and $(a,b) \in A \times B$. If $A = B$, we say that T is p-nonexpansive.*

Lemma 1. *[14] Let (X, Γ) be a complete Hausdorff locally $\mathbb{K}$-convex space if $T: X \to X$ is a p-contraction mapping then T has a unique fixed point $\bar{x}$ in X, and $T^k x \to \bar{x}$ for every $x \in X$.*

Proof. Let $y \in X$ and $k \ge 1$ we have

$$\begin{aligned} p\left(T^k y - y\right) &\le \max\left\{p\left(T^k y - T^{k-1}y\right), p\left(T^{k-1}y - T^{k-2}y\right), ..., p(Ty - y)\right\} \\ &\le \max\left\{\gamma^k p(Ty - y), \gamma^{k-1} p(Ty - y), ..., p(Ty - y)\right\} \end{aligned}$$

then $\max\left\{\gamma^k p(Ty - y), \gamma^{k-1} p(Ty - y), ..., p(Ty - y)\right\} = p(Ty - y)$, which implies that for all $x \in X$ and $k \ge 1$

$$p\left(T^k x - x\right) \leq p\left(Tx - x\right).$$

For every $p \in \Gamma$ and $k \geq 1$, Choose n sufficiently large. Then for $y = T^n x$, we have

$$\begin{aligned} p\left(T^{n+k}x - T^n x\right) &\leq p\left(T^{n+1}x - T^n x\right) \\ &\leq \gamma_p^n p\left(Tx - x\right) \end{aligned}$$

Since $0 \leq \gamma_p < 1$, $\gamma_p^n \to 0$ as $n \to \infty$, we get $p\left(T^{n+k}x - T^n x\right) \to 0$. Thus $\left\{T^k x\right\}$ is a p-Cauchy sequence and so it converges to a point $\bar{x}$ in X. Clearly $T\bar{x} = \bar{x}$ and uniqueness of the fixed point follows as usual since X is Hausdorff. □

Theorem 3. *Let (X, Γ) be a complete Hausdorff locally $\mathbb{K}$-convex space and (A, B) be two nonempty closed subsets of X. Assume that $T : A \cup B \to A \cup B$ is a relatively noncyclic mapping such that for some $\gamma_p \in (0, 1)$*

$$p\left(Tx - Ty\right) \leq \gamma_p p\left(a - b\right)$$

for all $p \in \Gamma$ and $(a, b) \in A \times B$ then $D_p(A, B) = 0$. Moreover, the mapping T has a fixed point in $A \cup B$ if and only if $A \cap B \neq \emptyset$.

Proof. Let $\{a_n\}$ and $\{b_n\}$ be two sequences in A and B respectively such that $p\left(a_n - b_n\right) \to D_p(A, B)$. Then

$$D_p(A, B) \leq p\left(Ta_n - Tb_n\right) \leq \gamma_p p\left(a_n - b_n\right).$$

Taking limit when n tends to infinity, we see that necessarily $D_p(A, B) = 0$. Suppose first that $A \cap B \neq \emptyset$. If we apply the Theorem 1 in $A \cap B$, there exists a fixed point of T that in fact is unique in $A \cap B$.

On the other hand, suppose that T has a fixed point $\bar{b}$ in $A \cup B$. Without loss of generality, suppose that $\bar{b} \in B$. Then, given a point $a_0 \in A$, if we denote $a_n = T^n a_0$ we have

$$p\left(a_n - \bar{b}\right) \leq \gamma_p p\left(a_{n-1} - \bar{b}\right) \leq \gamma_p^2 p\left(a_{n-2} - \bar{b}\right) \leq \cdots \leq \gamma_p^n p\left(a_0 - \bar{b}\right)$$

Since $0 \leq \gamma_p < 1$, $\gamma_p^n \to 0$ as $n \to \infty$, we get that $\{a_n\}$ converges to $\bar{b}$. Since A is closed, $\bar{a} \in A \cap B$ and the result follows. □

Theorem 4. *Let (X, Γ) be a complete Hausdorff locally $\mathbb{K}$-convex space and (A, B) be two nonempty closed subsets of X such that $A_0^p \neq \emptyset$. Assume that (A, B) satisfies the p-property. Let $T : A \cup B \to A \cup B$ be a relatively relatively noncyclic mapping that satisfies the conditions*

(i) $T_{|A}$ is p-contraction,
(ii) T is relatively p-nonexpansive.

Then the minimization problem (1) has a solution

Proof. Let $a \in A_0^p$ then exists $b \in B$ such that $p\left(a - b\right) = D_p(A, B)$. Since T is relatively p-nonexpansive; so

$$p\left(Ta - Tb\right) \leq p\left(a - b\right) = D_p(A, B)$$

Hence, $Ta \in A_0^p$, therefore $T\left(A_0^p\right) \subseteq A_0^p$. Now let $a_0 \in A_0^p$. By Lemma 1 if $a_{n+1} = Ta_n$, then $a_n \to \bar{a}$ where $\bar{a}$ is a fixed point of T in A. Since $a_0 \in A_0^p$, then exists $b_0 \in B$ such that $p(a_0 - b_0) = D_p(A, B)$. Again, since $a_1 = Ta_0 \in A_0^p$, then there exists $b_1 \in B$ such that $p(a_1 - b_1) = D_p(A, B)$.

Inductively, using this process for all $n \in \mathbb{N} \cup \{0\}$ we have a sequence $\{b_n\}$ in B such that

$$p(a_n - b_n) = D_p(A, B).$$

Since (A, B) has the p-property, we get that for all $n, m \in \mathbb{N} \cup \{0\}$

$$p(a_n - b_m) = p(a_n - b_m).$$

This implies that $\{b_n\}$ is a Cauchy sequence, and hence there exists $\bar{b} \in B$ such that $a_n \to \bar{b}$. We now have

$$p\left(\bar{a} - \bar{b}\right) = \lim_{n \to \infty} p(a_n - b_n) = D_p(A, B)$$

We know that T is relatively nonexpansive, so that

$$p\left(T\bar{a} - T\bar{b}\right) \leq p\left(\bar{a} - \bar{b}\right) = D_p(A, B)$$

Thus $p\left(\bar{a} - T\bar{b}\right) = p\left(\bar{a} - T\bar{b}\right)$, since (A, B) has property P. Hence $\left(\bar{a} - \bar{b}\right) \in A \times B$ is a solution of (1). □

Author Contributions: Conceptualization, E.M.; Supervision, A.M. and L.S.; Validation, A.M.; Writing—original draft, T.S. and A.B.

Funding: This research received no external funding.

Conflicts of Interest: Research was supported by a National Centre of Scientific and Technological Research grant. The authors would like to express their gratitude to the editor and the anonymous referees for their constructive comments and suggestions, which have improved the quality of the manuscript..

References

1. Monna, A.F. *Analyse Non-Archimedienne*; Springer: Berlin/Heidelberg, Germany; New York, NY, USA, 1970.
2. Roovij, A.C.M.V. *Non-Archimedean Functional Analysis*; Marcel Dekker: New York, NY, USA, 1978.
3. Van Tiel, J. Espaces localement K-convexes I–III. *Indag. Math.* **1965**, *27*, 249–289. [CrossRef]
4. Perez-Garcia, C.; Schikhof, W.H. Locally Convex Spaces over Non-Archimedean Valued Fields. In *Cambridge Studies in Advanced Mathematics*; Cambridge University Press: Cambridge, UK, 2010.
5. Ciric, L.B. A generalization of Banachscontraction principle. *Proc. Am. Math. Soc.* **1974**, *45*, 267–273.
6. Kirk, W.A.; Srinivasan, P.S.; Veeramani, P. Fixed points for mappings satisfying cyclical contractive conditions. *Fixed Point Theory* **2003**, *4*, 79–89.
7. Eldred, A.; Kirk, W.A.; Veeramani, P. Proximal normal structureand relatively nonexpansive mappings. *Stud. Math.* **2005**, *171*, 283–293. [CrossRef]
8. Sankar Raj, V. A best proximity point theorem for weakly contractive non-self-mappings. *Nonlinear Anal.* **2011**, *74*, 4804–4808. [CrossRef]
9. Al-Thagafi, M.A.; Shahzad, N. Convergence and existence results for best proximity points. *Nonlinear Anal.* **2009**, *70*, 3665–3671. [CrossRef]
10. Edraoui, M.; Aamri, M.; Lazaiz, S. Fixed Point Theorem in Locally K-Convex Space. *Int. J. Math. Anal.* **2018**, *12*, 485–490. [CrossRef]
11. Zaslavski, A.J. Two fixed point results for a class of mappings of contractive type. *J. Nonlinear Var. Anal.* **2018**, *2*, 113–119.

12. Park, S. Some general fixed point theorems on topological vector spaces. *Appl. Set-Valued Anal. Optim.* **2019**, *1*, 19–28.
13. Abkar, A.; Gabeleh, M. Global optimal solutions of noncyclic mappings in metric spaces. *J. Optim. Theory Appl.* **2012**, *153*, 298–305. [CrossRef]
14. Cain, G.; Nashed, M. Fixed points and stability for a sum of two operators in locally convex spaces. *Pac. J. Math.* **1971**, *39*, 581–592 [CrossRef]

© 2019 by the authors. Licensee MDPI, Basel, Switzerland. This article is an open access article distributed under the terms and conditions of the Creative Commons Attribution (CC BY) license (http://creativecommons.org/licenses/by/4.0/).

Article

On Fixed Point Results for Modified JS-Contractions with Applications

Vahid Parvaneh [1,*], Nawab Hussain [2], Aiman Mukheimer [3] and Hassen Aydi [4,5,*]

1. Department of Mathematics, Payame Noor University, P.O. Box 19395-3697, Tehran, Iran
2. Department of Mathematics, King Abdulaziz University P.O. Box 80203, Jeddah 21589, Saudi Arabia
3. Department of Mathematics and General Sciences, Prince Sultan University, Riyadh 11586, Saudi Arabia
4. Institut Supérieur d'Informatique et des Techniques de Communication, Université de Sousse, H. Sousse 4000, Tunisia
5. China Medical University Hospital, China Medical University, Taichung 40402, Taiwan
* Correspondence: zam.dalahoo@gmail.com (V.P.); hassen.aydi@isima.rnu.tn (H.A.)

Received: 8 June 2019; Accepted: 23 July 2019; Published: 24 July 2019

Abstract: In [Fixed Point Theory Appl., 2015 (2015):185], the authors introduced a new concept of modified contractive mappings, generalizing Ćirić, Chatterjea, Kannan, and Reich type contractions. They applied the condition (θ_4) (see page 3, Section 2 of the above paper). Later, in [Fixed Point Theory Appl., 2016 (2016):62], Jiang et al. claimed that the results in [Fixed Point Theory Appl., 2015 (2015):185] are not real generalizations. In this paper, by restricting the conditions of the control functions, we obtain a real generalization of the Banach contraction principle (BCP). At the end, we introduce a weakly JS-contractive condition generalizing the JS-contractive condition.

Keywords: metric space; fixed point; weakly JS-contraction

---

1. Introduction

The Banach contraction principle (BCP) [1] is one of the famous results in fixed point theory which has attracted many authors. Many extensions and generalizations have been appeared in literature by weakening the topology itself of the space or by considering different contractive conditions (for single and valued mappings). For more details, see ([2–23]).

Definition 1. *Given a mapping* $Y : X \to X$ *on a metric space* (X, d).

(a) *Such* Y *is a C-contraction if there is* $\mu \in \left(0, \frac{1}{2}\right)$ *such that for all* $\Omega, \omega \in X$, [24]

$$d(Y\Omega, Y\omega) \le \mu \left(d(\Omega, Y\omega) + d(\omega, Y\Omega) \right).$$

(b) *Such* Y *is a K-contraction if there is* $\mu \in \left(0, \frac{1}{2}\right)$ *such that for all* $Y\Omega \in X$, [25]

$$d(Y\Omega, Y\omega) \le \mu \left(d(\Omega, Y\Omega) + d(\omega, Y\omega) \right).$$

(c) *Such* Y *is a Reich contraction if there are* q, r *and* $s \ge 0$ *with* $q + r + s < 1$ *such that for all* $\Omega, \omega \in X$,

$$d(Y\Omega, Y\omega) < q \cdot d(\Omega, \omega) + r \cdot d(\Omega, Y\Omega) + s \cdot d(\omega, Y\omega).$$

Denote by Θ the set of functions $\theta : (0, \infty) \to (1, \infty)$ satisfying the following assertions:

(θ_1) θ is non-decreasing;

(θ_2) for each $\{h_k\} \subseteq (0, \infty)$, $\lim_{k \to \infty} \theta(h_k) = 1$ if and only if $\lim_{k \to \infty} h_k = 0$;

(θ_3) there are $m \in (0, 1)$ and $\tau \in (0, \infty]$ so that

$$\lim_{u \to 0^+} \frac{\theta(u) - 1}{u^m} = \tau;$$

(θ_4) $\theta(i + j) \leq \theta(i)\theta(j)$ for all $i, j > 0$.

By Δ we denote the class of functions $\theta \in \Theta$ without condition (θ_4).

Theorem 1. *([26, Corollary 2.1]) Let $Y : X \to X$ be a self-mapping on a complete metric space (X, d). Suppose there are $\theta \in \Delta$ and $\mu \in (0, 1)$ so that*

$$\Omega, \omega \in X, \quad d(Y\Omega, Y\omega) \neq 0 \quad \text{implies} \quad \theta\left(d(Y\Omega, Y\omega)\right) \leq \left(\theta\left(d(\Omega, \omega)\right)\right)^\mu.$$

Then T has a unique fixed point.

Note that the BCP comes immediately from Theorem 1. Motivated by [26], Hussain et al. [27] gave sufficient conditions for the existence of a fixed point of a class of generalized contractive mappings via a control function $\theta \in \Theta$ in the setting of complete metric spaces and b-complete b-metric spaces. Denote by Λ the set of functions $\theta : (0, \infty) \to (1, \infty)$ verifying (θ_1), (θ_2) and (θ_4). On the other hand, when considering (X, d) as a metric space and $\theta \in \Lambda$ (that is, the condition (θ_3) is omitted from Θ), Jiang et al. [28] proved that $D(x, y) = \ln(\theta(d(x, y)))$ defines itself a metric on X (see Lemma 1 in [28]) and proved that the results in [27] are not generalizations of Ćirić, Chatterjea, Kannan, and Reich results.

In this paper, we more restrict the conditions on the control function θ. For this, denote by Θ' the set of functions $\theta : (0, \infty) \to (1, \infty)$ so that

(θ_1) θ is continuous and strictly increasing;
(θ_2) for each $\{h_k\} \subseteq (0, \infty)$, $\lim_{k \to \infty} \theta(h_k) = 1$ if and only if $\lim_{k \to \infty} h_k = 0$.

Let (X, d) be a metric space. For $\theta \in \Theta'$ (that is, without the condition (θ_4)), note that $D(x, y) = \ln(\theta(d(x, y)))$ does not define a metric on X (we can not ensure the triangular inequality for a metric). Consequently, we are not in same direction as Jiang et al. [28]. Even for such restricted control function θ, we also obtain a real generalization of the Banach contraction principle. In fact, we will complete the work of Hussain et al. [27]. We refer the readers to Theorem 3 of [16].

2. Main Results

Definition 2. *Let $Y : X \to X$ be a self-mapping on a metric space (X, d). Such Y is said to be a $\mathcal{P}$-contraction, whenever there are $\theta \in \Theta'$ and $\tau_1, \tau_2, \tau_3, \tau_4 \geq 0$ with $\tau_1 + \tau_2 + \tau_3 + \tau_4 < 1$ such that the following holds:*

$$\theta\left(d(Y\Omega, Y\omega)\right) \leq \left(\theta\left(d(\Omega, \omega)\right)\right)^{\tau_1} \left(\theta\left(d(\Omega, Y\Omega)\right)\right)^{\tau_2} \left(\theta\left(d(\omega, Y\omega)\right)\right)^{\tau_3} \left(\theta\left(\frac{d(\Omega, Y\omega) + d(\omega, Y\Omega)}{2}\right)\right)^{\tau_4}, \quad (1)$$

for all $\Omega, \omega \in X$.

As a new generalization of the BCP, we have

Theorem 2. *Each $\mathcal{P}$-contraction mapping on a complete metric space has a unique fixed point.*

Proof. Let $\Omega_0 \in X$ be arbitrary. Define $\{\Omega_n\}$ by $\Omega_n = Y\Omega_{n-1}$, $n \geq 1$. If there is $\Omega_N = \Omega_{N+1}$ for some N, nothing is to prove. We assume that $\Omega_n \neq \Omega_{n+1}$ for each $n \geq 0$.

We claim that

$$\lim_{n \to \infty} d(\Omega_n, \Omega_{n+1}) = 0.$$

In view of (1), we have

$$\theta\left(d(\Omega_{n+1}, \Omega_n)\right) = \theta\left(d(Y\Omega_n, Y\Omega_{n-1})\right) \quad (2)$$
$$\leq \left(\theta\left(d(\Omega_n, \Omega_{n-1})\right)\right)^{\tau_1} \left(\theta\left(d(\Omega_n, Y\Omega_n)\right)\right)^{\tau_2}$$
$$\left(\theta\left(d(\Omega_{n-1}, Y\Omega_{n-1})\right)\right)^{\tau_3} \left(\theta\left(\frac{d(\Omega_n, Y\Omega_{n-1}) + d(\Omega_{n-1}, Y\Omega_n)}{2}\right)\right)^{\tau_4}$$
$$\leq \left(\theta\left(d(\Omega_n, \Omega_{n-1})\right)\right)^{\tau_1} \left(\theta\left(d(\Omega_n, \Omega_{n+1})\right)\right)^{\tau_2}$$
$$\left(\theta\left(d(\Omega_{n-1}, \Omega_n)\right)\right)^{\tau_3} \left(\theta\left(\frac{d(\Omega_{n-1}, \Omega_{n+1})}{2}\right)\right)^{\tau_4}$$
$$\leq \left(\theta\left(d(\Omega_n, \Omega_{n-1})\right)\right)^{\tau_1+\tau_3} \left(\theta\left(d(\Omega_n, \Omega_{n+1})\right)\right)^{\tau_2} \left(\theta\left(\max\{d(\Omega_{n-1}, \Omega_n), d(\Omega_n, \Omega_{n+1})\}\right)\right)^{\tau_4}.$$

If for some N, we have

$$d(\Omega_{N-1}, \Omega_N) < d(\Omega_N, \Omega_{N+1}),$$

then in view of (θ_1), we get that

$$\theta(d(\Omega_{N-1}, \Omega_N)) < \theta(d(\Omega_N, \Omega_{N+1})). \quad (3)$$

Using (2), we have

$$\theta\left(d(\Omega_{N+1}, \Omega_N)\right) \leq \left(\theta\left(d(\Omega_N, \Omega_{N-1})\right)\right)^{\tau_1+\tau_3} \left(\theta\left(d(\Omega_N, \Omega_{N+1})\right)\right)^{\tau_2+\tau_4}. \quad (4)$$

Therefore,

$$\theta\left(d(\Omega_{N+1}, \Omega_N)\right) \leq \left(\theta\left(d(\Omega_N, \Omega_{N-1})\right)\right)^{\frac{\tau_1+\tau_3}{1-\tau_2-\tau_4}} \leq \theta\left(d(\Omega_N, \Omega_{N-1})\right),$$

which is a contradiction with respect to (3).

Consequently, for all $n \geq 1$,

$$\max\left\{d(\Omega_{n-1}, \Omega_n), d(\Omega_n, \Omega_{n+1})\right\} = d(\Omega_{n-1}, \Omega_n),$$

which yields that

$$1 < \theta(d(\Omega_{n+1}, \Omega_n)) \leq (\theta(d(\Omega_1, \Omega_0)))^{\left[\frac{\tau_1+\tau_3+\tau_4}{1-\tau_2}\right]^n}.$$

At the limit, we have

$$\lim_{n \to \infty} \theta(d(\Omega_n, \Omega_{n+1})) = 1.$$

According to (θ_2), we get

$$\lim_{n \to \infty} d(\Omega_n, \Omega_{n+1}) = 0. \quad (5)$$

In order to show that $\{\Omega_n\}$ is a Cauchy sequence, suppose the contrary, i.e., there is $\varepsilon > 0$ for which we can find m_i and n_i so that

$$n_i > m_i > i, \quad d(\Omega_{m_i}, \Omega_{n_i}) \geq \varepsilon. \quad (6)$$

That is,

$$d(\Omega_{m_i}, \Omega_{n_i-1}) < \varepsilon. \quad (7)$$

From (6), one writes

$$d(\Omega_{m_i-1}, \Omega_{n_i-1}) \leq d(\Omega_{m_i-1}, \Omega_{m_i}) + d(\Omega_{m_i}, \Omega_{n_i-1}).$$

In view of (5) and (7), we get

$$\limsup_{i\to\infty} d(\Omega_{m_i-1}, \Omega_{n_i-1}) \leq \varepsilon. \qquad (8)$$

Analogously,

$$\limsup_{i\to\infty} d(\Omega_{m_i-1}, \Omega_{n_i}) \leq \varepsilon. \qquad (9)$$

On the other hand, we have

$$\theta\left(d(\Omega_{m_i}, \Omega_{n_i})\right) = \theta\left(d(Y\Omega_{m_i-1}, Y\Omega_{n_i-1})\right)$$
$$\leq \left(\theta\left(d(\Omega_{m_i-1}, \Omega_{n_i-1})\right)\right)^{\tau_1} \left(\theta\left(d(\Omega_{m_i-1}, Y\Omega_{m_i-1})\right)\right)^{\tau_2}$$
$$\left(\theta\left(d(\Omega_{n_i-1}, Y\Omega_{n_i-1})\right)\right)^{\tau_3} \left(\theta\left(\frac{d(\Omega_{m_i-1}, Y\Omega_{n_i-1}) + d(\Omega_{n_i-1}, Y\Omega_{m_i-1})}{2}\right)\right)^{\tau_4}$$
$$\leq \left(\theta\left(d(\Omega_{m_i-1}, \Omega_{n_i-1})\right)\right)^{\tau_1} \left(\theta\left(d(\Omega_{m_i-1}, \Omega_{m_i})\right)\right)^{\tau_2}$$
$$\left(\theta\left(d(\Omega_{n_i-1}, \Omega_{n_i})\right)\right)^{\tau_3} \left(\theta\left(\frac{d(\Omega_{m_i-1}, \Omega_{n_i}) + d(\Omega_{n_i-1}, \Omega_{m_i})}{2}\right)\right)^{\tau_4}.$$

Using now (θ_1) and (5)–(8), we have

$$\theta(\varepsilon) \leq \theta\left(\limsup_{i\to\infty} d(\Omega_{m_i}, \Omega_{n_i})\right)$$
$$\leq \left(\theta\left(\limsup_{i\to\infty} d(\Omega_{m_i-1}, \Omega_{n_i-1})\right)\right)^{\tau_1} \left(\theta\left(\limsup_{i\to\infty} d(\Omega_{m_i-1}, \Omega_{m_i})\right)\right)^{\tau_2}$$
$$\left(\theta\left(\limsup_{i\to\infty} d(\Omega_{n_i-1}, \Omega_{n_i})\right)\right)^{\tau_3} \left(\theta\left(\limsup_{i\to\infty} \frac{d(\Omega_{m_i-1}, \Omega_{n_i}) + d(\Omega_{n_i-1}, \Omega_{m_i})}{2}\right)\right)^{\tau_4}$$
$$\leq (\theta(\varepsilon))^{\tau_1} (\theta(\varepsilon))^{\tau_4}.$$

This implies that

$$1 < \theta(\varepsilon) \leq (\theta(\varepsilon))^{\tau_1 + \tau_4},$$

which is a contradiction. Thus, $\{\Omega_n\}$ is a Cauchy sequence. The completeness of X implies that there is $\Omega \in X$ so that $\Omega_n \to \Omega$ as $n \to \infty$. On the other hand,

$$\theta\left(d(\Omega_n, Y\Omega)\right) = \theta\left(d(Y\Omega_{n-1}, Y\Omega)\right)$$
$$\leq \left(\theta\left(d(\Omega_{n-1}, \Omega)\right)\right)^{\tau_1} \left(\theta\left(d(\Omega_{n-1}, Y\Omega_{n-1})\right)\right)^{\tau_2}$$
$$\left(\theta\left(d(\Omega, Y\Omega)\right)\right)^{\tau_3} \left(\theta\left(\frac{d(Y\Omega, \Omega_{n-1}) + d(\Omega, Y\Omega_{n-1})}{2}\right)\right)^{\tau_4}$$
$$\leq \left(\theta\left(d(\Omega_{n-1}, \Omega)\right)\right)^{\tau_1} \left(\theta\left(d(\Omega_{n-1}, \Omega_n)\right)\right)^{\tau_2}$$
$$\left(\theta\left(d(\Omega, Y\Omega)\right)\right)^{\tau_3} \left(\theta\left(\frac{d(Y\Omega, \Omega_{n-1}) + d(\Omega, \Omega_n)}{2}\right)\right)^{\tau_4}.$$

Taking $n \to \infty$ and using (θ_1) and (5), we have

$$\theta(d(\Omega, Y\Omega)) \leq (\theta(d(\Omega, \Omega)))^{\tau_1} (\theta(d(\Omega, \Omega)))^{\tau_2}$$
$$(\theta(d(\Omega, Y\Omega)))^{\tau_3} (\theta(d(\Omega, Y\Omega)))^{\tau_4}$$
$$= (\theta(d(\Omega, Y\Omega)))^{\tau_3 + \tau_4}.$$

We deduce that $\Omega = Y\Omega$, so Ω is a fixed point.

Let there are two points Ω, ω which are two different fixed points of Y. So,

$$\theta\left(d(Y\Omega, Y\omega)\right) \leq \left(\theta\left(d(\Omega, \omega)\right)\right)^{\tau_1} \left(\theta\left(d(\Omega, Y\Omega)\right)\right)^{\tau_2}$$
$$\left(\theta\left(d(\omega, Y\omega)\right)\right)^{\tau_3} \left(\theta\left(d(\Omega, Y\Omega)\right)\right)^{\tau_4}$$
$$= \left(\theta\left(d(\Omega, Y\Omega)\right)\right)^{\tau_3+\tau_4}.$$

We deduce that $\Omega = Y\Omega$, so Ω is a fixed point.

Let Ω, ω be two distinct fixed points of Y. We have

$$\theta\left(d(\Omega, \omega)\right) = \theta\left(d(Y\Omega, Y\omega)\right) \leq \left(\theta\left(d(\Omega, \omega)\right)\right)^{\tau_1} \left(\theta\left(d(\Omega, \Omega)\right)\right)^{\tau_2}$$
$$\left(\theta\left(d(\omega, \omega)\right)\right)^{\tau_3} \left(\theta\left(d(\Omega, \omega)\right)\right)^{\tau_4}$$
$$= \left(\theta\left(d(\Omega, \omega)\right)\right)^{\tau_1+\tau_4} < \theta\left(d(\Omega, \omega)\right),$$

which is a contradiction. So, Ω has a unique fixed point. □

Remark 1. *In Theorem 2, we can substitute the continuity of θ by the continuity of Y.*

By setting $\theta(t) = e^{\sqrt{t}}$, we have

Corollary 1. *Let $Y : X \to X$ be a mapping on a complete metric space (X, d) such that the following holds:*

$$\sqrt{d(Y\Omega, Y\omega)} \leq \tau_1 \sqrt{d(\Omega, \omega)} + \tau_2 \sqrt{d(\Omega, Y\Omega)} + \tau_3 \sqrt{d(\omega, Y\omega)} + \tau_4 \sqrt{\frac{d(\Omega, Y\omega) + d(\omega, Y\Omega)}{2}},$$

for all $\Omega, y \in X$, where $\theta \in \mathcal{P}$ and $\tau_1, \tau_2, \tau_3, \tau_4 \geq 0$ so that $\tau_1 + \tau_2 + \tau_3 + \tau_4 < 1$. Then Y has a unique fixed point.

Remark 2. *Taking $\tau_1 = \tau_4 = 0$ in the Corollary 1, we get Theorem 2.6 of [27].*
Taking $\tau_4 = 0$ in Theorem 1, we get Theorem 2.8 of [27].

Setting $\theta(t) = e^{\sqrt[n]{t}}$ in Theorem 2, we have

Corollary 2. *Let (Ω, d) be a complete metric space and let $Y : X \to X$ be such that the following holds:*

$$\sqrt[n]{d(Y\Omega, Y\omega)} \leq \tau_1 \sqrt[n]{d(\Omega, \omega)} + \tau_2 \sqrt[n]{d(\Omega, Y\Omega)} + \tau_3 \sqrt[n]{d(\omega, Y\omega)} + \tau_4 \sqrt[n]{\frac{d(\Omega, Y\omega) + d(\omega, Y\Omega)}{2}},$$

for all $\Omega, \omega \in X$, where $\theta \in \mathcal{P}$ and $\tau_1, \tau_2, \tau_3, \tau_4 \geq 0$ such that $\tau_1 + \tau_2 + \tau_3 + \tau_3 < 1$. Then Y has a unique fixed point.

Remark 3 ([12]). *Other examples of functions in the set $\mathcal{P}$ are*

$$f(t) = \cosh t, \quad f(t) = e^{te^t}, \quad f(t) = e^{\sqrt{te^{\sqrt{t}}}},$$

$$f(t) = \frac{2\cosh t}{1+\cosh t}, \quad f(t) = \frac{2e^{te^t}}{1+e^{te^t}}, \quad f(t) = \frac{2e^{\sqrt{te}\sqrt{t}}}{1+e^{\sqrt{te}\sqrt{t}}},$$

$$f(t) = 1 + \ln(1+t), \quad f(t) = e^{\sqrt{te^t}},$$

$$f(t) = \frac{2+2\ln(1+t)}{2+\ln(1+t)}, \quad f(t) = \frac{2e^{\sqrt{te^t}}}{1+e^{\sqrt{te^t}}},$$

for all $t > 0$.

By setting $\theta(t) = e^{te^t}$, we have

Corollary 3. *Let* $Y : X \to X$ *be a continuous mapping on a complete metric space* (X, d). *Suppose that there are* $\tau_1, \tau_2, \tau_3, \tau_4 \geq 0$ *with* $\tau_1 + \tau_2 + \tau_3 + \tau_4 < 1$ *such that the following holds:*

$$d(Y\Omega, Y\omega)e^{d(Y\Omega, Y\omega)} \leq \tau_1 d(\Omega, \omega)e^{d(\Omega, \omega)} + \tau_2 d(\Omega, Y\Omega)e^{d(\Omega, Y\Omega)}$$
$$+ \tau_3 d(\omega, Y\omega)e^{d(\omega, Y\omega)} + \tau_4 d(\omega, Y\Omega)e^{\frac{d(\Omega, Y\omega) + d(\omega, Y\Omega)}{2}},$$

for all $\Omega, \omega \in X$. *Then there is a unique fixed point of* Y.

Corollary 4. *Let* $Y : X \to X$ *be a continuous mapping on a complete metric space* (X, d). *Suppose that there are* $\tau_1, \tau_2, \tau_3, \tau_4 \geq 0$ *with* $\tau_1 + \tau_2 + \tau_3 + \tau_4 < 1$ *such that the following holds:*

$$\frac{2e^{d(Y\Omega, Y\omega)}e^{d(Y\Omega, Y\omega)}}{1 + e^{d(Y\Omega, Y\omega)}e^{d(Y\Omega, Y\omega)}} \leq \left[\frac{2e^{d(\Omega, \omega)}e^{d(\Omega, \omega)}}{1 + e^{d(\Omega, \omega)}e^{d(\Omega, \omega)}}\right]^{\tau_1} \left[\frac{2e^{d(\Omega, Y\Omega)}e^{d(\Omega, Y\Omega)}}{1 + e^{d(\Omega, Y\Omega)}e^{d(\Omega, Y\Omega)}}\right]^{\tau_2}$$
$$\left[\frac{2e^{d(\omega, Y\omega)}e^{d(\omega, Y\omega)}}{1 + e^{d(\omega, Y\omega)}e^{d(\omega, Y\omega)}}\right]^{\tau_3} \left[\frac{2e^{\frac{d(\Omega, Y\omega) + d(\omega, Y\Omega)}{2}}e^{\frac{d(\Omega, Y\omega) + d(\omega, Y\Omega)}{2}}}{1 + e^{\frac{d(\Omega, Y\omega) + d(\omega, Y\Omega)}{2}}e^{\frac{d(\Omega, Y\omega) + d(\omega, Y\Omega)}{2}}}\right]^{\tau_4},$$

for all $\Omega, \omega \in X$. *Then there is a unique fixed point of* Y.

Corollary 5. *Let* $Y : X \to X$ *be a continuous mapping on a complete metric space* (X, d). *Suppose that there are* $\tau_1, \tau_2, \tau_3, \tau_4 \geq 0$ *with* $\tau_1 + \tau_2 + \tau_3 + \tau_4 < 1$ *such that the following holds:*

$$1 + \ln(1 + d(Y\Omega, Y\omega)) \leq [1 + \ln(1 + d(\Omega, \omega))]^{\tau_1} [1 + \ln(1 + d(\Omega, Y\Omega))]^{\tau_2}$$
$$[1 + \ln(1 + d(\omega, Y\omega))]^{\tau_3} \left[1 + \ln\left(1 + \frac{d(\Omega, Y\omega) + d(\omega, Y\Omega)}{2}\right)\right]^{\tau_4},$$

for all $\Omega, \omega \in X$. *Then* Y *has a unique fixed point.*

Example 1. Let $X = [0, 5]$ be endowed with the metric $d(\Omega, \omega) = |\Omega - \omega|$ for all $\Omega, \omega \in X$. Define $Y : X \to X$ and $\theta : (0, \infty) \to (1, \infty)$ by

$$Y\Omega = \begin{cases} \frac{2}{3\pi}\Omega \arctan \Omega, & \text{if } \Omega \in [0, \alpha], \\ \frac{1}{3}\sinh^{-1}\Omega & \text{if } \Omega \in [\alpha, +5], \end{cases}$$

where $\alpha \, (\simeq 2.06)$ is the positive solution of the equation

$$\frac{2}{3\pi}\Omega \arctan \Omega = \frac{1}{3}\sinh^{-1}\Omega.$$

Take $\theta(t) = e^{te^t}$. Choose $\tau_1 = \frac{37}{100}$ and $\tau_i = \frac{1}{5}$ for $i = 2, 3, 4$.
Let $\Omega, \omega \in X = [0, 5]$. We have the following cases:
Case 1: $\Omega, \omega \in [0, \alpha]$. According to the mean value Theorem for $t \longmapsto g(t) := \frac{2}{3\pi}t \arctan t$ on the interval $J = (\min(\omega, \Omega), \max(\omega, \Omega)) \subset [0, \alpha]$, there is some $c \in J$ such that

$$d(Y\Omega, Y\omega) = \left|\frac{2}{3\pi}\Omega \arctan \Omega - \frac{2}{3\pi}\omega \arctan \omega\right| \leq g'(c)d(\Omega, \omega),$$

where
$$g'(c) = \frac{2}{3\pi}\arctan c + \frac{2}{3\pi}\frac{c}{1+c^2} \leq \frac{2}{3\pi}\frac{6}{5} + \frac{2}{3\pi}\frac{1}{2} \leq \frac{17}{15\pi} \leq \frac{37}{100},$$

because that $\arctan c \leq \frac{6}{5}$, for each $c \in [0, \alpha]$, and $\frac{c}{1+c^2} \leq \frac{1}{2}$, for each $c \geq 0$.

Therefore,

$$\theta(d(Y\Omega, Y\omega)) = e^{d(Y\Omega, Y\omega)}e^{d(Y\Omega, Y\omega)}$$
$$= e^{d(\frac{2}{3\pi}\Omega \arctan \Omega, \frac{2}{3\pi}\omega \arctan \omega)}e^{d(\frac{2}{3\pi}\Omega \arctan \Omega, \frac{2}{3\pi}\omega \arctan \omega)}$$
$$\leq \left[e^{d(\Omega,\omega)}e^{d(\Omega,\omega)}\right]^{\frac{37}{100}}$$
$$\leq \left[e^{d(\Omega,\omega)}e^{d(\Omega,\omega)}\right]^{\frac{37}{100}} \cdot \left[e^{d(\Omega,Y\Omega)}e^{d(\Omega,Y\Omega)}\right]^{\frac{20}{100}} \cdot$$
$$\left[e^{d(\omega,Y\omega)}e^{d(\omega,Y\omega)}\right]^{\frac{20}{100}} \cdot \left[e^{\frac{d(\Omega,Y\omega)+d(\omega,Y\Omega)}{2}}e^{\frac{d(\Omega,Y\omega)+d(\omega,Y\Omega)}{2}}\right]^{\frac{20}{100}}.$$

Case 2: $\Omega \in [0, \alpha]$ and $\omega \in [\alpha, 5]$. Here,

$$\frac{2}{3\pi}\omega \arctan \omega \geq \frac{1}{3}\sinh^{-1}\omega$$

for all $\omega \in [\alpha, 5]$. Using the mean value Theorem on the function $t \to \frac{2}{3\pi}t\arctan t$ on the interval $[\Omega, \omega]$, we have

$$d(Y\Omega, Y\omega) = \left|\frac{2}{3\pi}\Omega \arctan \Omega - \frac{1}{3}\sinh^{-1}\omega\right| = \frac{1}{3}\sinh^{-1}\omega - \frac{2}{3\pi}\Omega \arctan \Omega$$
$$\leq \frac{2}{3\pi}\omega \arctan \omega - \frac{2}{3\pi}\Omega \arctan \Omega$$
$$\leq \frac{37}{100}d(\Omega, \omega),$$

Therefore, as in case 1,

$$\theta(d(Y\Omega, Y\omega)) = e^{d(Y\Omega, Y\omega)}e^{d(Y\Omega, Y\omega)}$$
$$\leq \left[e^{d(\Omega,\omega)}e^{d(\Omega,\omega)}\right]^{\frac{37}{100}} \cdot \left[e^{d(\Omega,Y\Omega)}e^{d(\Omega,Y\Omega)}\right]^{\frac{20}{100}} \cdot$$
$$\left[e^{d(\omega,Y\omega)}e^{d(\omega,Y\omega)}\right]^{\frac{20}{100}} \cdot \left[e^{\frac{d(\Omega,Y\omega)+d(\omega,Y\Omega)}{2}}e^{\frac{d(\Omega,Y\omega)+d(\omega,Y\Omega)}{2}}\right]^{\frac{20}{100}}.$$

Case 3: $\omega \in [0, \alpha]$ and $\Omega \in [\alpha, 5]$. It is similar to case 2.

Case 4: $\Omega, \omega \in [\alpha, 5]$. Here, one writes

$$d(Y\Omega, Y\omega) = \left|\frac{1}{3}\sinh^{-1}\Omega - \frac{1}{3}\sinh^{-1}\omega\right| \leq \frac{37}{100}d(\Omega, \omega).$$

Similarly,

$$\theta(d(Y\Omega, Y\omega)) \leq \left[e^{d(\Omega,\omega)}e^{d(\Omega,\omega)}\right]^{\frac{37}{100}} \cdot \left[e^{d(\Omega,Y\Omega)}e^{d(\Omega,Y\Omega)}\right]^{\frac{20}{100}} \cdot$$
$$\left[e^{d(\omega,Y\omega)}e^{d(\omega,Y\omega)}\right]^{\frac{20}{100}} \cdot \left[e^{\frac{d(\Omega,Y\omega)+d(\omega,Y\Omega)}{2}}e^{\frac{d(\Omega,Y\omega)+d(\omega,Y\Omega)}{2}}\right]^{\frac{20}{100}}.$$

Hence, Y is a P-contraction. Thus all the conditions of Theorem 2 hold and Y has a fixed point ($\Omega = 0$).

3. Weak-JS Contractive Conditions

Let Φ be the class of functions $\phi : [1, \infty) \to [0, \infty)$ satisfying the following properties:

(ϕ_1) ϕ is continuous;
(ϕ_2) $\phi(1) = 0$;
(ϕ_3) or each $\{b_n\} \subseteq (1, \infty)$, $\lim_{n \to \infty} \phi(b_n) = 0$ iff $\lim_{n \to \infty} b_n = 1$.

Remark 4. *It is clear that $Y(t) = t - \sqrt[n]{t}$ ($n \geq 1$) belongs to Φ. Other examples are $Y(t) = e^{t-1} - 1$ and $Y(t) = \ln t$.*

Definition 3. *Let (X, d) be a metric space and let Y be a self-mapping on X.*
We say that Y is a weakly JS-contraction if for all $\Omega, \omega \in X$ with $d(Y\Omega, Y\omega) > 0$, we have

$$\theta\big(d(Y\Omega, Y\omega)\big) \leq \theta\big(d(\Omega, \omega)\big) - \phi\big(\theta(d(\Omega, \omega))\big) \tag{10}$$

where $\phi \in \Phi$ and $\theta \in \Theta'$.

Theorem 3. *Let (X, d) be a complete metric space. Let Y be a self-mapping on X so that*

(i) *Y is a weakly JS-contraction;*
(ii) *Y is continuous.*

Then Y has a unique fixed point.

Proof. Let $\Omega_0 \in X$ be arbitrary. Define $\{\Omega_n\}$ by $\Omega_n = Y^n \Omega_0 = Y\Omega_{n-1}$. Without loss of generality, assume that $\Omega_n \neq \Omega_{n+1}$ for each $n \geq 0$. Since Y is a weakly JS-contraction, we derive

$$\theta\big(d(\Omega_n, \Omega_{n+1})\big) = \theta\big(d(Y\Omega_{n-1}, Y\Omega_n)\big) \leq \theta\big(d(\Omega_{n-1}, \Omega_n)\big) - \phi\big(\theta(d(\Omega_{n-1}, \Omega_n))\big). \tag{11}$$

So, we deduce that $\{\theta(d(\Omega_n, \Omega_{n+1}))\}$ is decreasing, and so there is $r \geq 1$ so such $\lim_{n \to \infty} \theta\big(d(\Omega_n, \Omega_{n+1})\big) = r$. We will prove that $r = 1$.

Taking $n \to \infty$, we have

$$r - \phi(r) = r. \tag{12}$$

So,

$$\lim_{n \to \infty} \phi\big(\theta(d(\Omega_{n-1}, \Omega_n))\big) = 0. \tag{13}$$

That is,

$$\lim_{n \to \infty} \theta(d(\Omega_{n-1}, \Omega_n)) = 1, \tag{14}$$

i.e.,

$$\lim_{n \to \infty} d(\Omega_{n-1}, \Omega_n) = 0. \tag{15}$$

We claim that $\{\Omega_n\}$ is a Cauchy sequence.
We argue by contradiction, i.e., there is $\varepsilon > 0$ for which there are $\{\Omega_{m_i}\}$ and $\{\Omega_{n_i}\}$ of $\{\Omega_n\}$ so that

$$n_i > m_i > i \text{ and } d(\Omega_{m_i}, \Omega_{n_i}) \geq \varepsilon. \tag{16}$$

From (16) and using the triangular inequality, we get

$$\varepsilon \leq d(\Omega_{m_i}, \Omega_{n_i})$$
$$\leq d(\Omega_{m_i}, \Omega_{m_i+1}) + d(\Omega_{m_i+1}, \Omega_{n_i})$$
$$\leq d(\Omega_{m_i}, \Omega_{m_i+1}) + d(\Omega_{m_i+1}, \Omega_{n_i+1}) + d(\Omega_{n_i+1}, \Omega_{n_i}).$$

Taking $i \to \infty$, and using (15), we get

$$\varepsilon \leq \limsup_{i \to \infty} d(\Omega_{m_i+1}, \Omega_{n_i+1}). \tag{17}$$

Also,

$$d(\Omega_{n_i}, \Omega_{m_i}) \leq d(\Omega_{n_i}, \Omega_{n_i-1}) + d(\Omega_{n_i-1}, \Omega_{m_i}).$$

Then, from (15),

$$\limsup_{i \to \infty} d(\Omega_{n_i}, \Omega_{m_i}) \leq \varepsilon. \tag{18}$$

As $d(Y\Omega_{m_i}, Y\Omega_{n_i}) > 0$, we may apply (10) to get that

$$\theta(d(\Omega_{m_i+1}, \Omega_{n_i+1})) = \theta(d(Y\Omega_{m_i}, Y\Omega_{n_i}))$$
$$\leq \theta(d(\Omega_{m_i}, \Omega_{n_i})) - \phi(\theta(d(\Omega_{m_i}, \Omega_{n_i}))).$$

Now, taking $i \to \infty$ and using (θ1), (17) and (18), we have

$$\theta(\varepsilon) \leq \theta(\limsup_{i \to \infty} d(\Omega_{m_i+1}, \Omega_{n_i+1}))$$
$$\leq \theta(\limsup_{i \to \infty} d(\Omega_{m_i}, \Omega_{n_i})) - \liminf_{i \to \infty} \phi(\theta(d(\Omega_{m_i}, \Omega_{n_i})))$$
$$\leq \theta(\varepsilon) - \liminf_{i \to \infty} \phi(\theta(d(\Omega_{m_i}, \Omega_{n_i}))).$$

This implies that

$$\liminf_{i \to \infty} d(\Omega_{m_i}, \Omega_{n_i}) = 0,$$

which is a contradiction with respect to (16).

Thus, $\{\Omega_n\}$ is a Cauchy sequence in the complete metric space (Ω, d), so there is some $\Omega \in X$ such that $\lim_{n \to \infty} d(\Omega_n, \Omega) = 0$.

Now, since Y is continuous, we get that $\Omega_{n+1} = Y\Omega_n \to Y\Omega$ as $n \to \infty$. That is, $\Omega = Y\Omega$. Thus, Y has a fixed point.

Let $\Omega, \omega \in Fix(T)$ so that $\Omega \neq \omega$. Consider

$$\theta(d(\Omega, \omega)) = \theta(d(Y\Omega, Y\omega)) \leq \theta(d(\Omega, y)) - \phi(\theta(d(\Omega, \omega))).$$

Thus,

$$\phi(\theta(d(\Omega, \omega))) = 0.$$

which is a contradiction. Hence, $\Omega = \omega$. □

One can obtain many other contractive conditions by substituting suitable values of θ and ϕ in (10).

Taking $\phi(t) = t - t^\alpha$ for all $t \geq 1$ and $\alpha \in [0, 1)$, we obtain the JS-contractive condition.

Without the continuity assumption of Y, we have

Theorem 4. *Let (X,d) be a complete metric space. Let $Y : X \to X$ be a mapping. Suppose that*

$$\theta(d(Y\Omega, Y\omega)) \leq \theta(d(\Omega, \omega)) - \phi(\theta(d(\Omega, \omega))), \tag{19}$$

for all $\Omega, \omega \in X$, where $\theta \in \Theta'$ and $\phi \in \Phi$. Then Y has a unique fixed point.

Proof. For $\Omega_0 \in X$, let $\{\Omega_n\}$ be defined by $\Omega_{n+1} = Y\Omega_n$ for $n \geq 0$. Note that there is $\Omega \in X$ such that

$$\lim_{n \to \infty} d(\Omega_n, \Omega) = 0.$$

We also have
$$d(\Omega, Y\Omega) \leq d(\Omega, Y\Omega_n) + d(Y\Omega_n, Y\Omega). \tag{20}$$

From (19),
$$1 \leq \theta(d(Y\Omega_n, Y\Omega)) \leq \theta(d(\Omega_n, \Omega)) - \phi(\theta(d(\Omega_n, \Omega))), \tag{21}$$

Hence, we get that $\lim_{n \to \infty} \theta(d(Y\Omega_n, Y\Omega)) = 1$. Thus, we have $\lim_{n \to \infty} d(Y\Omega_n, Y\Omega) = 0$ which by (20), implies that $Y\Omega = \Omega$. □

Example 2. *Let $\Omega = [2, \infty)$. Take the metric*

$$d(\rho, \varrho) = |\rho - \varrho|$$

for all $\rho, \varrho \in \Omega$. Define $Y : \Omega \to \Omega$, $\varphi : [1, \infty) \to [0, \infty)$ and $\theta : [0, \infty) \to [1, \infty)$ by

$$Y\rho = \ln(100 + \rho),$$

$$\varphi(\rho) = \ln(\rho),$$

and $\theta(t) = e^t$. Note that for all $x \geq 0$, one has $e^{\frac{x}{100}} \leq e^x - x$. Now, for all $\rho, \varrho \in \Omega$, we have

$$\theta(d(Y\rho, Y\varrho)) = e^{d(Y\rho, Y\varrho)}$$
$$= e^{(|\ln(100+\rho) - \ln(100+\varrho)|)}$$
$$\leq e^{\frac{|\rho-\varrho|}{100}}$$
$$\leq e^{|\rho-\varrho|} - |\rho - \varrho|$$
$$= e^{d(\rho,\varrho)} - d(\rho, \varrho)$$
$$= \theta(d(\rho, \varrho)) - \varphi(\theta(d(\rho, \varrho))).$$

Thus, Y is a weakly JS-contraction. All hypotheses of Theorem 3 are verified, so Y has a unique fixed point, which is, $u \simeq \frac{4651}{1000}$.

4. Application to Nonlinear Integral Equations

Consider the following nonlinear integral equation

$$\Omega(t) = \phi(t) + \int_a^b \chi(t, s, \Omega(s)) ds, \tag{22}$$

where $a, b \in \mathbb{R}$, $\Omega \in C[a,b]$ (the set of continuous functions from $[a,b]$ to $\mathbb{R}$), $\phi : [a,b] \to \mathbb{R}$ and $\chi : [a,b] \times [a,b] \times \mathbb{R} \to \mathbb{R}$ are given functions.

Theorem 5. *Assume that*

(i) $\chi : [a,b] \times [a,b] \times \mathbb{R} \to \mathbb{R}$ *is continuous and there is $\theta \in \Theta$ so that $\theta(\sup_{t \in [a,b]} f(t)) \le \sup_{t \in [a,b]} \theta(f(t))$ for arbitrary function f with*

$$\theta\left(\int_a^b |(\chi(t,s,\Omega(s))ds - \chi(t,s,\omega(s))|ds\right) \le \int_a^b \theta(|\chi(t,s,\Omega(s)) - \chi(t,s,\omega(s))|)ds;$$

(ii) *there is $\tau_i \in (0,1)$ so that*

$$\theta(|\chi(t,s,\Omega(s)) - \chi(t,s,\omega(s))|)$$
$$\le \frac{[\theta(|\Omega(t) - y(t)|)]^{\tau_1}[\theta(|\Omega(t) - \int_a^b \chi(t,s,\Omega(s)ds)|)]^{\tau_2}[\theta(|\omega(t) - \int_a^b \chi(t,s,\omega(s))ds|)]^{\tau_3}}{b-a}$$
$$[\theta(|\omega(t) - \int_a^b \chi(t,s,\Omega(s))ds|)]^{\tau_4}$$

for all $\Omega, \omega \in C[a,b]$ and $t,s \in [a,b]$.

Then (22) has a unique solution.

Proof. Let $X = C[a,b]$. Define the metric d on X by $d(\Omega,\omega) = \sup_{t \in [a,b]} |\Omega(t) - \omega(t)|$. Then (X,d) is a complete metric space. Consider $Y : X \to X$ by $Y\Omega(t) = \phi(t) + \int_a^t \chi(t,s,\Omega(s))ds$. Let $\Omega, \omega \in X$ and $t \in [a,b]$. We have

$$\theta(|Y\Omega(t) - Y\omega(t)|)$$
$$= \theta(|\int_a^t \chi(t,s,\Omega(s))ds - \int_a^t \chi(t,s,\omega(s))ds|)$$
$$\le \int_a^b \theta(|\chi(t,s,\Omega(s)) - \chi(t,s,\omega(s))|)ds$$
$$\le \int_a^b \frac{[\theta(|\Omega(t) - \omega(t)|)]^{\tau_1}[\theta(|\Omega(t) - \int_a^b \chi(t,s,\Omega(s)ds)|)]^{\tau_2}[\theta(|\omega(t) - \int_a^b \chi(t,s,\omega(s))ds|)]^{\tau_3}}{b-a}$$
$$[\theta(|\omega(t) - \int_a^b \chi(t,s,\Omega(s))ds|)]^{\tau_4}ds$$
$$\le \frac{1}{b-a}\int_a^b [\theta(d(\Omega,\omega))]^{\tau_1}[\theta(d(\Omega,Y\Omega))]^{\tau_2}[\theta(d(\omega,Y\omega))]^{\tau_3}[\theta(d(\omega,Y\Omega))]^{\tau_4}ds$$
$$= [\theta(d(\Omega,\omega))]^{\tau_1}[\theta(d(\Omega,Y\Omega))]^{\tau_2}[\theta(d(\omega,Y\omega))]^{\tau_3}[\theta(d(\omega,Y\Omega))]^{\tau_4}.$$

Thus Y is a $\mathcal{P}$-contraction. All the conditions of Theorem 2 hold, and so Y has a unique fixed point, that is, (22) has a unique solution. $\square$

5. Conclusions

In this paper, we restricted the conditions on the control function θ (with respect to the ones given in [27,28]) and we obtained a real generalization of the Banach contraction principle (BCP). We also initiated a weakly JS-contractive condition that generalizes its corresponding of Jleli and Samet [26], and we provided some related fixed point results.

Author Contributions: All authors contributed equally and significantly in writing this article. All authors read and approved the final manuscript.

Funding: This research received no external funding.

Acknowledgments: The third author would like to thank Prince Sultan University for funding this work through research group Nonlinear Analysis Methods in Applied Mathematics (NAMAM) group number RG-DES-2017-01-17.

Conflicts of Interest: The authors declare that they have no competing interests regarding the publication of this paper.

References

1. Banach, S. Sur Les Operations Dans Les Ensembles Abstraits et Leur Application Aux Equations Integrales. *Fund. Math.* **1922**, *3*, 133–181. [CrossRef]
2. Aydi, H.; Abbas, M.; Vetro, C. Partial Hausdorff metric and Nadler's fixed point theorem on partial metric spaces. *Topol. Appl.* **2012**, *159*, 3234–3242. [CrossRef]
3. Aydi, H.; Bota, M.F.; Karapinar, E.; Moradi, S. A common fixed point for weak ϕ-contractions on b-metric spaces. *Fixed Point Theory* **2012**, *13*, 337–346.
4. Aydi, H.; Karapinar, E.; Samet, B. Fixed point Theorems for various classes of cyclic mappings. *J. Appl. Math.* **2012**, *2012*, 867216. [CrossRef]
5. Aydi, H.; Shatanawi, W.; Vetro, C. On generalized weakly G-contraction mapping in G-metric spaces. *Comput. Math. Appl.* **2011**, *62*, 4222–4229. [CrossRef]
6. Boyd, D.W.; Wong, J.S.W. On nonlinear contractions. *Proc. Am. Math. Soc.* **1969**, *20*, 458–464. [CrossRef]
7. Ćirić, L.B. A generalization of Banach's contraction principle. *Proc. Am. Math. Soc.* **1974**, *45*, 267–273. [CrossRef]
8. Ćirić, L.B. Generalized contractions and fixed-point theorems. *Publ. Inst. Math.* **1971**, *12*, 19–26.
9. Ćirić, L.B. Multi-valued nonlinear contraction mappings. *Nonlinear Anal.* **2009**, *71*, 2716–2723. [CrossRef]
10. Ćirić, L.B.; Samet, B.; Aydi, H.; Vetro, C. Common fixed points of generalized contractions on partial metric spaces and an application. *Appl. Math. Comput.* **2011**, *218*, 2398–2406. [CrossRef]
11. Geraghty, M. On contractive mappings. *Proc. Am. Math. Soc.* **1973**, *40*, 604–608. [CrossRef]
12. Hussain, N.; Dorić, D.; Kadelburg, Z.; Radenović, S. Suzuki-type fixed point results in metric type spaces. *Fixed Point Theory Appl.* **2012**, *2012*, 126. [CrossRef]
13. Karapinar, E.; Alqahtani, O.; Aydi, H. On interpolative Hardy-Rogers type contractions. *Symmetry* **2018**, *11*, 8. [CrossRef]
14. Meir, A.; Keeler, E. A theoremon contraction mapping. *J. Math. Anal. Appl.* **1969**, *28*, 326–329. [CrossRef]
15. Mizoguchi, N.; Takahashi, W. Fixed point theorems for multivalued mappings on complete metric spaces. *J. Math. Anal. Appl.* **1989**, *141*, 177–188. [CrossRef]
16. Mustafa, Z.; Parvaneh, V.; Jaradat, M.M.M.; Kadelburg, Z. Extended rectangular b-metric spaces and some fixed point theorems for contractive mappings. *Symmetry* **2019**, *11*, 594. [CrossRef]
17. Patle, P.; Patel, D.; Aydi, H.; Radenović, S. On H^+-type multivalued contractions and applications in symmetric and probabilistic spaces. *Mathematics* **2019**, *7*, 144. [CrossRef]
18. Reich, S. Some remarks concerning contraction mappings. *Can. Math. Bull.* **1971**, *14*, 121–124. [CrossRef]
19. Wardowski, D. Fixed points of a new type of contractive mappings in complete metric spaces. *Fixed Point Theory Appl.* **2012**, *2012*, 94. [CrossRef]
20. Lohawech, P.; Kaewcharoen, A. Fixed point theorems for generalized JS-quasi-contractions in complete partial b-metric spaces. *J. Nonlineart Sci. Appl.* **2019**, *12*, 728–739. [CrossRef]
21. Mlaiki, N.; Kukić, K.; Gardasević-Filipović, M.; Aydi, H. On almost b-metric spaces and related fixed point results. *Axioms* **2019**, *8*, 70. [CrossRef]
22. Dosenović, T.; Radenović, S. A comment on "Fixed point theorems of JS-quasi-contractions". *Indian J. Math. Dharma Prakash Gupta Meml.* **2018**, *60*, 141–152.
23. Kadelburg, Z.; Radenović, S.; Shukla, S. Boyd-Wong and Meir-Keeler type theorems in generalized metric spaces. *J. Adv. Math. Stud.* **2016**, *9*, 83–93.
24. Chatterjea, S.K. Fixed Point Theorems. *C. R. Acad. Bulgare Sci.* **1972**, *25*, 727–730. [CrossRef]
25. Kannan, R. Some results on fixed points. *Bull. Calc. Math. Soc.* **1968**, *60*, 71–76.
26. Jleli, M.; Samet, B. A new generalization of the Banach contraction principle. *J. Inequal. Appl.* **2014**, *2014*, 38. [CrossRef]

27. Hussain, N.; Parvaneh, V.; Samet, B.; Vetro, C. Some fixed point theorems for generalized contractive mappings in complete metric spaces. *Fixed Point Theory Appl.* **2015**, *2015*, 185. [CrossRef]
28. Jiang, S.; Li, Z.; Damjanović, B. A note on Some fixed point theorems for generalized contractive mappings in complete metric spaces. *Fixed Point Theory Appl.* **2016**, *2016*, 62. [CrossRef]

© 2019 by the authors. Licensee MDPI, Basel, Switzerland. This article is an open access article distributed under the terms and conditions of the Creative Commons Attribution (CC BY) license (http://creativecommons.org/licenses/by/4.0/).

Article
Best Proximity Point Results for Geraghty Type $\mathcal{Z}$-Proximal Contractions with an Application

Hüseyin Işık [1,2,*], Hassen Aydi [3], Nabil Mlaiki [4] and Stojan Radenović [5]

1. Nonlinear Analysis Research Group, Ton Duc Thang University, Ho Chi Minh City 700000, Vietnam
2. Faculty of Mathematics and Statistics, Ton Duc Thang University, Ho Chi Minh City 700000, Vietnam
3. Institut Supérieur d'Informatique et des Techniques de Communication, Université de Sousse, H. Sousse 4000, Tunisia
4. Department of Mathematics and General Sciences, Prince Sultan University, Riyadh 11586, Saudi Arabia
5. Department of Mathematics, College of Science, King Saud University, Riyadh 11451, Saudi Arabia
* Correspondence: huseyin.isik@tdtu.edu.vn

Received: 18 June 2019; Accepted: 11 July 2019; Published: 18 July 2019

Abstract: In this study, we establish the existence and uniqueness theorems of the best proximity points for Geraghty type $\mathcal{Z}$-proximal contractions defined on a complete metric space. The presented results improve and generalize some recent results in the literature. An example, as well as an application to a variational inequality problem are also given in order to illustrate the effectiveness of our generalizations.

Keywords: best proximity point; $\mathcal{Z}$-contraction; geraghty type contraction; simulation function; admissible mapping; variational inequality

MSC: 47H10; 54H25

1. Introduction

Numerous problems in science and engineering defined by nonlinear functional equations can be solved by reducing them to an equivalent fixed-point problem. In fact, an operator equation

$$Gx = 0 \qquad (1)$$

may be expressed as a fixed-point equation $\mathcal{T}x = x$. Accordingly, the Equation (1) has a solution if the self-mapping $\mathcal{T}$ has a fixed point. However, for a non-self mapping $\mathcal{T} : P \to Q$, the equation $\mathcal{T}x = x$ does not necessarily admit a solution. Here, it is quite natural to find an approximate solution x^* such that the distance $d(x^*, \mathcal{T}x^*)$ is minimum, in which case x^* and $\mathcal{T}x^*$ are in close proximity to each other. Herein, the optimal approximate solution x^*, for which $d(x^*, \mathcal{T}x^*) = d(P,Q)$, is called a best proximity point of $\mathcal{T}$. The main aim of the best proximity point theory is to give sufficient conditions for finding the existence of a solution to the nonlinear programming problem,

$$\min_{\zeta \in P} d(\zeta, \mathcal{T}\zeta). \qquad (2)$$

Moreover, a best proximity point generates to a fixed point if the mapping under consideration is a self-mapping. For more details on this research subject, see [1–15].

In 2015, Khojasteh et al. [16] presented the notion of $\mathcal{Z}$-contraction involving a new class of mappings—namely, simulation functions, and proved new fixed-point theorems via different methods to others in the literature. For more details, see [17–20].

Definition 1 ([16]). *A simulation function is a mapping $\zeta : [0,\infty) \times [0,\infty) \to \mathbb{R}$ so that:*

(ζ_1) $\zeta(0,0) = 0$;
(ζ_2) $\zeta(\mu, \eta) < \eta - \mu$ for all $\mu, \eta > 0$;
(ζ_3) *If $(\mu_n), (\eta_n)$ are sequences in $(0, \infty)$ so that $\lim_{n\to\infty} \mu_n = \lim_{n\to\infty} \eta_n > 0$, then*

$$\limsup_{n\to\infty} \zeta(\mu_n, \eta_n) < 0. \tag{3}$$

Theorem 1 ([16]). *Let (M,d) be a complete metric space and $\mathcal{T} : M \to M$ be a $\mathcal{Z}$-contraction with respect to $\zeta \in \mathcal{Z}$—that is,*

$$\zeta(d(\mathcal{T}\xi, \mathcal{T}\omega), d(\xi, \omega)) \geq 0, \quad \text{for all } \xi, \omega \in M.$$

Then, $\mathcal{T}$ admits a unique fixed point (say $\tau \in X$) and, for each $\xi_0 \in M$, the Picard sequence $\{\mathcal{T}^n \xi_0\}$ is convergent to τ.

In this study, we will consider simulation functions satisfying only the condition (ζ_2). For the sake of convenience, we identify the set of all simulation functions satisfying only the condition (ζ_2) by $\mathcal{Z}$.

The main concern of the paper is to establish theorems on the existence and uniqueness of best proximity points for Geraghty type $\mathcal{Z}$-proximal contractions in complete metric spaces. The obtained results complement and extend some known results from the literature. An example, as well as an application to a variational inequality problem, is also given in order to illustrate the effectiveness of our generalizations.

2. Preliminaries

Let P and Q be two non-empty subsets of a metric space, (M, d). Consider:

$$d(P, Q) := \inf \{d(\rho, \nu) : \rho \in P, \nu \in Q\};$$
$$P_0 := \{\rho \in P : d(\rho, \nu) = d(P, Q) \text{ for some } \nu \in Q\};$$
$$Q_0 := \{\nu \in Q : d(\rho, \nu) = d(P, Q) \text{ for some } \rho \in P\}.$$

Denote by

$$B_{est}(\mathcal{T}) = \{u \in P : d(u, \mathcal{T}u) = d(P, Q)\},$$

the set of all best proximity points of a non-self-mapping $\mathcal{T} : P \to Q$. In the study [5], Caballero et al. familiarized the notion of Geraghty contraction for non-self-mappings as follows:

Definition 2 ([5]). *Let P, Q be two non-empty subsets of a metric space, (M, d). A mapping $\mathcal{T} : P \to Q$ is called a Geraghty contraction if there is $\beta \in \Sigma$, so that for all $\xi, \omega \in P$*

$$d(\mathcal{T}\xi, \mathcal{T}\omega) \leq \beta(d(\xi, \omega)) \cdot d(\xi, \omega), \tag{4}$$

where the class Σ is the set of functions $\beta : [0, \infty) \to [0, 1)$, satisfying

$$\beta(t_n) \to 1 \implies t_n \to 0.$$

In the paper [10], Jleli and Samet initiated the concepts of α-ψ-proximal contractive and α-proximal admissible mappings. They provided related best-proximity-point results. Subsequently, Hussain et al. [7] modified the aforesaid notions and substantiated certain best-proximity-point theorems.

Definition 3 ([10]). *Let $\mathcal{T} : P \to Q$ and $\alpha : P \times P \to [0, \infty)$ be given mappings. Then, $\mathcal{T}$ is called α-proximal admissible if*

$$\left. \begin{array}{l} \alpha(u_1, u_2) \geq 1 \\ d(p_1, \mathcal{T} u_1) = d(P, Q) \\ d(p_2, \mathcal{T} u_2) = d(P, Q) \end{array} \right\} \implies \alpha(p_1, p_2) \geq 1,$$

for all $u_1, u_2, p_1, p_2 \in P$.

Definition 4 ([7]). *Let $\mathcal{T} : P \to Q$ and $\alpha, \eta : P \times P \to [0, \infty)$ be given mappings. Such $\mathcal{T}$ is said to be (α, η)-proximal admissible if*

$$\left. \begin{array}{l} \alpha(u_1, u_2) \geq \eta(u_1, u_2) \\ d(p_1, \mathcal{T} u_1) = d(P, Q) \\ d(p_2, \mathcal{T} u_2) = d(P, Q) \end{array} \right\} \implies \alpha(p_1, p_2) \geq \eta(p_1, p_2),$$

for all $u_1, u_2, p_1, p_2 \in P$.

Note that if $\eta(u, v) = 1$ for all $u, v \in P$, then Definition 4 corresponds to Definition 3.
Very recently, Tchier et al. in [14] initiated the concept of $\mathcal{Z}$-proximal contractions.

Definition 5 ([14]). *Let P and Q be two non-empty subsets of a metric space, (M, d). A non-self-mapping $\mathcal{T} : P \to Q$ is called a $\mathcal{Z}$-proximal contraction if there is a simulation function ζ so that*

$$\left. \begin{array}{l} d(\rho, \mathcal{T} u) = d(P, Q) \\ d(v, \mathcal{T} v) = d(P, Q) \end{array} \right\} \implies \zeta(d(\rho, v), d(u, v)) \geq 0, \tag{5}$$

for all $\rho, v, u, v \in P$.

Now, we introduce a new concept which will be efficiently used in our results.

Definition 6. *Let $\mathcal{T} : P \to Q$ and $\alpha, \eta : P \times P \to [0, \infty)$ be given mappings. Then, $\mathcal{T}$ is said to be triangular (α, η)-proximal admissible, if*

(1) $\mathcal{T}$ *is (α, η)-proximal admissible;*
(2) $\alpha(u, v) \geq \eta(u, v)$ *and* $\alpha(v, z) \geq \eta(v, z)$ *implies that* $\alpha(u, z) \geq \eta(u, z)$, *for all $u, v, z \in P$.*

Now, we describe a new class of contractions for non-self-mappings which generalize the concept of Geraghty-contractions.

Definition 7. *Let P and Q be two non-empty subsets of a metric space (M, d), $\zeta \in \mathcal{Z}$ and $\alpha, \eta : P \times P \to [0, \infty)$ and $\beta \in \Sigma$. A non-self-mapping $\mathcal{T} : P \to Q$ is said to be a Geraghty type $\mathcal{Z}$-proximal contraction, if for all $u, v, \rho, v \in P$, the following implication holds:*

$$\left. \begin{array}{l} \alpha(u, v) \geq \eta(u, v) \\ d(\rho, \mathcal{T} u) = d(P, Q) \\ d(v, \mathcal{T} v) = d(P, Q) \end{array} \right\} \implies \zeta(d(\rho, v), \beta(d(u, v))d(u, v)) \geq 0. \tag{6}$$

Remark 1. *If $\mathcal{T} : P \to Q$ is a Geraghty type $\mathcal{Z}$-proximal contraction, then by (ζ_2) and Definition 7, the following implication holds for all $u, v, \rho, v \in P$ with $u \neq v$:*

$$\left. \begin{array}{l} \alpha(u, v) \geq \eta(u, v) \\ d(\rho, \mathcal{T} u) = d(P, Q) \\ d(v, \mathcal{T} v) = d(P, Q) \end{array} \right\} \implies d(\rho, v) < \beta(d(u, v))d(u, v). \tag{7}$$

3. Main Results

Our first result is as follows.

Theorem 2. *Let (P, Q) be a pair of non-empty subsets of a complete metric space (M, d) so that P_0 is non-empty, $T : P \to Q$ and $\alpha, \eta : P \times P \to [0, \infty)$ be given mappings. Suppose that:*

(i) *P is closed and $T(P_0) \subseteq Q_0$;*
(ii) *T is triangular (α, η)-proximal admissible;*
(iii) *There are $u_0, u_1 \in P_0$ so that $d(u_1, Tu_0) = d(P, Q)$ and $\alpha(u_0, u_1) \geq \eta(u_0, u_1)$;*
(iv) *T is a continuous Geraghty type $\mathcal{Z}$-proximal contraction.*

Then, T has a best proximity point in P. If $\alpha(u, v) \geq \eta(u, v)$ for all $u, v \in B_{est}(T)$, then T has a unique best proximity point $u^* \in P$. Moreover, for every $u \in P$, $\lim_{n \to \infty} T^n u = u^*$.

Proof. From the condition (iii), there are $u_0, u_1 \in P_0$ so that

$$d(u_1, Tu_0) = d(P, Q) \quad \text{and} \quad \alpha(u_0, u_1) \geq \eta(u_0, u_1).$$

Since $T(P_0) \subseteq Q_0$, there is $u_2 \in P_0$ so that

$$d(u_2, Tu_1) = d(P, Q).$$

Thus, we get

$$\alpha(u_0, u_1) \geq \eta(u_0, u_1),$$
$$d(u_1, Tu_0) = d(P, Q),$$
$$d(u_2, Tu_1) = d(P, Q).$$

Since T is (α, η)-proximal admissible, we get $\alpha(u_1, u_2) \geq \eta(u_1, u_2)$. Now, we have

$$d(u_2, Tu_1) = d(P, Q) \quad \text{and} \quad \alpha(u_1, u_2) \geq \eta(u_1, u_2).$$

Again, since $T(P_0) \subseteq Q_0$, there exists $u_3 \in P_0$ such that

$$d(u_3, Tu_2) = d(P, Q),$$

and thus,

$$\alpha(u_1, u_2) \geq \eta(u_1, u_2),$$
$$d(u_2, Tu_1) = d(P, Q),$$
$$d(u_3, Tu_2) = d(P, Q).$$

Since T is (α, η)-proximal admissible, this implies that $\alpha(u_2, u_3) \geq \eta(u_2, u_3)$. Thus, we have

$$d(u_3, Tu_2) = d(P, Q) \quad \text{and} \quad \alpha(u_2, u_3) \geq \eta(u_2, u_3).$$

By repeating this process, we build a sequence $\{u_n\}$ in $P_0 \subseteq P$ so that

$$d(u_{n+1}, Tu_n) = d(P, Q) \quad \text{and} \quad \alpha(u_n, u_{n+1}) \geq \eta(u_n, u_{n+1}), \tag{8}$$

for all $n \in \mathbb{N} \cup \{0\}$. If there is n_0 so that $u_{n_0} = u_{n_0+1}$, then

$$d(u_{n_0}, Tu_{n_0}) = d(u_{n_0+1}, Tu_{n_0}) = d(P, Q).$$

That is, u_{n_0} is a best proximity point of T. We should suppose that $u_n \neq u_{n+1}$, for all n.

From (8), for all $n \in \mathbb{N}$, we get

$$\alpha(u_{n-1}, u_n) \geq \eta(u_{n-1}, u_n),$$
$$d(u_n, Tu_{n-1}) = d(P, Q),$$
$$d(u_{n+1}, Tu_n) = d(P, Q).$$

On the grounds that T is a Geraghty type $\mathcal{Z}$-proximal contraction, by utilizing Remark 1, we deduce that

$$d(u_n, u_{n+1}) < \beta(d(u_{n-1}, u_n))d(u_{n-1}, u_n), \tag{9}$$

which requires that $d(u_n, u_{n+1}) < d(u_{n-1}, u_n)$, for all n. Therefore, the sequence $\{d(u_n, u_{n+1})\}$ is decreasing, and so there is $\lambda \geq 0$ so that $\lim_{n \to \infty} d(u_n, u_{n+1}) = \lambda$. Now, we shall show that $\lambda = 0$. On the contrary, assume that $\lambda > 0$. Then, taking into account (9), for any $n \in \mathbb{N}$,

$$d(u_n, u_{n+1}) < \beta(d(u_{n-1}, u_n))d(u_{n-1}, u_n) < d(u_{n-1}, u_n).$$

This yields, for any $n \in \mathbb{N}$,

$$0 < \frac{d(u_n, u_{n+1})}{d(u_{n-1}, u_n)} < \beta(d(u_{n-1}, u_n)) < 1.$$

Taking $n \to \infty$, we find that

$$\lim_{n \to \infty} \beta(d(u_{n-1}, u_n)) = 1,$$

and since $\beta \in \Sigma$, $\lim_{n \to \infty} d(u_{n-1}, u_n) = 0$. This contradicts our assumption $\lim_{n \to \infty} d(u_{n-1}, u_n) = \lambda > 0$. Therefore, we get

$$\lim_{n \to \infty} d(u_{n-1}, u_n) = 0, \quad \text{for all } n \in \mathbb{N}. \tag{10}$$

We shall prove that $\{u_n\}$ is Cauchy in P. By contradiction, suppose that $\{u_n\}$ is not a Cauchy sequence, so there is an $\varepsilon > 0$ for which we can find $\{u_{m_k}\}$ and $\{u_{n_k}\}$ of $\{u_n\}$ such that n_k is the smallest index for which $n_k > m_k > k$ and

$$d(u_{m_k}, u_{n_k}) \geq \varepsilon \quad \text{and} \quad d(u_{m_k}, u_{n_k-1}) < \varepsilon. \tag{11}$$

We have

$$\varepsilon \leq d(u_{m_k}, u_{n_k}) \leq d(u_{m_k}, u_{n_k-1}) + d(u_{n_k-1}, u_{n_k})$$
$$< \varepsilon + d(u_{n_k-1}, u_{n_k}).$$

Taking $k \to \infty$, by (10), we get

$$\lim_{k \to \infty} d(u_{m_k}, u_{n_k}) = \varepsilon. \tag{12}$$

By triangular inequality,

$$\left| d(u_{m_k+1}, u_{n_k+1}) - d(u_{m_k}, u_{n_k}) \right| \leq d(u_{m_k+1}, u_{m_k}) + d(u_{n_k}, u_{n_k+1}),$$

which yields that

$$\lim_{k \to \infty} d(x_{m_k+1}, x_{n_k+1}) = \varepsilon. \tag{13}$$

Since T is triangular (α, η)-proximal admissible, by using (8), we infer

$$\alpha(u_m, u_n) \geq \eta(u_m, u_n), \quad \text{for all } n, m \in \mathbb{N} \text{ with } m < n. \tag{14}$$

Combining (8) and (14), for all $k \in \mathbb{N}$, we have

$$\alpha(u_{m_k}, u_{n_k}) \geq \eta(u_{m_k}, u_{n_k}),$$
$$d(u_{m_k+1}, \mathcal{T}u_{m_k}) = d(P, Q),$$
$$d(u_{n_k+1}, \mathcal{T}u_{n_k}) = d(P, Q).$$

Regarding the fact that $\mathcal{T}$ is a Geraghty type $\mathcal{Z}$-proximal contraction, from Remark 1, we deduce that

$$d(u_{m_k+1}, u_{n_k+1}) < \beta(d(u_{m_k}, u_{n_k}))d(u_{m_k}, u_{n_k}) < d(u_{m_k}, u_{n_k}).$$

Taking the limit as k tends to ∞ on both sides of the last inequality, and using the Equations (12) and (13), we get

$$\varepsilon \leq \lim_{k \to \infty} \beta(d(u_{m_k}, u_{n_k}))\varepsilon \leq \varepsilon,$$

which implies that $\lim_{k \to \infty} \beta(d(u_{m_k}, u_{n_k})) = 1$, and so $\lim_{k \to \infty} d(u_{m_k}, u_{n_k}) = 0$ which contradicts $\varepsilon > 0$. Hence, $\{u_n\}$ is a Cauchy sequence in P. Since P is a closed subset of the complete metric space (M, d), there is $p \in P$ so that

$$\lim_{n \to \infty} d(u_n, p) = 0. \tag{15}$$

Since $\mathcal{T}$ is continuous, we have

$$\lim_{n \to \infty} d(\mathcal{T}u_n, \mathcal{T}p) = 0. \tag{16}$$

Combining (8), (15), and (16), we get

$$d(P, Q) = \lim_{n \to \infty} d(u_{n+1}, \mathcal{T}u_n) = d(p, \mathcal{T}p).$$

Therefore, $u \in P$ is a best proximity point of $\mathcal{T}$. Finally, we shall show that the set $B_{est}(\mathcal{T})$ is a singleton. Suppose that r is another best proximity point of $\mathcal{T}$, that is, $d(r, \mathcal{T}r) = d(P, Q)$. Then, by the hypothesis, we have $\alpha(p, r) \geq \eta(p, r)$—that is,

$$\alpha(p, r) \geq \eta(p, r),$$
$$d(p, \mathcal{T}p) = d(P, Q),$$
$$d(r, \mathcal{T}r) = d(P, Q).$$

Then, from Remark 1, we deduce

$$d(p, r) < \beta(d(p, r))d(p, r) < d(p, r),$$

which is a contradiction. Hence, we have a unique best proximity point of $\mathcal{T}$. □

Let us consider the following assertion in order to remove the continuity on the operator $\mathcal{T}$ in the next theorem.

(C) If a sequence $\{u_n\}$ in P is convergent to $u \in P$ so that $\alpha(u_n, u_{n+1}) \geq \eta(u_n, u_{n+1})$, then $\alpha(u_n, u) \geq \eta(u_n, u)$ for all $n \in \mathbb{N}$.

Theorem 3. *Let (P, Q) be a pair of non-empty subsets of a complete metric space (M, d) so that P_0 is non-empty, $\mathcal{T} : P \to Q$ and $\alpha, \eta : P \times P \to [0, \infty)$ be given mappings. Suppose that:*

(i) P is closed and $\mathcal{T}(P_0) \subseteq Q_0$;

(ii) $\mathcal{T}$ is triangular (α, η)-proximal admissible;
(iii) there are $u_0, u_1 \in P_0$ so that $d(u_1, \mathcal{T}u_0) = d(P, Q)$ and $\alpha(u_0, u_1) \geq \eta(u_0, u_1)$;
(iv) the condition (C) holds and $\mathcal{T}$ is a Geraghty type $\mathcal{Z}$-proximal contraction.

Then, $\mathcal{T}$ has a best proximity point in P. If $\alpha(u,v) \geq \eta(u,v)$ for all $u, v \in B_{est}(\mathcal{T})$, then $\mathcal{T}$ has a unique best proximity point $u^* \in P$. Moreover, for each $u \in P$, we have $\lim_{n\to\infty} \mathcal{T}^n u = u^*$.

Proof. Following the proof of Theorem 2, there exists a Cauchy sequence $\{u_n\} \subset P_0$ satisfying (8) and $u_n \to p$. On account of (i), P_0 is closed, and so $p \in P_0$. Also, since $\mathcal{T}(P_0) \subseteq Q_0$, there is $z \in P_0$ so that

$$d(z, \mathcal{T}p) = d(P, Q). \tag{17}$$

Taking (C) and (8) into account, we infer

$$\alpha(u_n, p) \geq \eta(u_n, p), \quad \text{for all } n \in \mathbb{N}.$$

Since $\mathcal{T}$ is (α, η)-proximal admissible and

$$\begin{aligned} \alpha(u_n, p) &\geq \eta(u_n, p), \\ d(u_{n+1}, \mathcal{T}u_n) &= d(P, Q), \\ d(z, \mathcal{T}p) &= d(P, Q), \end{aligned} \tag{18}$$

so, we conclude that

$$\alpha(u_{n+1}, z) \geq \eta(u_{n+1}, z), \quad \text{for all } n \in \mathbb{N}. \tag{19}$$

Considering (18), (19) and Remark 1, we have

$$d(u_{n+1}, z) < \beta(d(u_n, p))d(u_n, p) < d(u_n, p),$$

which implies that $\lim_{n\to\infty} d(u_{n+1}, z) = 0$. By the uniqueness of the limit, we obtain $z = p$. Thus, by (17), we deduce that $d(p, \mathcal{T}p) = d(P, Q)$. Uniqueness of the best proximity point follows from the proof of Theorem 2. □

Example 1. Let $M = \mathbb{R}^2$ be endowed with the Euclidean metric, $P = \{(0, u) : u \geq 0\}$ and $Q = \{(1, u) : u \geq 0\}$. Note that $d(P, Q) = 1$, $P_0 = P$ and $Q_0 = Q$. Let

$$\begin{cases} \beta(t) = \frac{1}{1+t}, & \text{if } t > 0 \\ \beta(t) = \frac{1}{2}, & \text{otherwise}. \end{cases}$$

Then, $\beta \in \Sigma$. Define $\mathcal{T} : P \to Q$ and $\alpha : P \times P \to [0, \infty)$ by

$$\mathcal{T}(0, u) = \begin{cases} (1, \frac{u}{9}), & \text{if } 0 \leq u \leq 1, \\ (1, u^2), & \text{if } u > 1, \end{cases}$$

and

$$\alpha((0, u), (0, v)) = \begin{cases} 2\eta((0, u), (0, v)), & \text{if } u, v \in [0, 1], \text{ or } u = v \\ 0, & \text{otherwise}. \end{cases}$$

Choose $\zeta(t,s) = \frac{2}{3}s - t$ for all $t,s \in [0,\infty)$. Let $u,v,p,q \geq 0$ be such that

$$\begin{cases} \alpha((0,u),(0,v)) \geq \eta((0,u),(0,v)) \\ d((0,p),T(0,u)) = d(P,Q) = 1 \\ d((0,q),T(0,v)) = d(P,Q) = 1. \end{cases}$$

Then, $u,v \in [0,1]$ or $u = v$.

$u,v \in [0,1]$. Here, $T(0,u) = (1, \frac{u}{9})$ and $T(0,v) = (1, \frac{v}{9})$. Also,

$$\sqrt{1 + (p - \frac{u}{9})^2} = \sqrt{1 + (q - \frac{v}{9})^2} = 1,$$

that is, $p = \frac{u}{9}$ and $q = \frac{v}{9}$. So, $\alpha((0,p),(0,q)) \geq d((0,p),(0,q))$. Moreover,

$$\zeta(d((0,p),(0,q)), \beta(d((0,u),(0,v)))d((0,u),(0,v)))$$
$$= \frac{2}{3}\beta(d((0,u),(0,v)))d((0,u),(0,v)) - d((0,\frac{u}{9}),(0,\frac{v}{9}))$$
$$= \frac{2}{3}\beta(|u-v|)|u-v| - \frac{|u-v|}{9}.$$

If $u = v$, then $\beta(|u-v|) = \frac{1}{2}$ and the right-hand side of the above inequality is equal to 0.
If $u \neq v$, we have

$$\zeta(d((0,p),(0,q)), \beta(d((0,u),(0,v)))d((0,u),(0,v)))$$
$$= \frac{2}{3}\frac{|u-v|}{1+|u-v|} - \frac{|u-v|}{9} \geq 0.$$

$u = v > 1$. Here, $T(0,u) = (1,u^2)$ and $T(0,v) = (1,v^2)$. Similarly, we get that $p = q = u^2 = v^2$. So, $\alpha((0,p),(0,q)) = 0 = \eta((0,p),(0,q))$.
Also, $\zeta(d((0,p),(0,q)), \beta(d((0,u),(0,v)))d((0,u),(0,v))) \geq 0$.
In each case, we get that T is an (α,η)-proximal admissible. It is also easy to see that T is triangular (α,η)-proximal admissible. Also, T is a Geraghty type $\mathcal{Z}$-proximal contraction. Also, if $\{u_n = (0,p_n)\}$ is a sequence in P such that $\alpha(u_n, u_{n+1}) \geq \eta(u_n, u_{n+1})$ for all n and $u_n = (0,p_n) \to u = (0,p)$ as $n \to \infty$, then $p_n \to p$. We have $p_n, p_{n+1} \in [0,1]$ or $p_n = p_{n+1}$. We get that $p \in [0,1]$ or $p_n = p$. This implies that $\alpha(u_n, u) \geq \eta(u_n, u)$ for all n.
Moreover, there is $(u_0, u_1) = ((0,1), (0, \frac{1}{9})) \in P_0 \times P_0$ so that

$$d(u_1, Tu_0) = 1 = d(P,Q) \text{ and } \alpha(u_0, u_1) \geq d(u_0, u_1).$$

Consequently, all conditions of Theorem 3 are satisfied. Therefore, T has a unique best proximity point in P, which is $(0,0)$. On the other side, we indicate that (4) is not satisfied. In fact, for $u = (0,2), v = (0,3)$, we have

$$d(Tu, Tv) = d(T(0,2), T(0,3)) = d((0,4),(0,9))$$
$$= 5 > \frac{1}{2} = \beta(d((0,2),(0,3)))d((0,2),(0,3))$$
$$= \beta(d(u,v))d(u,v).$$

Corollary 1. Let (P, Q) be a pair of non-empty subsets of a complete metric space (M, d), such that P_0 is non-empty. Suppose that $T : P \to Q$ is a Geraghty-proximal contraction—that is, the following implication holds for all $u, v, \rho, \nu \in P$:

$$\left. \begin{array}{l} d(\rho, Tu) = d(P, Q) \\ d(\nu, Tv) = d(P, Q) \end{array} \right\} \Longrightarrow \zeta(d(\rho, \nu), \beta(d(u, v))d(u, v)) \geq 0.$$

Also, assume that P is closed and $T(P_0) \subseteq Q_0$. Then, T has a unique best proximity point $u^* \in P$. Moreover, for each $u \in P$, we have $\lim_{n \to \infty} T^n u = u^*$.

Proof. We take $\alpha(\sigma, \varsigma) = \eta(\sigma, \varsigma) = 1$ in the proof of Theorem 2 (resp. Theorem 3). □

4. Some Consequences

In this section we give new fixed-point results on a metric space endowed with a partial ordering/graph by using the results provided in the previous section. Define

$$\alpha, \eta : M \times M \to [0, \infty), \quad \alpha(u, v) = \begin{cases} \eta(u, v), & \text{if } u \preceq v, \\ 0, & \text{otherwise.} \end{cases}$$

Definition 8. Let $(M, \preceq, d)$ be a partially ordered metric space, (P, Q) be a pair of non-empty subsets of M, and $T : P \to Q$ be a given mapping. Such T is said to be $\preceq$-proximal increasing if

$$\left. \begin{array}{l} u_1 \preceq u_2 \\ d(p_1, Tu_1) = d(P, Q) \\ d(p_2, Tu_2) = d(P, Q) \end{array} \right\} \Longrightarrow p_1 \preceq p_2,$$

for all $u_1, u_2, p_1, p_2 \in P$.

Then, the following result is a direct consequence of Theorem 2 (resp. Theorem 3).

Theorem 4. Let (P, Q) be a pair of non-empty subsets of a complete ordered metric space $(M, \preceq, d)$ so that P_0 is non-empty and $T : P \to Q$ be a given non-self-mapping. Suppose that:

(i) P is closed and $T(P_0) \subseteq Q_0$;
(ii) T is $\preceq$-proximal increasing;
(iii) There are $u_0, u_1 \in P_0$ so that $d(u_1, Tu_0) = d(P, Q)$ and $u_0 \preceq u_1$;
(iv) T is continuous or, for every sequence $\{u_n\}$ in P is convergent to $u \in P$ so that $u_n \preceq u_{n+1}$, we have $u_n \preceq u$ for all $n \in \mathbb{N}$;
(v) There exist $\zeta \in \mathcal{Z}$ and $\beta \in \Sigma$, such that for all $u, v, \rho, \nu \in P$,

$$\left. \begin{array}{l} u \preceq v \\ d(\rho, Tu) = d(P, Q) \\ d(\nu, Tv) = d(P, Q) \end{array} \right\} \Longrightarrow \zeta(d(\rho, \nu), \beta(d(u, v))d(u, v)) \geq 0. \quad (20)$$

Then, T has a best proximity point in P. If $u \preceq v$ for all $u, v \in \text{Best}(T)$, then T has a unique best proximity point $u^* \in P$. Moreover, for every $u \in P$, $\lim_{n \to \infty} T^n u = u^*$.

Now, we present the existence of the best proximity point for non-self mappings from a metric space M, endowed with a graph, into the space of non-empty closed and bounded subsets of the metric space. Consider a graph G, such that the set $V(G)$ of its vertices coincides with M and the set

$E(G)$ of its edges contains all loops; that is, $E(G) \supseteq \Delta$, where $\Delta = \{(u,u) : u \in M\}$. We assume G has no parallel edges, so we can identify G with the pair $(V(G), E(G))$.

Define

$$\alpha, \eta: M \times M \to [0, +\infty), \quad \alpha(u,v) = \begin{cases} \eta(u,v), & \text{if } (u,v) \in E(G), \\ 0, & \text{otherwise.} \end{cases}$$

Definition 9. *Let (M,d) be a complete metric space endowed with a graph G and (P,Q) be a pair of non-empty subsets of M and $\mathcal{T}: P \to Q$ be a given mapping. Such $\mathcal{T}$ is said to be triangular G-proximal, if*

(1) *for all $u_1, u_2, p_1, p_2 \in P$,*

$$\left. \begin{array}{r} (u_1, u_2) \in E(G) \\ d(p_1, \mathcal{T} u_1) = d(P,Q) \\ d(p_2, \mathcal{T} u_2) = d(P,Q) \end{array} \right\} \implies (p_1, p_2) \in E(G);$$

(2) *$(u,v) \in E(G)$ and $(v,z) \in E(G)$ implies that $(u,z) \in E(G)$, for all $u,v,z \in P$.*

for all $u_1, u_2, p_1, p_2 \in P$.

The following result is a direct consequence of Theorem 2 (resp. Theorem 3).

Theorem 5. *Let (M,d) be a complete metric space endowed with a graph G and (P,Q) be a pair of non-empty subsets of M so that P_0 is non-empty and $\mathcal{T}: P \to Q$ be a given non-self mapping. Suppose that:*

(i) *P is closed and $\mathcal{T}(P_0) \subseteq Q_0$;*
(ii) *$\mathcal{T}$ is triangular G-proximal;*
(iii) *There are $u_0, u_1 \in P_0$ so that $d(u_1, \mathcal{T} u_0) = d(P,Q)$ and $(u_0, u_1) \in E(G)$;*
(iv) *$\mathcal{T}$ is continuous or, for every sequence $\{u_n\}$ in P is convergent to $u \in P$ so that $(u_n, u_{n+1}) \in E(G)$, we have $(u_n, u) \in E(G)$ for all $n \in \mathbb{N}$;*
(v) *There exist $\zeta \in \mathcal{Z}$ and $\beta \in \Sigma$ such that for all $u, v, \rho, \nu \in P$,*

$$\left. \begin{array}{r} (u,v) \in E(G) \\ d(\rho, \mathcal{T} u) = d(P,Q) \\ d(\nu, \mathcal{T} v) = d(P,Q) \end{array} \right\} \implies \zeta(d(\rho, \nu), \beta(d(u,v))d(u,v)) \geq 0. \qquad (21)$$

Then, $\mathcal{T}$ has a best proximity point in P. If $(u,v) \in E(G)$ for all $u,v \in B_{est}(\mathcal{T})$, then $\mathcal{T}$ has a unique best proximity point $u^* \in P$. Moreover, for every $u \in P$, $\lim_{n \to \infty} \mathcal{T}^n u = u^*$.

5. A Variational Inequality Problem

Let C be a non-empty, closed, and convex subset of a real Hilbert space H, with inner product $\langle \cdot, \cdot \rangle$ and a norm $\|\cdot\|$. A variational inequality problem is given in the following:

$$\text{Find } u \in C \text{ so that } \langle Su, v-u \rangle \geq 0 \text{ for all } v \in C, \qquad (22)$$

where $S: H \to H$ is a given operator. The above problem can be seen in operations research, economics, and mathematical physics, especially in calculus of variations associated with the minimization of infinite-dimensional functionals. See [21] and the references therein. It appears in variant problems of nonlinear analysis, such as complementarity and equilibrium problems, optimization, and finding fixed points; see [21–23]. To solve problem (22), we define the metric projection operator $P_C: H \to C$. Note that for every $u \in H$, there is a unique nearest point $P_C u \in C$ so that

$$\|u - P_C u\| \leq \|u - v\|, \quad \text{for all } v \in C.$$

The two lemmas below correlate the solvability of a variational inequality problem to the solvability of a special fixed-point problem.

Lemma 1 ([24]). *Let $z \in H$. Then, $u \in C$ is such that $\langle u - z, y - u \rangle \geq 0$, for all $y \in C$ iff $u = P_C z$.*

Lemma 2 ([24]). *Let $S : H \to H$. Then, $u \in C$ is a solution of $\langle Su, v - u \rangle \geq 0$, for all $v \in C$, if $u = P_C(u - \lambda Su)$, with $\lambda > 0$.*

The main theorem of this section is:

Theorem 6. *Let C be a non-empty, closed, and convex subset of a real Hilbert space H. Assume that $S : H \to H$ is such that $P_C(I - \lambda S) : C \to C$ is a Geraghty-proximal contraction. Then, there is a unique element $u^* \in C$, such that $\langle Su^*, v - u^* \rangle \geq 0$ for all $v \in C$. Also, for any $u_0 \in C$, the sequence $\{u_n\}$ given as $u_{n+1} = P_C(u_n - \lambda Su_n)$ where $\lambda > 0$ and $n \in \mathbb{N} \cup \{0\}$, is convergent to u^*.*

Proof. We consider the operator $\mathcal{T} : C \to C$ defined by $\mathcal{T}x = P_C(x - \lambda Sx)$ for all $x \in C$. By Lemma 2, $u \in C$ is a solution of $\langle Su, v - u \rangle \geq 0$ for all $v \in C$, if $u = \mathcal{T}u$. Now, $\mathcal{T}$ verifies all the hypotheses of Corollary 1 with $P = Q = C$. Now, from Corollary 1, the fixed-point problem $u = \mathcal{T}u$ possesses a unique solution $u^* \in C$. □

Author Contributions: H.I. analyzed and prepared/edited the manuscript, H.A. analyzed and prepared/edited the manuscript, N.M. analyzed and prepared the manuscript, S.R. analyzed and prepared the manuscript.

Funding: This research received no external funding.

Acknowledgments: The third author would like to thank Prince Sultan University for funding this work through the research group Nonlinear Analysis Methods in Applied Mathematics (NAMAM) group number RG-DES-2017-01-17.

Conflicts of Interest: The authors declare that they have no competing interests regarding the publication of this paper.

References

1. Abkar, A.; Gabeleh, M. Best proximity points for cyclic mappings in ordered metric spaces. *J. Optim. Theory Appl.* **2011**, *150*, 188–193. [CrossRef]
2. Al-Thagafi, M.A.; Shahzad, N. Best proximity pairs and equilibrium pairs for Kakutani multimaps. *Nonlinear Anal.* **2009**, *70*, 1209–1216. [CrossRef]
3. Aydi, H.; Felhi, A. On best proximity points for various α-proximal contractions on metric-like spaces. *J. Nonlinear Sci. Appl.* **2016**, *9*, 5202–5218. [CrossRef]
4. Aydi, H.; Felhi, A. Best proximity points for cyclic Kannan-Chatterjea-Ćirić type contractions on metric-like spaces. *J. Nonlinear Sci. Appl.* **2016**, *9*, 2458–2466. [CrossRef]
5. Caballero, J.; Harjani, J.; Sadarangani, K. A best proximity point theorem for Geraghty-contractions. *Fixed Point Theory Appl.* **2012**, *2012*, 231. [CrossRef]
6. Eldred, A.A.; Veeramani, P. Existence and convergence of best proximity points. *J. Math. Anal. Appl.* **2006**, *323*, 1001–1006. [CrossRef]
7. Hussain, N.; Kutbi, M.A.; Salimi, P. Best proximity point results for modified α-ψ-proximal rational contractions. *Abstr. Appl. Anal.* **2013**, *2013*, 927457. [CrossRef]
8. Hussain, N.; Latif, A.; Salimi, P. New fixed point results for contractive maps involving dominating auxiliary functions. *J. Nonlinear Sci. Appl.* **2016**, *9*, 4114–4126. [CrossRef]
9. Işık, H.; Sezen, M.S.; Vetro, C. φ-Best proximity point theorems and applications to variational inequality problems. *J. Fixed Point Theory Appl.* **2017**, *19*, 3177–3189. [CrossRef]
10. Jleli, M.; Samet, B. Best proximity points for α-ψ-proximal contractive type mappings and application. *Bull. Sci. Math.* **2013**, *137*, 977–995. [CrossRef]
11. Basha, S.S.; Veeramani, P. Best proximity pair theorems for multifunctions with open fibres. *J. Approx. Theory* **2000**, *103*, 119–129. [CrossRef]

12. Sahmim, S.; Felhi, A.; Aydi, H. Convergence Best Proximity Points for Generalized Contraction Pairs. *Mathematics* **2019**, *7*, 176. [CrossRef]
13. Souyah, N.; Aydi, H.; Abdeljawad, T.; Mlaiki, N. Best proximity point theorems on rectangular metric spaces endowed with a graph. *Axioms* **2019**, *8*, 17. [CrossRef]
14. Tchier, F.; Vetro, C.; Vetro, F. Best approximation and variational inequality problems involving a simulation function. *Fixed Point Theory Appl.* **2016**, *2016*, 26. [CrossRef]
15. Samet, B.; Vetro, C.; Vetro, P. Fixed point theorems for α-ψ-contractive type mappings. *Nonlinear Anal.* **2012**, *75*, 2154–2165. [CrossRef]
16. Khojasteh, F.; Shukla, S.; Radenović, S. A new approach to the study of fixed point theorems via simulation functions. *Filomat* **2015**, *29*, 1189–1194. [CrossRef]
17. Argoubi, H.; Samet, B.; Vetro, C. Nonlinear contractions involving simulation functions in metric space with a partial order. *J. Nonlinear Sci. Appl.* **2015**, *8*, 1082–1094. [CrossRef]
18. Işık, H.; Gungor, N.B.; Park, C.; Jang, S.Y. Fixed point theorems for almost $\mathcal{Z}$-contractions with an application. *Mathematics* **2018**, *6*, 37. [CrossRef]
19. Nastasi, A.; Vetro, P. Fixed point results on metric and partial metric spaces via simulation functions. *J. Nonlinear Sci. Appl.* **2015**, *8*, 1059–1069. [CrossRef]
20. Radenovic, S.; Vetro, F.; Vujakovic, J. An alternative and easy approach to fixed point results via simulation functions. *Demonstr. Math.* **2017**, *50*, 223–230. [CrossRef]
21. Kinderlehrer, D.; Stampacchia, G. *An Introduction to Variational Inequalities and Their Applications*; Academic Press: New York, NY, USA, 1980.
22. Fang, S.C.; Petersen, E.L. Generalized variational inequalities. *J. Optim. Theory Appl.* **1982**, *38*, 363–383. [CrossRef]
23. Todd, M.J. *The Computations of Fixed Points and Applications*; Springer: Berlin/Heidelberg, Germany, 1976.
24. Deutsch, F. *Best Approximation in Inner Product Spaces*; Springer: New York, NY, USA, 2001.

© 2019 by the authors. Licensee MDPI, Basel, Switzerland. This article is an open access article distributed under the terms and conditions of the Creative Commons Attribution (CC BY) license (http://creativecommons.org/licenses/by/4.0/).

Article

Recent Advances on the Results for Nonunique Fixed in Various Spaces

Erdal Karapınar

Department of Medical Research, China Medical University, Taichung 40402, Taiwan; karapinar@mail.cmuh.org.tw

Received: 21 March 2019; Accepted: 27 May 2019; Published: 5 June 2019

Abstract: In this short survey, we aim to underline the importance of the non-unique fixed point results in various abstract spaces. We recall a brief background on the topic and we combine, collect and unify several existing non-unique fixed points in the literature. Some interesting examples are considered.

Keywords: non-unique fixed point; contractions; partial metric; simulation function; Branciari distance; b-Branciari distance

MSC: 47H10; 54H25

---

1. Introduction

It is very common to consider to existing a fixed point of a certain mapping while presuming it is unique. This is true, considering a solution of a fixed point problem $G(x) = Fx - x = 0$ is unique. On the other hand, in the real world, in particular in nonlinear systems, the solution need to be unique. In such case, non-unique or periodic solutions also have worth for understanding the corresponding phenomena.

The first known result for finding nonunique fixed points for certain operators was proposed by Ćirić [1]. In this well-known paper, Ćirić [1] emphasized the worth and importance of the notion of the non-unique fixed points (also, the periodic fixed points)in the setting of complete metric spaces. Inspired by this initial report of Ćirić [1], several significant results has been released on nonunique fixed point theorems for various fixed point problems, see e.g., [1–12].

This survey can be considered as a continuation of the recent paper [13].

2. Preliminaries

This section is devoted to collecting and recalling the basic notions and fundamental results without considering the proofs. On the other hand, in the following sections, we show how to derive these basic results from the upcoming theorems that we state.

From now on, we preserve the letters $\mathbb{R}_0^+$, to denote the set of non-negative real numbers. In addition, $\mathbb{N}_0$ present the set of positive integer numbers with zero.

The first definition is orbitally continuous, and has a key role in the non-unique fixed point results.

Definition 1. *(see [1]) Let F be a self-map on a metric space (S, δ).*

(*i*) *F is said to be an orbitally continuous mapping if*

$$\lim_{i \to \infty} F^{n_i} x = z \tag{1}$$

implies

$$\lim_{i \to \infty} FF^{n_i}x = Fz \qquad (2)$$

for each $x \in S$.

(ii) *If every Cauchy (fundamental) sequence of type $\{F^{n_i}x\}_{i \in \mathbb{N}}$ converges, then metric space (S, δ) is orbitally complete*

Throughout this section, the letter F is reserved for presenting a self-mapping on a non-empty set which is endowed a standard metric δ. Moreover, the pair (S, δ) represents standard metric space. We presume also that (S, δ) is orbitally complete in all upcoming theorems, corollaries, lemmas and propositions. A point z is called a periodic point of a function F of period m if $F^m(z) = z$, where $F^0(x) = x$ and $F^m(x)$ is iteratively defined by $F^m(x) = T(F^{m-1}(x))$. The set $Fix_S(F)$ indicate the set of all fixed point of F on S.

Theorem 1. [**Non-unique fixed point theorem of Ćirić** [1]] *If there is $k \in [0, 1)$ such that*

$$\min\{\delta(Fx, Fy), \delta(x, Fx), \delta(y, Fy)\} - \min\{\delta(x, Fy), \delta(Fx, y)\} \leq k\delta(x, y),$$

for all $x, y \in S$, then the mapping F possesses a fixed point in S. Indeed, for an arbitrary initial point $x_0 \in S$ the recursive sequence $\{F^n x_0\}_{n \in \mathbb{N}}$ converges to a fixed point of F.

Theorem 2. [**Nonunique fixed point of Achari** [2]] *If there exists $k \in [0, 1)$ such that for all $x, y \in S$,*

$$\frac{P(x,y) - Q(x,y)}{R(x,y)} \leq k\delta(x, y), \qquad (3)$$

where

$$\begin{aligned}
P(x, y) &= \min\{\delta(Fx, Fy)\delta(x, y), \delta(x, Fx)\delta(y, Fy)\}, \\
Q(x, y) &= \min\{\delta(x, Fx)\delta(x, Fy), \delta(y, Fy)\delta(Fx, y)\}, \\
R(x, y) &= \min\{\delta(x, Fx), \delta(y, Fy)\}.
\end{aligned}$$

with $R(x, y) \neq 0$. Then, the mapping F possesses a fixed point in S. Indeed, for an arbitrary initial point $x_0 \in S$ the recursive sequence $\{F^n x_0\}_{n \in \mathbb{N}}$ converges to a fixed point of F.

Theorem 3. [**Nonunique fixed point of Pachpatte** [11]] *Suppose that there exists $k \in [0, 1)$ such that*

$$m(x, y) - n(x, y) \leq k\delta(x, Fx)\delta(y, Fy), \qquad (4)$$

for all $x, y \in S$, where

$$\begin{aligned}
m(x, y) &= \min\{[\delta(Fx, Fy)]^2, \delta(x, y)\delta(Fx, Fy), [\delta(y, Fy)]^2\}, \\
n(x, y) &= \min\{\delta(x, Fx)\delta(y, Fy), \delta(x, Fy)\delta(y, Fx)\}.
\end{aligned}$$

Then, the mapping F possesses a fixed point in S. Indeed, for an arbitrary initial point $x_0 \in S$ the recursive sequence $\{F^n x_0\}_{n \in \mathbb{N}}$ converges to a fixed point of F.

Theorem 4. [**Nonunique fixed point of Ćirić-Jotić** [14]] *If there exists $k \in [0, 1)$ and $a \geq 0$ such that*

$$J(x, y) - aI(x, y) \leq kL(x, y), \qquad (5)$$

for all distinct $x, y \in S$ where

$$J(x,y) = \min\left\{\begin{array}{c} \delta(Fx,Fy), \delta(x,y), \delta(x,Fx), \delta(y,Fy), \frac{\delta(x,Fx)[1+\delta(y,Fy)]}{1+\delta(x,y)}, \\ \frac{\delta(y,Fy)[1+\delta(x,Fx)]}{1+\delta(x,y)}, \frac{\min\{d^2(Fx,Fy), d^2(x,Fx), d^2(y,Fy)\}}{\delta(x,y)} \end{array}\right\},$$

$$I(x,y) = \min\{\delta(x,Fy), \delta(y,Fx)\},$$

$$L(x,y) = \max\{\delta(x,y), \delta(x,Fx)\}.$$

Then, the mapping F possesses a fixed point in S. Indeed, for an arbitrary initial point $x_0 \in S$ the recursive sequence $\{F^n x_0\}_{n \in \mathbb{N}}$ converges to a fixed point of F.

Theorem 5. [**Nonunique fixed point of Karapınar** [15]] *If there exist real numbers a_1, a_2, a_3, a_4, a_5 and a self mapping $F : S \to S$ satisfies the conditions*

$$0 \leq \frac{a_4 - a_2}{a_1 + a_2} < 1, \quad a_1 + a_2 \neq 0, \quad a_1 + a_2 + a_3 > 0 \text{ and } 0 \leq a_3 - a_5 \tag{6}$$

$$E(x,y) \leq a_4 \delta(x,y) + a_5 \delta(x, F^2 x) \tag{7}$$

where

$$E(x,y) := a_1 \delta(Fx,Fy) + a_2 [\delta(x,Fx) + \delta(y,Fy)] + a_3 [\delta(y,Fx) + \delta(x,Fy)],$$

hold for all $x, y \in S$. Then, the mapping F possesses a fixed point in S. Indeed, for an arbitrary initial point $x_0 \in S$ the recursive sequence $\{F^n x_0\}_{n \in \mathbb{N}}$ converges to a fixed point of F.

Our aim is mainly to get the corresponding nonunique fixed point theorems in the setting of various abstract spaces, such as, partial metric spaces, Branciari distance.

In what follows, we express the definition of a comparison function. This notion was considered first by Browder [16] and later by Rus [17] and many others. We say that a function $\varphi : [0, \infty) \to [0, \infty)$ is a comparison function [16,17] if it is not only nondecreasing but also $\varphi^n(t) \to 0$ as $n \to \infty$ for every $t \in [0, \infty)$, where φ^n is the n-th iterate of φ. A simple example of such mappings is $\psi(t) = \frac{kt}{n}$ where $k \in [0,1)$ and $n \in \{2, 3, \cdots\}$.

Let Ψ denote the set of all functions $\psi : [0, \infty) \to [0, \infty)$ such that

(Ψ_1) ψ is nondecreasing;
(Ψ_2) $\sum_{n=1}^{+\infty} \psi^n(t) < \infty$ for all $t > 0$.

A function $\psi \in \Psi$ is named as (c)-comparison.

For more details and examples of both comparison and (c)-comparison functions, we refer to e.g., [17].

Lemma 1 ([17]). *Suppose that $\phi : [0, \infty) \to [0, \infty)$ is a comparison function. Then, we have*

1. *ϕ is continuous at 0;*
2. *each iterate ϕ^k of ϕ, $k \geq 1$, is also a comparison function;*
3. *$\phi(t) < t$ for all $t > 0$.*

It is clear that if ϕ is a (c)-comparison function is a comparison function. Hence, the properties above are also valid for (c)-comparison functions.

Definition 2. *A function $\zeta : [0, \infty) \times [0, \infty) \to \mathbb{R}$ is named* simulation *if*

(ζ_1) $\zeta(t,s) < s - t$ for all $t, s > 0$;

(ζ_2) if $\{t_n\}, \{s_n\}$ are sequences in $(0, \infty)$ such that $\lim\limits_{n \to \infty} t_n = \lim\limits_{n \to \infty} s_n > 0$, then

$$\limsup_{n \to \infty} \zeta(t_n, s_n) < 0. \qquad (8)$$

In the original definition, given in [18], there is a condition, $\zeta(0,0) = 0$. This condition is superfluous and hence it was dropped, see e.g., Argoubi et al. [19]. Let $\mathcal{Z}$ denote the family of all simulation functions $\zeta : [0, \infty) \times [0, \infty) \to \mathbb{R}$, i.e., verifying ($\zeta_1$) and ($\zeta_2$).

Due to (ζ_1), we deduce

$$\zeta(t, t) < 0 \text{ for all } t > 0. \qquad (9)$$

The following example is derived from [18,20,21].

Example 1. Let $\mu_i : \mathbb{R}_0^+ \to \mathbb{R}_0^+$ be continuous functions such that $\mu_i(t) = 0$ if and only if, $t = 0$. For $i = 1, 2, 3, 4, 5, 6$, we define the mappings $\zeta_i : \mathbb{R}_0^+ \times \mathbb{R}_0^+ \to \mathbb{R}$, as follows

(i) $\zeta_1(t,s) = \mu_1(s) - \mu_2(t)$ for all $t, s \in [0, \infty)$, where $\mu_1, \mu_2 : \mathbb{R}_0^+ \to \mathbb{R}_0^+$ are two continuous functions such that $\mu_1(t) = \mu_2(t) = 0$ if and only if $t = 0$ and $\mu_1(t) < t \le \mu_2(t)$ for all $t > 0$.

(ii) $\zeta_2(t,s) = s - \dfrac{f(t,s)}{g(t,s)} t$ for all $t, s \in [0, \infty)$, where $f, g : [0, \infty)^2 \to (0, \infty)$ are two continuous functions with respect to each variable such that $f(t,s) > g(t,s)$ for all $t, s > 0$.

(iii) $\zeta_3(t,s) = s - \mu_3(s) - t$ for all $t, s \in [0, \infty)$.

(iv) $\zeta_4(t,s) = s\varphi(s) - t$ for all $s, t \in [0, \infty)$, where $\varphi : [0, \infty) \to [0, 1)$ is a function such that $\limsup\limits_{t \to r^+} \varphi(t) < 1$ for all $r > 0$.

(v) $\zeta_5(t,s) = \eta(s) - t$ for all $s, t \in [0, \infty)$, where $\eta : \mathbb{R}_0^+ \to \mathbb{R}_0^+$ is an upper semi-continuous mapping such that $\eta(t) < t$ for all $t > 0$ and $\eta(0) = 0$.

(vi) $\zeta_6(t,s) = s - \int_0^t \mu(u) du$ for all $s, t \in [0, \infty)$, where $\mu : [0, \infty) \to [0, \infty)$ is a function such that $\int_0^\varepsilon \mu(u) du$ exists and $\int_0^\varepsilon \mu(u) du > \varepsilon$, for each $\varepsilon > 0$.

It is clear that each function ζ_i ($i = 1, 2, 3, 4, 5, 6$) forms a simulation function.

3. Nonunique Fixed Point Results in Partial Metric Space

In this section, we start with recollecting the definition of a partial metric that is one of the most significant generalization of a metric concept. The main difference between a partial metric from the standard metric is on the self-distance axiom. Despite a standard distance function in partial metric, offered by Matthews [22], self-distance is not necessarily equal to zero. From the mathematical point of view, it seems that the definition of a partial metric is inconsistent, even if it seems fallacious. By contrast with the expectations and knowledge, zero self-distance is quite logical and rational the framework of computer sciences. Indeed, we put the notion of partial across to reader by examining the following classical example:

Let $\mathcal{S}$ be the union of the set of all finite sequence ($\mathcal{S}_F$) with the set of all infinite sequence ($\mathcal{S}_i$). We shall propose a distance function in the following way:

$$\delta : \mathcal{S} \times \mathcal{S} \to [0, \infty) \text{ such that } \delta(x, y) = 2^{-\sup\{n | \forall i < n \text{ such that } x_i = y_i\}}. \qquad (10)$$

It is easy to check that all metric axioms are fulfilled on the restriction of the domain of δ to $\mathcal{S}_I$. On the other hand, in case of the restriction of the domain $\mathcal{S}$ to $\mathcal{S}_F$, the function δ fails to self-distance axioms. More precisely, taking finite sequences into account, in particular, for the finite sequence $x = (x_1, x_2, \cdots, x_m)$, for some positive integer m, the self-distance $\rho(x, y) = \dfrac{1}{2^m} \ne 0$. This simple example indicate that the idea of non-zero distance has a logic and worthy. In computer science programming, usage of the finite sequences are more reasonable and affective in case of taking the termination of the program into account. Roughly speaking, one can declare that programming with infinite sequence may leads to infinite loops in running and has a problem of termination and hence getting an output.

Another simple but effective example [22,23]) can be given by using the maximum operator. To put a finer point on it, consider set of all non-negative real numbers with maximum operator, i.e.,

$$\rho : [0,\infty) \times [0,\infty) \to [0,\infty) \text{ such that } \rho(r_1,r_2) = \max\{r_1,r_2\}. \tag{11}$$

In particular, $\rho(3,3) = 3 \neq 0$.

After the intuitive introduction of partial metric, now, we shall state the formal definition of it as follows:

Definition 3. *(See e.g., [22,23]) A function $\rho : S \times S \to \mathbb{R}_0^+$ on a (non-empty) set S is named as a partial metric if the following axioms are fulfilled*

(P1) $z = w \Leftrightarrow \rho(z,z) = \rho(w,w) = \rho(z,w)$,
(P2) $\rho(z,z) \leq \rho(z,w)$,
(P3) $\rho(z,w) = \rho(w,z)$,
(P4) $\rho(z,w) \leq \rho(z,v) + \rho(v,w) - \rho(v,v)$,

for all $z,w,v \in S$. Here, the coupled letter (S,ρ) is said to be a partial metric space.

Despite the fact that the self-distance is not necessarily zero, we derive, from (P1) and (P2), that $\rho(x,y) = 0$ yields the reflexivity $x = y$.

Hereafter, the pair (S,δ) present a standard metric space and the pair (S,ρ) indicate a partial metric space. For avoiding so many repetitions, we shall not put these presumes in all statements in the upcoming definitions, theorems and corollaries.

Example 2. *(See e.g., [24,25]) Functions $\sigma_i : S \times S \to \mathbb{R}_0^+$ ($i \in \{1,2,3\}$) are defined by*

$$\begin{aligned} \sigma_1(z,w) &= \delta(z,w) + C, \\ \sigma_2(z,w) &= \delta(z,w) + \max\{\gamma(z),\gamma(w)\}, \\ \sigma_3(z,w) &= \delta(z,w) + \rho(z,w). \end{aligned}$$

It clear that all three functions, defined above, form partial metrics on S, where $\gamma : S \to \mathbb{R}_0^+$ is an arbitrary function and $C \geq 0$.

Example 3. *(See [22,23]) Let $S = \{[q,r] : q,b \in \mathbb{R}, q \leq r\}$ and define $\rho([q,r],[s,t]) = \max\{r,t\} - \min\{q,s\}$. Then (S,ρ) forms a partial metric space.*

Example 4. *(See [22]) Let $\rho : S \times S \to \mathbb{R}_0^+$, where $S = [0,1] \cup [2,3]$.*
Define $\rho(q,r) = \begin{cases} \max\{q,r\} & \text{if } \{q,r\} \cap [2,3] \neq \emptyset, \\ |q-r| & \text{if } \{q,r\} \subset [0,1]. \end{cases}$
Then (S,ρ) is a partial metric space.

The topology τ_ρ, induced by a partial metric ρ defined on a non-empty set S, is classified as T_0 with a base of the family of open ρ-balls $\{O_\rho(x,\epsilon) : q \in S, \epsilon > 0\}$ where

$$O_\rho(q,\epsilon) = \{r \in S : \rho(q,r) < \rho(r,r) + \epsilon\}$$

for all $q \in S$ and $\epsilon > 0$.

A sequence $\{x_n\}_{n \in \mathbb{N}}$ in a partial metric space (S,ρ) converges to a point $x \in S$ (in brief, $x_n \to x$,) if and only if $\rho(x,x) = \lim_{n \to \infty} \rho(x,x_n)$.

Regarding the following example, we shall underline the fact that the limit of a sequence is not necessarily unique in partial metric space. It can be easily observed an example by regarding the partial metric space considered in Example 11. If we take the sequence $\{\frac{1}{n^3+1}\}_{n\in\mathbb{N}}$ into account, we derive that

$$\rho(1,1) = \lim_{n\to\infty} \rho(1, \frac{1}{n^3+1}) \quad \text{and} \quad \rho(2,2) = \lim_{n\to\infty} \rho(2, \frac{1}{n^3+1}).$$

On the other hand, the limit of a sequence is unique, under certain additional conditions. In particular, the following lemma was proposed for the uniqueness of the limit.

Lemma 2. *(See e.g., [24,25]) Consider a sequence $\{x_n\}_{n\in\mathbb{N}}$ in (S,ρ) with $x_n \to x$ and $x_n \to y$. If*

$$\lim_{n\to\infty} \rho(x_n, x_n) = \rho(x,x) = \rho(y,y),$$

then $x = y$.

It is quite natural to expect a close connection between the notions of the standard metric and partial metric. Indeed, a function $\delta_\rho : S \times S \to \mathbb{R}_0^+$ defined as

$$\delta_\rho(x,y) = 2\rho(x,y) - \rho(x,x) - \rho(y,y), \tag{12}$$

forms a standard metric on S, see e.g., [23]. In addition, the functions $\delta_0, \delta_m^\rho : S \times S \to [0,\infty)$ defined by

$$\delta_0(x,y) = \begin{cases} 0 & \text{if } x = y \\ \rho(x,y) & \text{otherwise}. \end{cases} \tag{13}$$

and
$$\begin{aligned}\delta_m^\rho(x,y) &= \rho(x,y) - \min\{\rho(x,x), \rho(y,y)\} \\ &= \max\{\rho(x,y) - \rho(x,x), \rho(x,y) - \rho(y,y)\}\end{aligned}$$

form metrics on S (see e.g., [26], respectively). Moreover, we have $\tau_\rho \subseteq \tau_{\delta_\rho} = \tau_{\delta_m^\rho} \subseteq \tau_{\delta_0}$. In particular, both δ_ρ and δ_ρ^m are the Euclidean metric on S which are based on the partial metric space (S,ρ) of Example 11.

In what follows we give the definition of fundamental topological concepts as follows:

Definition 4. *(See e.g., [6,22,23,27]) Let (S,ρ) be a partial metric space.*

1. *A sequence $\{x_n\}_{n\in\mathbb{N}}$ in S converges to $x^* \in S$ if*

$$\lim_{n\to\infty} \delta_\rho(x^*, x_n) = 0 \Leftrightarrow \rho(x^*, x^*) = \lim_{n\to\infty} \rho(x^*, x_n) = \lim_{n,m\to\infty} \rho(x_n, x_m). \tag{14}$$

2. *A sequence $\{x_n\}_{n\in\mathbb{N}}$ in S is called a fundamental (or, Cauchy) sequence in (S,ρ) if $\lim_{n,m\to\infty} \rho(x_n, x_m)$ exists and is finite, that is,*
   (*) for each $\varepsilon > 0$ there is $n_0 \in \mathbb{N}$ such that $\rho(x_n, x_m) - \rho(x_n, x_n) < \varepsilon$ whenever $n_0 \leq n \leq m$.
3. *(S,ρ) is called complete if every Cauchy sequence $\{x_n\}_{n\in\mathbb{N}}$ converges to a point $x^* \in S$ such that $\rho(x^*, x^*) = \lim_{n,m\to\infty} \rho(x_n, x_m)$.*

In the sequel, the following characterizations of topological concepts shall be used efficiently.

Lemma 3. *(See [23])*

1. *A partial metric space (S,ρ) is complete if and only if the corresponding metric space (S,δ_ρ) is complete.*
2. *A sequence $\{x_n\}_{n\in\mathbb{N}}$ in (S,ρ) is a fundamental if and only if it forms a fundamental sequence in the corresponding metric space (S,δ_ρ).*

We underline that the partial metric spaces considered in Example 11, Example 3 and Example 4 are complete.

Lemma 4. *Let (S, ρ) be a partial metric space and let $\{x_n\}_{n \in \mathbb{N}}$ and $\{y_n\}_{n \in \mathbb{N}}$ be sequences in S such that $x_n \to x^*$ and $y_n \to y^*$ with respect to τ_{δ_ρ}. Then*

$$\lim_{n \to \infty} \rho(x_n, y_n) = \rho(x^*, y^*).$$

For our purposes, we need to recall the following notion which is an adaptation of Definition 1 in the context of partial metric spaces.

Definition 5. *(cf. [1])*

1. A self-mapping F, defined on a partial metric space (S, ρ), is said to be an orbitally continuous if

$$\lim_{i,j \to \infty} \rho(F^{n_i} x, F^{n_j} x) = \lim_{i \to \infty} \rho(F^{n_i} x, x^*) = \rho(x^*, x^*), \qquad (15)$$

implies

$$\lim_{i,j \to \infty} \rho(FF^{n_i} x, FF^{n_j} x) = \lim_{i \to \infty} \rho(FF^{n_i} x, Fx^*) = \rho(Fx^*, Fx^*), \qquad (16)$$

for each $x \in S$.

Equivalently, F is orbitally continuous provided that if $F^{n_i} x \to z$ with respect to τ_{δ_ρ}, then $F^{n_i+1} x \to Fz$ with respect to τ_{δ_ρ}, for each $x \in S$.

2. A partial metric space (S, ρ) is said to be an orbitally complete if each fundamental sequence of type $\{F^{n_i} x\}_{i \in \mathbb{N}}$ converges with respect to τ_{δ_ρ}, that is, if there is $z \in S$ such that

$$\lim_{i,j \to \infty} \rho(F^{n_i} x, F^{n_j} x) = \lim_{i \to \infty} \rho(F^{n_i} x, z) = \rho(z, z). \qquad (17)$$

In the following lines in this section, we focus on non-unique fixed points of certain mappings in the framework of partial metric spaces that are successors results in the direction of a renowned Ćirić [1] result. The presented results in this section not only extend but also enrich several earlier results on the topic in the literature, in particular the pioneer works [1,2,11,28]). We also present examples to emphasize the advantages of the usage of partial metric spaces rather than standard metric spaces.

Throughout this section, we presume that F is an orbitally continuous self-map of an orbitally complete partial metric space (S, ρ).

3.1. Ćirić Type Non-Unique Fixed Points on Partial Metric Spaces

The first result is the following one.

Theorem 6. *If $\phi \in \Phi$ such that*

$$C(x, y) \leq \phi(\rho(x, y)), \qquad (18)$$

where

$$C(x, y) := \min\{\rho(Fx, Fy), \rho(x, Fx), \rho(y, Fy)\} - \min\{\delta_m^\rho(x, Fy), \delta_m^\rho(Fx, y)\}, \qquad (19)$$

for all $x, y \in S$, then, for each $x_0 \in S$, the sequence $\{F^n x_0\}_{n \in \mathbb{N}_0}$ converges with respect to τ_{δ_ρ} to a fixed point of F.

Proof. We construct an iterative sequence $\{x_n\}_{n \in \mathbb{N}_0}$, by starting an arbitrary initial point $x_0 \in S$, as follows:

$$x_{n+1} = F x_n, \quad n \in \mathbb{N}_0.$$

If there exists $n_0 \in \mathbb{N}_0$ such that $x_{n_0} = x_{n_0+1}$, then x_{n_0} forms a fixed point of F and hence the proof is completed trivially. Accordingly, by avoiding the simplicity case, we assume then that $x_n \neq x_{n+1}$ for each $n \in \mathbb{N}_0$.

Substituting $x = x_n$ and $y = x_{n+1}$ in (18) we find the inequality

$$C(x_n, x_{n+1}) \leq \phi(\rho(x_n, x_{n+1})),$$

which is equal to

$$\min\{\rho(x_{n+1}, x_{n+2}), \rho(x_n, x_{n+1}), \rho(x_{n+1}, x_{n+2})\}$$
$$- \min\{\delta_m^\rho(x_n, x_{n+2}), \delta_m^\rho(x_{n+1}, x_{n+1})\}$$
$$\leq \phi(\rho(x_n, x_{n+1})).$$

Attendantly, we observe that

$$\min\{\rho(x_n, x_{n+1}), \rho(x_{n+1}, x_{n+2})\} \leq \phi(\rho(x_n, x_{n+1})). \tag{20}$$

Suppose $\rho(x_{n_0}, x_{n_0+1}) \leq \rho(x_{n_0+1}, x_{n_0+2})$ for some $n_0 \in \mathbb{N}_0$. Then, from the preceding inequalities we observe that

$$\rho(x_{n_0}, x_{n_0+1}) \leq \phi(\rho(x_n, x_{n+1})) < \rho(x_{n_0}, x_{n_0+1}),$$

which is a contradiction.

Therefore $\rho(x_n, x_{n+1}) > \rho(x_{n+1}, x_{n+2})$ for all $n \in \mathbb{N}_0$.

Hence, by (20) we get

$$\rho(x_{n+1}, x_{n+2}) \leq \phi(\rho(x_n, x_{n+1})) \leq \cdots \leq \phi^{n+1}(\rho(x_0, x_1)), \tag{21}$$

for all $n \in \mathbb{N}_0$.

In what follows, we indicate that the constructed sequence $\{x_n\}_{n\in\mathbb{N}}$ is fundamental (Cauchy) in (S, ρ). For this goal, take $n, m \in \mathbb{N}_0$ with $n < m$ and employ (21) and (P4), as follows:

$$\begin{aligned}\rho(x_n, x_m) &\leq \rho(x_n, x_{n+1}) + \cdots + \rho(x_{m-1}, x_m) - \sum_{k=n}^{m-1} \rho(x_k, x_k)\\ &\leq \phi^n(\rho(x_0, x_1)) \cdots + \phi^{m-1}(\rho(x_0, x_1))\\ &\leq \sum_{k=n}^{m-1} \phi^k(\rho(x_0, x_1)) \to 0 \text{ as } n \to \infty.\end{aligned}$$

Consequently, $\{x_n\}_{n\in\mathbb{N}_0}$ is a fundamental sequence in (S, ρ). Since $x_n = F^n x_0$ for all n, and (S, ρ) is F-orbitally complete, there is $x^* \in S$ such that $x_n \to x^*$ with respect to τ_{δ_ρ}. Moreover, we have

$$\rho(x^*, x^*) = \lim_{n \to \infty} \rho(x^*, x_n) = \lim_{n,m \to \infty} \rho(x_n, x_m) = 0.$$

By the orbital continuity of F, we deduce that $x_n \to Fx^*$ with respect to τ_{δ_ρ}. Hence $x^* = Fx^*$. □

Definition 6. *The self-mapping $F : S \to S$ is called Ćirić type simulated if there exists $k \in (0, 1)$ and $\zeta \in \mathcal{Z}$ such that*

$$\zeta(m_F(x, y), c_F(x, y)) \geq 0 \tag{22}$$

for all $x, y \in S$, where

$$m_F(x, y) := \min\{\rho(Fx, Fy), \rho(x, Fx), \rho(y, Fy)\} - \min\{\delta_m^\rho((x, Fy), \delta_m^\rho((Fx, y)\}.$$

$$c_F(x,y) := k(\rho(x,y) - \rho(x,x)) + \rho(y,y),$$

Theorem 7. *If F is a Ćirić type simulated mapping, then for each $x_0 \in S$ the sequence $\{F^n x_0\}_{n \in \mathbb{N}_0}$ converges to a fixed point of F.*

Proof. We construct a recursive sequence $\{x_n\}_{n \in \mathbb{N}_0}$, by taking an arbitrary point $x_0 \in S$, as follows:

$$x_{n+1} = F x_n, \quad n \in \mathbb{N}_0.$$

We presume that $x_n \neq x_{n+1}$ for each $n \in \mathbb{N}_0$. Indeed, if there exists non-negative integer n_0 such that $x_{n_0} = x_{n_0+1}$, then x_{n_0} forms a fixed point of F that terminate the proof.

Substituting $x = x_n$ and $y = x_{n+1}$ in (22) we obtain

$$0 \leq \zeta(m_F(x_n, y), c_F(x_n, y)) < c_F(x_n, y) - m_F(x_n, y)$$

where

$$\begin{aligned} m_F(x_n, x_{n+1}) &= \min\{\rho(F x_n, F x_{n+1}), \rho(x_n, F x_n), \rho(x_{n+1}, F x_{n+1})\} \\ &\quad - \min\{\delta_m^\rho((x_n, F x_{n+1}), \delta_m^\rho((F x_n, x_{n+1})\}. \end{aligned}$$

and

$$c_F(x_n, x_{n+1}) = k(\rho(x_n, x_{n+1}) - \rho(x_n, x_n)) + \rho(x_{n+1}, x_{n+1}),$$

A simple evaluation yields that

$$\begin{aligned} &\min\{\rho(x_{n+1}, x_{n+2}), \rho(x_n, x_{n+1}), \rho(x_{n+1}, x_{n+2})\} \\ &\quad - \min\{\delta_m^\rho(x_n, x_{n+2}), \delta_m^\rho(x_{n+1}, x_{n+1})\} \\ &\leq k(\rho(x_n, x_{n+1}) - \rho(x_n, x_n)) + \rho(x_{n+1}, x_{n+1}). \end{aligned}$$

Consequently, we get that

$$\begin{aligned} &\min\{\rho(x_n, x_{n+1}), \rho(x_{n+1}, x_{n+2})\} \\ &\leq k(\rho(x_n, x_{n+1}) - \rho(x_n, x_n)) + \rho(x_{n+1}, x_{n+1}), \end{aligned} \tag{23}$$

Substituting $x = x_{n+1}$ and $y = x_n$, with a revising order, in (22), we get

$$0 \leq \zeta(m_F(x_{n+1} x_n), c_F(x_{n+1} x_n)) < c_F(x_{n+1} x_n) - m_F(x_{n+1} x_n)$$

where

$$\begin{aligned} m_F(x_{n+1} x_n) &= \min\{\rho(F x_{n+1} F x_n), \rho(x_{n+1} F x_{n+1}), \rho(x_n, F x_n)\} \\ &\quad - \min\{\delta_m^\rho((x_{n+1} F x_n), \delta_m^\rho((F x_{n+1} x_n)\}. \end{aligned}$$

and

$$c_F(x_{n+1} x_n) := k(\rho(x_{n+1} x_n) - \rho(x_{n+1} x_{n+1})) + \rho(x_n, x_n),$$

By a simple calculation, we derive that

$$\begin{aligned} &\min\{\rho(x_{n+2}, x_{n+1}), \rho(x_{n+1}, x_{n+2}), \rho(x_n, x_{n+1})\} \\ &\quad - \min\{\delta_m^\rho(x_{n+1}, x_{n+1}), \delta_m^\rho(x_{n+2}, x_n)\} \\ &\leq k(\rho(x_{n+1}, x_n) - \rho(x_{n+1}, x_{n+1})) + \rho(x_n, x_n), \end{aligned}$$

which imply that

$$\begin{aligned} &\min\{\rho(x_n, x_{n+1}), \rho(x_{n+1}, x_{n+2})\} \\ &\leq k(\rho(x_n, x_{n+1}) - \rho(x_{n+1}, x_{n+1})) + \rho(x_n, x_n). \end{aligned} \tag{24}$$

Suppose $\rho(x_{n_0}, x_{n_0+1}) \leq \rho(x_{n_0+1}, x_{n_0+2})$ for some $n_0 \in \mathbb{N}_0$. Then, on account of two inequalities (23) and (24), we obtain that

$$(1-k)\rho(x_{n_0}, x_{n_0+1}) \leq \min\{\rho(x_{n_0+1}, x_{n_0+1}) - kp(x_{n_0}, x_{n_0}),$$
$$\rho(x_{n_0}, x_{n_0}) - kp(x_{n_0+1}, x_{n_0+1})\}.$$

If, for instance, $\rho(x_{n_0+1}, x_{n_0+1}) \leq \rho(x_{n_0}, x_{n_0})$, we have

$$(1-k)\rho(x_{n_0}, x_{n_0+1}) \leq \rho(x_{n_0+1}, x_{n_0+1}) - kp(x_{n_0}, x_{n_0})$$
$$\leq (1-k)\rho(x_{n_0+1}, x_{n_0+1})$$
$$\leq (1-k)\rho(x_{n_0}, x_{n_0}),$$

so, by using (P2), $\rho(x_{n_0}, x_{n_0+1}) = \rho(x_{n_0}, x_{n_0}) = \rho(x_{n_0+1}, x_{n_0+1})$, and hence $x_{n_0} = x_{n_0+1}$, a contradiction.

Therefore $\rho(x_n, x_{n+1}) > \rho(x_{n+1}, x_{n+2})$ for all $n \in \mathbb{N}_0$.

Hence, by (23) we get

$$\rho(x_{n+1}, x_{n+2}) - \rho(x_{n+1}, x_{n+1}) \leq k(\rho(x_n, x_{n+1}) - \rho(x_n, x_n))$$
$$\leq k^2(\rho(x_{n-1}, x_n) - \rho(x_{n-1}, x_{n-1})) \quad (25)$$
$$\leq \ldots \leq k^{n+1}((\rho(x_0, x_1) - \rho(x_0, x_0)),$$

for all $n \in \mathbb{N}_0$.

As a next step, we indicate that the sequence $\{x_n\}_{n \in \mathbb{N}}$ is fundamental in (S, ρ). For this aim, we let $n, m \in \mathbb{N}_0$ with $n < m$ and by using (25) and (P4), we find

$$\rho(x_n, x_m) - \rho(x_n, x_n) \leq \rho(x_n, x_{n+1}) + \cdots + \rho(x_{m-1}, x_m) - \sum_{k=n}^{m-1} \rho(x_k, x_k)$$
$$\leq (k^n + \cdots + k^{m-1})\rho(x_0, x_1).$$

Attendantly, the sequence $\{x_n\}_{n \in \mathbb{N}_0}$ fulfills the condition $(*)$ of Definition 4 and hence $\{x_n\}_{n \in \mathbb{N}_0}$ is a fundamental sequence in (S, ρ). On account of that (S, ρ) is F-orbitally complete and keeping $x_n = F^n x_0$ for all n, in mind, we deduce that there is $x^* \in S$ such that $x_n \to x^*$. By the orbital continuity of F, we conclude that $x_n \to Fx^*$. Accordingly, we have $x^* = Fx^*$ which concludes the proof. □

Regarding Example 1 (i), we conclude the following result from Theorem 7.

Theorem 8. *If there is $k \in (0,1)$ such that*

$$\min\{\rho(Fx, Fy), \rho(x, Fx), \rho(y, Fy)\} - \min\{\delta^\rho_m(x, Fy), \delta^\rho_m(Fx, y)\}$$
$$\leq k(\rho(x,y) - \rho(x,x)) + \rho(y,y), \quad (26)$$

for all $x, y \in S$, then, the mapping F possesses a fixed point in S. Indeed, for an arbitrary initial point $x_0 \in S$ the recursive sequence $\{F^n x_0\}_{n \in \mathbb{N}}$ converges to a fixed point of F.

Regarding that the class of metric functions are contained in the class of partial metric, we deduce the renowned result of Ćirić [1].

Corollary 1. *[1] Theorem 1. Let F be an orbitally continuous self-map of a F-orbitally complete metric space (S, δ). If there is $k \in (0,1)$ such that*

$$\min\{\delta(Fx, Fy), \delta(x, Fx), \delta(y, Fy)\} - \min\{\delta(x, Fy), \delta(Fx, y)\}$$
$$\leq k\delta(x,y), \quad (27)$$

for all $x,y \in \mathcal{S}$, then for each $x_0 \in \mathcal{S}$ the sequence $\{F^n x_0\}_{n \in \mathbb{N}_0}$ converges to a fixed point of F.

In what follows we put two illustrative examples to show that Theorem 8 is a genuine extension of Corollary 1 for the metrics δ_ρ and δ_m^ρ, and δ_0, respectively.

Example 5 ([6]). *Consider the set $S = \{0, 1, 2\}$ equipped with a partial metric $\rho : S \times S \to \mathbb{R}_0^+$ with a definition $\rho(x,y) = \max\{x,y\}$ for all $x,y \in \mathcal{S}$. We set a self-mapping $F : S \to S$ in a way that $F0 = F1 = 0$ and $F2 = 1$. Notice that the completeness of a partial metric space (S, ρ) yields that it is also F-orbitally complete. Note also that F is orbitally continuous. An elementary evaluation yields that*

$$\min\{\rho(Fx, Fy), \rho(x, Fx), \rho(y, Fy)\} - \min\{\delta_m^\rho(x, Fy), \delta_m^\rho(Fx, y)\}$$
$$\leq \tfrac{1}{2}(\rho(x,y) - \rho(x,x)) + \rho(y,y),$$

for all $x,y \in \mathcal{S}$. Thus, we conclude that all hypotheses of Theorem 8 are fulfilled. On the other hand,

$$\min\{\delta_\rho(T1, T2), \delta_\rho(1, T1), \delta_\rho(2, T2)\} - \min\{\delta_\rho(1, T2), \delta_\rho(T1, 2)\}$$
$$= 1 - 0 = 1 > k = kd_p(1,2),$$

for any $k \in (0,1)$. As a result, Corollary 1 cannot be applied to the complete metric space (S, δ_ρ). In fact, it cannot be applied to (X, δ_m^ρ), because $\delta_m^\rho = \delta_\rho$, in this case.

Example 6 ([6]). *Consider the set $S = [1, \infty)$ equipped with a partial metric $\rho : S \times S \to \mathbb{R}_0^+$ with a definition $\rho(x,y) = \max\{x,y\}$ for all $x,y \in \mathcal{S}$. We set a self-mapping $F : S \to S$ in a way that $Fx = (x+1)/2$ for all $x \in \mathcal{S}$. As it is mentioned in Example 5, (S, ρ) is F-orbitally complete since it is already complete. In addition, F is continuous with respect to τ_{δ_ρ}, and hence it is orbitally continuous.*

In what follows we shall prove that F fulfills the contraction condition (55) for any $k \in (0,1)$. We consider two distinct cases for $x, y \in \mathcal{S}$ as follows:
Case 1. If $x = y$ then

$$\min\{\rho(Fx, Fy), \rho(x, Fx), \rho(y, Fy)\} - \min\{\delta_m^\rho(x, Fy), \delta_m^\rho(Fx, y)\}$$
$$= \min\{\tfrac{x+1}{2}, x, x\} - (x - \tfrac{x+1}{2}) = 1$$
$$\leq x = \rho(x,x) = k((\rho(x,y) - \rho(x,x)) + \rho(y,y).$$

Case 2. Suppose now $x \neq y$. Regarding the analogy, we presume only $x > y$. (Please note that the case $x < y$ is observed by verbatim.) We shall examine this case in two steps.
Step 1. If $Fx \geq y$, then

$$\min\{\rho(Fx, Fy), \rho(x, Fx), \rho(y, Fy)\} - \min\{\delta_m^\rho(x, Fy), \delta_m^\rho(Fx, y)\}$$
$$= \min\{\tfrac{x+1}{2}, x, y\} - \min\{x - \tfrac{y+1}{2}, \tfrac{x+1}{2} - y\}$$
$$= y - (\tfrac{x+1}{2} - y) = 2y - \tfrac{x+1}{2}$$
$$\leq y = \rho(y,y) = k((\rho(x,y) - \rho(x,x)) + \rho(y,y).$$

Step 2. If $Fx < y$, we have

$$\min\{\rho(Fx, Fy), \rho(x, Fx), \rho(y, Fy)\} - \min\{\delta_m^\rho(x, Fy), \delta_m^\rho(Fx, y)\}$$
$$= \min\{\tfrac{x+1}{2}, x, y\} - \min\{x - \tfrac{y+1}{2}, y - \tfrac{x+1}{2}\}$$
$$= \tfrac{x+1}{2} - (y - \tfrac{x+1}{2}) = x + 1 - y$$
$$< y = \rho(y,y) = k((\rho(x,y) - \rho(x,x)) + \rho(y,y).$$

Consequently, all hypotheses of Theorem 8 are satisfied. In fact F possesses a (unique) fixed point, namely, $x = 1$.

Now, we shall indicate that Corollary 1 cannot be applied to the self-map F and the complete metric space (S, δ_0). Indeed, given $k \in (0,1)$, choose $x > 1$ such that $x + 1 > 2kx$, and let $y = Fx$. Then

$$\min\{\delta_0(Fx, Fy), \delta_0(x, Fx), \delta_0(y, Fy)\} - \min\{\delta_0(x, Fy), \delta_0(Fx, y)\}$$
$$= \min\{\frac{x+1}{2}, x\} - \min\{x, 0\} = \frac{x+1}{2} > kx = k\rho_0(x, y).$$

As a result, the contraction condition (27) is not fulfilled.

The following theorem characterize Theorem 3 [1] in the setting of partial metric spaces.

Theorem 9. *Suppose that F satisfies the inequality*

$$\min\{\rho(Fx, Fy), \rho(x, Fx), \rho(y, Fy)\} - \min\{\delta_m^\rho(x, Fy), \delta_m^\rho(Fx, y)\} < \rho(x, y) - \rho(x, x) + \rho(y, y), \quad (28)$$

for all $x, y \in S$ with $x \neq y$. If for some $x_0 \in S$ the sequence $\{F^n x_0\}_{n \in \mathbb{N}_0}$ has a cluster point $z \in S$ with respect to τ_{δ_ρ}, then z is a fixed point of F.

Proof. We shall construct a sequence by starting with an point $x_0 \in S$ so that the sequence $\{x_{n+1} =: F^n x_0\}_{n \in \mathbb{N}_0}$ has a cluster point $x^* \in S$ with respect to τ_{δ_ρ}.

If there is a non-negative integer n_0 so that $x_{n_0} = x_{n_0+1}$, then x_{n_0} forms a fixed point of F. Thus, we presume then that $x_n \neq x_{n+1}$ for each $n \in \mathbb{N}_0$.

By verbatim in the corresponding lines in Theorem 8, by substituting $x = x_n$ and $y = x_{n+1}$ in (28) we derive

$$\min\{\rho(x_n, x_{n+1}), \rho(x_{n+1}, x_{n+2})\} < \rho(x_n, x_{n+1}) - \rho(x_n, x_n) + \rho(x_{n+1}, x_{n+1}),$$

and substituting $x = x_{n+1}$ and $y = x_n$ in (28), we obtain

$$\min\{\rho(x_n, x_{n+1}), \rho(x_{n+1}, x_{n+2})\} < \rho(x_n, x_{n+1}) - \rho(x_{n+1}, x_{n+1}) + \rho(x_n, x_n).$$

If $\rho(x_{n_0}, x_{n_0+1}) \leq \rho(x_{n_0+1}, x_{n_0+2})$ for some $n_0 \in \mathbb{N}_0$, then, on account of the preceding two inequalities we get $\rho(x_{n_0}, x_{n_0}) < \rho(x_{n_0+1}, x_{n_0+1})$ and $\rho(x_{n_0+1}, x_{n_0+1}) < \rho(x_{n_0}, x_{n_0})$, respectively. It is a contradiction.

Consequently $\rho(x_n, x_{n+1}) > \rho(x_{n+1}, x_{n+2})$ for all $n \in \mathbb{N}_0$, and thus the sequence $\{\rho(F^n x_0, F^{n+1} x_0)\}_{n \in \mathbb{N}_0}$ is convergent. Since $\{F^n x_0\}_{n \in \mathbb{N}_0}$ has a cluster point $x^* \in X$ with respect to τ_{δ_ρ}, then there is a subsequence $\{F^{n_i} x_0\}_{i \in \mathbb{N}_0}$ of $\{F^n x_0\}_{n \in \mathbb{N}_0}$ which converges to x^*. By the orbital continuity of F we have $F^{n_i+1} x_0 \to Fx^*$, so by Lemma 4,

$$\lim_{i \to \infty} \rho(F^{n_i} x_0, F^{n_i+1} x_0) = \rho(x^*, Fx^*). \quad (29)$$

Therefore

$$\lim_{n \to \infty} \rho(F^n x_0, F^{n+1} x_0) = \rho(x^*, Fx^*). \quad (30)$$

Again, by the orbital continuity of F we have $F^{n_i+2} x_0 \to F^2 z$ with respect to τ_{δ_ρ} and hence

$$\lim_{n \to \infty} \rho(F^{n+1} x_0, F^{n+2} x_0) = \rho(Fx^*, F^2 x^*),$$

so

$$\rho(Fx^*, F^2 x^*) = \rho(x^*, Fx^*). \quad (31)$$

Assume $Fx^* \neq x^*$, i.e., $\rho(x^*, Fx^*) > 0$. So, one can substitute x and y with x^* and Fx^*, respectively, in (28) to deduce that
$$\min\{\rho(x^*, Fx^*), \rho(Fx^*, F^2 x^*)\} < \rho(x^*, Fx^*),$$
which yields that $\rho(Fx^*, F^2 x^*) < \rho(x^*, Fx^*)$. This contradicts the equality (31). Consequently we have $Fx^* = x^*$. □

3.2. Pachpatte Type Non-Unique Fixed Points on Partial Metric Spaces

Inspired from the renowned Ćirić's theorems [1], Pachpatte proved in Theorem 1 [11] that if a self-mapping F is an orbitally continuous on a F-orbitally complete metric space (S, δ) such that there is $k \in (0, 1)$ with

$$\begin{aligned}\min\{[\delta(Fx, Fx)]^2, \delta(x,y)\delta(Fx, Fy), [\delta(Fy, y)]^2\} \\ - \min\{\delta(x, Fx)\delta(y, Fy), \delta(x, Fy)\delta(y, Fx)\} \leq k\delta(x, Fx)\delta(Fy, y)\end{aligned} \quad (32)$$

for all $x, y \in S$, then for each $x_0 \in S$ the sequence $\{F^n x_0\}_{n \in \mathbb{N}_0}$ converges to a fixed point of F.

On the other hand, Pachpatte's theorem does not yield a good framework for a possible application. Indeed, under its conditions, if we denote a fixed point of F by x^*, it follows that for each $y \in S$, we have either $Ty = x^*$ or $Ty = y$. Indeed, let $y \neq x^*$ and suppose $Ty \neq x^*$. Then from

$$\begin{aligned}\min\{[\delta(Fx^*, Fy)]^2, \delta(x^*, y)\delta(Fx^*, Fy), [\delta(y, Fy)]^2\} \\ - \min\{\delta(x^*, Fx^*)\delta(y, Fy), \delta(x^*, Fy)\delta(y, Fx^*)\} \\ \leq k\delta(x^*, Fx^*)\delta(y, Fy),\end{aligned}$$

it follows
$$\min\{[\delta(x^*, Fy)]^2, \delta(x^*, y)\delta(x^*, Fy), [\delta(y, Fy)]^2\} = 0.$$

Hence $\delta(y, Fy) = 0$, i.e., $y = Ty$.

In what follows, we repair the contraction condition (32) so that the inconvenient case, pointed above, is removed.

The function ρ' defined on $S \times S$ by $\rho'(x, y) = \rho(x, y) - \rho(x, x)$ for all $x, y \in S$, where ρ is a partial metric on a set S. Please note that $\rho' = \rho$, whenever ρ is a metric on S.

Definition 7. *Let (S, ρ) be a partial metric space. The self-mapping $F : S \to S$ is called Pachpatte type simulated if there exists $k \in (0, 1)$ and $\zeta \in \mathcal{Z}$ such that*

$$\zeta(J_F(x, y) - I_F(x, y), K_F(x, y)) \geq 0 \quad (33)$$

for all $x, y \in S$, where

$$\begin{aligned}J_F(x, y) &= \min\{[\rho'(x, Fx)]^2, \rho'(x, y)\rho'(Fx, Fy), [\rho'(y, Fy)]^2\} \\ I_F(x, y) &= \{\delta_m^\rho(x, Fx)\delta_m^\rho(y, Fy), \delta_m^\rho(x, Fy)\delta_m^\rho(y, Fx)\} \\ K_F(x, y) &= k\min\{\rho'(x, Fx)\rho'(y, Fy), [\rho'(x, y)]^2\},\end{aligned}$$

Theorem 10. *If F is a Pachpatte type simulated mapping, then for each $x_0 \in S$ the sequence $\{F^n x_0\}_{n \in \mathbb{N}_0}$ converges with respect to τ_{δ_ρ} to a fixed point of F.*

Proof. As usual, we fix an arbitrary initial point $x_0 \in S$ and construct an recursive sequence $\{x_n\}_{n \in \omega}$ as $x_{n+1} = Fx_n$, $n \in \mathbb{N}_0$.

If there exists $n_0 \in \mathbb{N}_0$ such that $x_{n_0} = x_{n_0+1}$, then x_{n_0} is a fixed point of F. Assume then that $x_n \neq x_{n+1}$ for each $n \in \mathbb{N}_0$.

Substituting $x = x_n$ and $y = x_{n+1}$ in (33) we find the inequality

$$0 \leq \zeta(J_F(x_n, x_{n+1}) - I_F(x_n, x_{n+1}), K_F(x_n, x_{n+1}))$$
$$< K_F(x_n, x_{n+1}) - [J_F(x_n, x_{n+1}) - I_F(x_n, x_{n+1})],$$

where

$$\begin{aligned} J_F(x_n, x_{n+1}) &= \min\{[\rho'(x_n, Fx_n)]^2, \rho'(x_n, x_{n+1})\rho'(Fx_n, Fx_{n+1}), [p'(x_{n+1}, Fx_{n+1})]^2\} \\ I_F(x_n, x_{n+1}) &= \{\delta_m^\rho(x_n, Fx_n)\delta_m^\rho(x_{n+1}, Fx_{n+1}), \delta_m^\rho(x_n, Fx_{n+1})\delta_m^\rho(x_{n+1}, Fx_n)\} \\ K_F(x_n, x_{n+1}) &= k\min\{\rho'(x_n Fx_n)\rho'(x_{n+1}, Fx_{n+1}), [p'(x_n, x_{n+1})]^2\}, \end{aligned}$$

By a simple evaluation, we find that

$$\min\{[\rho'(x_n, x_{n+1})]^2, \rho'(x_n, x_{n+1})p'(x_{n+1}, x_{n+2}), [\rho'(x_{n+1}, x_{n+2})]^2\}$$
$$\leq k\min\{\rho'(x_n, x_{n+1})p'(x_{n+1}, x_{n+2}), [\rho'(x_n, x_{n+1})]^2\}. \tag{34}$$

By (34) we deduce that

$$\min\{[\rho'(x_n, x_{n+1})]^2, p'(x_n, x_{n+1})\rho'(x_{n+1}, x_{n+2}), [p'(x_{n+1}, x_{n+2})]^2\}$$
$$= [\rho'(x_{n+1}, x_{n+2})]^2,$$

and hence

$$\rho'(x_{n+1}, x_{n+2}) \leq k\rho'(x_n, x_{n+1}),$$

for all $n \in \mathbb{N}_0$. Accordingly, we find

$$\rho(x_n, x_{n+1}) - \rho(x_n, x_n) \leq k^n(\rho(x_0, x_1) - \rho(x_0, x_0)),$$

for all $n \in \mathbb{N}$. By verbatim of Theorem 8, we conclude that $\{x_n\}_{n\in\mathbb{N}_0}$ is a fundamental sequence in (S, ρ). Since (S, ρ) is F-orbitally complete and $x_n = F^n x_0$ for all n, there is $x^* \in S$ such that $x_n \to x^*$ with respect to τ_{δ_ρ}. On account of the orbital continuity of F, we derive that $x_n \to Fx^*$. As a result $x^* = Fx^*$ which concludes the proof. □

Regarding Example 1 (i), we conclude the following result from Theorem 10.

Theorem 11. *If there is $k \in (0, 1)$ such that*

$$J_F(x, y) - I_F(x, y) \leq K_F(x, y) \tag{35}$$

for all $x, y \in S$, where

$$\begin{aligned} J_F(x, y) &= \min\{[\rho'(x, Fx)]^2, \rho'(x, y)\rho'(Fx, Fy), [p'(y, Fy)]^2\} \\ I_F(x, y) &= \{\delta_m^\rho(x, Fx)\delta_m^\rho(y, Fy), \delta_m^\rho(x, Fy)\delta_m^\rho(y, Fx)\} \\ K_F(x, y) &= k\min\{\rho'(x, Fx)\rho'(y, Fy), [p'(x, y)]^2\}, \end{aligned}$$

then for each $x_0 \in S$ the sequence $\{F^n x_0\}_{n\in\mathbb{N}_0}$ converges with respect to τ_{δ_ρ} to a fixed point of F.

Corollary 2. *If there is $k \in (0, 1)$ such that*

$$\min\{[\delta(x, Fx)]^2, \delta(x, y)\delta(Fx, Fy), [\delta(y, Fy)]^2\}$$
$$- \min\{\delta(x, Fx)\delta(y, Fy), \delta(x, Fy)\delta(y, Fx)\} \tag{36}$$
$$\leq k\min\{\delta(x, Fx)\delta(y, Fy), [\delta(x, y)]^2\},$$

for all $x, y \in S$, then the iterative sequence $\{F^n x_0\}_{n\in\mathbb{N}_0}$, initiated by an arbitrary point $x_0 \in S$, converges to a fixed point of F.

Remark 1. Consider an orbitally continuous self-map F defined on a complete partial metric space $(S = \mathbb{R}_0^+, \rho)$ with $\rho(x,y) := \max\{x,y\}$. If $Fx \leq x$ for all $x \in S$, then it possesses a fixed point Notice that a mapping F with $Fx \leq x$ yields $\rho'(x, Fx) = 0$ for all $x \in S$. Accordingly, the condition (35) in Theorem 11, is fulfilled trivially.

In what follows we state an illustrative example where Theorem 11 can be applied but not Corollary 2 for any of the metrics δ_ρ, δ_m^ρ and δ_0.

Example 7. Suppose that F is an orbitally continuous self-map defined on a complete partial metric space $(S = \mathbb{R}_0^+, \rho)$ with $\rho(x,y) := \max\{x,y\}$. Consider $F : S \to S$ by $Fx = 0$ if $x < 2$ and $Fx = x - 1$ if $x \geq 2$. Please note that F is orbitally continuous. Indeed, for each $x \in S$, the sequence $F^n x \to 0$ with respect to τ_{δ_ρ}, and $F0 = 0$. In addition, on account of Remark 1 the inequality (35) is fulfilled. Consequently, all hypotheses of Theorem 11 are held.

Consider $x \geq 3$ and $y = Fx$. Thus, we have $x - y = 1$, and $y \geq 2$. Accordingly we find

$$\min\{[\delta_\rho(x, Fx)]^2, \delta_\rho(x,y)\delta_\rho(Fx, Fy), [\delta_\rho(y, Fy)]^2\}$$
$$- \min\{\delta_\rho(x, Fx)\delta_\rho(y, Fy), \delta_\rho(x, Fy)\delta_\rho(y, Fx)\}$$
$$= \min\{1, (x-y)^2, 1\} - 0 = 1$$
$$= \min\{\delta_\rho(x, Fx)\delta_\rho(y, Fy), [\delta_\rho(x, y)]^2\}.$$

As a result, condition (36) is not held for any $k \in (0,1)$, so we cannot apply Corollary 2 to (S, δ_ρ) (and thus to (X, δ_m^ρ) and the self-map F.

As a final step, for $k \in (0,1)$, choose $x \geq 3$ with $x > 1/(1-k)$, and $y = Fx$. Then

$$\min\{[\delta_0(x, Fx)]^2, \delta_0(x,y)\delta_0(Fx, Fy), [\delta_0(y, Fy)]^2\}$$
$$- \min\{\delta_0(x, Fx)\delta_0(y, Fy), \delta_0(x, Fy)\delta_0(y, Fx)\}$$
$$= \min\{x^2, x(x-1), (x-1)^2\} - 0 = (x-1)^2$$
$$> kx(x-1)$$
$$= k\min\{\delta_0(x, Fx)\delta_0(y, Fy), [\delta_0(x, y)]^2\}.$$

Consequently, we cannot apply Corollary 2 to (S, δ_0) and the self-map F (note that, in fact, F is orbitally continuous for (X, δ_0)).

4. Non Unique Fixed Points on b-Branciari Distance Space

In this section, we shall consider a distance function which is not a generalization of a metric. Indeed, when Branciari [29] suggested a new distance function by replacing the axiom of the triangle inequality in a standard metric definition with another variant, the axiom of the quadrilateral inequality, he aimed at getting an extension of a standard metric. As it can be seen in the upcoming lines, Branciari distance is completely different and incomparable with metric.

For the sake of completeness, we recollect the definition of a Branciari distance here.

Definition 8. (See e.g., [30]) For a nonempty set S we define a function $b : S \times S \longrightarrow [0, \infty)$

(b1) $b(z, w) = 0$ if and only if $z = w$(selfdistance/indistancy)
(b2) $b(z, w) = b(w, z)$(symmetry) (37)
(b3) $b(z, w) \leq b(z, u) + b(u, v) + b(v, w)$ (quadrilateral inequality),

for all $z, w \in S$ and all distinct $u, v \in S \setminus \{x, y\}$. We say that b is a Branciari distance (or rectangular metric, or generalized metric, or Branciari metric). The pair (S, b) is called a Branciari distance space and abbreviated as "BDS".

Notice that in some publication, Branciari distance space was named as "generalized metric space". However the phrase "generalized metric" was used to identify several extensions of the standard metric (see e.g., [29,31–44]). Based on this discussion, we shall use "Branciari distance" to avoid the confusion.

In what follows we recollect the basic topological concepts in the framework of Branciari distance spaces.

Definition 9. *(See e.g., [30])*

1. A sequence $\{x_n\}$ in a Branciari distance space $(\mathcal{S}, b)$ converges to a limit x^* if and only if $b(x_n, x^*) \to 0$ as $n \to \infty$.
2. we say that a sequence $\{x_n\}$, in a Branciari distance space $(\mathcal{S}, b)$, is fundamental if and only if for any given $\varepsilon > 0$ there exists positive integer $N(\varepsilon)$ such that $b(x_n, x_m) < \varepsilon$ for all $n > m > N(\varepsilon)$.
3. We say that a Branciari distance space $(\mathcal{S}, b)$ is complete whenever each fundamental sequence in $\mathcal{S}$ is convergent.
4. A mapping $H : (X, b) \to (X, b)$ is continuous if for any sequence $\{x_n\}$ in $\mathcal{S}$ such that $b(x_n, x) \to 0$ as $n \to \infty$, we have $b(Hx_n, Hx) \to 0$ as $n \to \infty$.

We underline the fact that despite the high similarity in the definitions of the basic topological in the framework of Branciari distance space, the topology of Branciari distance space is not compatible with topology of the standard metric space. These difference shall be indicated in the following example.

Example 8. *(cf. [37,45])* Let z_1, z_2, z_3 be distinct real numbers such that $z_1, z_2, z_3 > 2$. Set $S = Y \cup Z$ where $Z = \{0, z_1, z_2, z_3\}$ and $Y = \{\frac{1}{n^2+1} : n \in \mathbb{N}\}$. We investigate the function $b : \mathcal{S} \times \mathcal{S} \to [0, \infty)$ which is defined by

$$b(x, y) = \begin{cases} 0, & \text{if } x = y, \\ 1, & \text{if } x \neq y \text{ and } [\{x, y\} \subset Y \text{ or } \{x, y\} \subset Z], \\ y, & \text{if } x \in Y, y \in Z. \end{cases}$$

We have $b(y, z) = b(z, y) = z$ whenever $y \in Y$ and $z \in Z$. and $(\mathcal{S}, b)$ is a complete Branciari distance space. Notice that the statements $(P1)$–$(P4)$ are fulfilled:

(p1) Since $\lim_{n \to \infty} \frac{1}{n^2+1} = 0$, we have $\lim_{n \to \infty} b(\frac{1}{n^2+1}, \frac{1}{5}) \neq b(0, \frac{1}{5})$. Thus, the function b is not continuous:
(p2) There is no $r > 0$ such that $B_r(0) \cap B_r(z_i) = \emptyset$ for $i = 1, 2, 3$ and hence it is not Hausdorff.
(p3) It is clear that the ball $B_{\frac{3}{5}}(\frac{1}{5}) = \{0, \frac{1}{5}, z_1, z_2, z_3\}$ since there is no $r > 0$ such that $B_r(0) \subset B_{\frac{3}{5}}(\frac{1}{5})$, i.e., open balls may not be an open set.
(p4) The sequence $\{\frac{1}{n^2+1} : n \in \mathbb{N}\}$ converges to $0, z_1, z_2, z_3$ and hence not fundamental.

It is easily concluded that the differences between quadrilateral inequality and the triangle inequality lead to these significant differences between the topologies of the standard metric space and Branciari distance space. In brief, the following statements express the weakness of the structure of Branciari distance topology:

(p1) Branciari distance is not continuous, (see e.g., Example 8)
(p2) The limit in a Branciari distance space is not necessarily unique (i.e., it is not a Haussdorf, see e.g., Example 8)
(p3) open ball need not to open set, (see e.g., Example 8)
(p4) a convergent sequence in Branciari distance space needs not to be fundamental. (see e.g., Example 8)
(p5) the mentioned topologies are incompatible (see e.g., Example 7 in [44]).

Lemma 5. *(See e.g., [36,37])* Let $\{x_n\}$ be a fundamental sequence in a Branciari distance space $(\mathcal{S}, b)$. If $x_m \neq x_n$ whenever $m \neq n$, then the sequence $\{x_n\}$ converges to at most one point.

Later, regarding the well-known b-metric, defined by Czerwik [46] the notion of Branciari distance is refined as b-Branciari distance (See e.g., [47]).

Definition 10. *For a nonempty set S, we consider a function $\sigma : S \times S \longrightarrow [0, \infty)$ so that*

(b1) $\sigma(x, y) = 0$ *if and only if* $x = y$(indistancy)
(b2) $\sigma(x, y) = \sigma(y, x)$(symmetry) (38)
(b3) $\sigma(x, y) \leq s[\sigma(x, u) + \sigma(u, v) + \sigma(v, y)]$ *(modified quadrilateral inequality),*

for all $x, y \in S$ and all distinct $u, v \in S \setminus \{x, y\}$. Then, we say that σ is a b-Branciari distance (or b-rectangular metric, or b-Branciari metric, or b-generalized metric). In addition, the pair (S, σ) is named as a b-Branciari distance space and abbreviated as "b-BDS".

In what follows, we derive the characterization of fundamental topological notions (that we need in the sequel) in context of b-Branciari distance spaces (See e.g., [8]).

Definition 11.

1. A sequence $\{x_n\}$ in a b-Branciari distance space (S, σ) is convergent to a limit x if and only if $\sigma(x_n, x) \to 0$ as $n \to \infty$.
2. A sequence $\{x_n\}$ in a b-Branciari distance space (S, σ) is fundamental (or, Cauchy) if and only if for every $\varepsilon > 0$ there exists positive integer $N(\varepsilon)$ such that $\sigma(x_n, x_m) < \varepsilon$ for all $n > m > N(\varepsilon)$.
3. A b-Branciari distance space (S, σ) is called complete if every fundamental sequence in S is b-Branciari distance space convergent.
4. A mapping $H : (X, \sigma) \to (X, \sigma)$ is continuous if for any sequence $\{x_n\}$ in S such that $\sigma(x_n, x) \to 0$ as $n \to \infty$, we have $\sigma(Hx_n, Hx) \to 0$ as $n \to \infty$.

As is mentioned above, the topology of Branciari distance space has difficulties (p1)–(p5), and these weakness are hereditarily valid for the topology of b-Branciari distance space. It is easy to see that Example 8 can be modified for b-Branciari distance space to indicate that the same problems holds for the topology of b-Branciari distance space (see e.g., [47]).

Now, we propose the following proposition that helps to simplify the upcoming proofs.

Lemma 6 ([8]). *If a sequence $\{x_n\}$ in (S, σ) is Cauchy with $x_m \neq x_n$ whenever $m \neq n$, then the sequence $\{x_n\}$ can converge to at most one point.*

We consider the characterization of some basic but crucial topological notions in the context of b-BDS.

Definition 12. *Let (S, σ) be a b-Branciari distance space and H be a self-map of S.*

1. *H is called orbitally continuous if*

$$\lim_{i \to \infty} H^{n_i} x = z \qquad (39)$$

implies

$$\lim_{i \to \infty} H H^{n_i} x = Hz \qquad (40)$$

for each $x \in S$.
2. *(S, σ) is called orbitally complete if every Cauchy sequence of type $\{H^{n_i} x\}_{i \in \mathbb{N}}$ converges with respect to τ_σ.*

We say that x^* is a periodic point of a function H of period m if $H^m(x^*) = x^*$, where $H^m(x) = H(H^{m-1}(x))$ for $m \in \mathbb{N}$ and $H^0(x) = x$.

In the following lines, we examine some non-unique fixed point results in the context of b-BDS. The presented results not only improve, extend several results in the corresponding literature, but also enrich them.

Henceforward, the couple (S, σ) represent b-Branciari metric space. The letter H be an orbitally continuous self-map on b-Branciari metric space- (S, σ) with $s \geq 1$. In all upcoming result, we assume that b-Branciari metric space- (S, σ) is orbitally complete. Avoiding from the repetitions, we shall not indicate the above assumptions to all theorems, corollaries and lemmas.

4.1. Ćirić Type Non-Unique Fixed Point Results

Definition 13. *A self-mapping $H : S \to S$ is called ψ-Ćirić type simulated if there exist $\zeta \in \mathcal{Z}$ and $\psi \in \Psi$ such that*

$$P_H(x, y) \leq \psi(\sigma(x, y)), \tag{41}$$

for all $x, y \in S$, where

$$P_H(x, y) := \min\{\sigma(Hx, Hy), \sigma(x, Hx), \sigma(y, Hy)\} - \min\{\sigma(x, Hy), \sigma(Hx, y)\}$$

Theorem 12. *If a mappings H is ψ-Ćirić type simulated, then for each $x_0 \in S$ the sequence $\{H^n x_0\}_{n \in \mathbb{N}}$ converges to a fixed point of H.*

Proof. Starting from an arbitrary point $x \in S$, we shall built an iterative sequence $\{x_n\}$ in the following way:

$$x_0 := x \text{ and } x_n = Hx_{n-1} \text{ for all } n \in \mathbb{N}. \tag{42}$$

We suppose that

$$x_n \neq x_{n-1} \text{ for all } n \in \mathbb{N}. \tag{43}$$

Indeed, if for some $n \in \mathbb{N}$ we have the inequality $x_n = Hx_{n-1} = x_{n-1}$, then, the proof is completed.

By substituting $x = x_{n-1}$ and $y = x_n$ in the inequality (44), we derive that

$$P_H(x_{n-1}, x_n) \leq \psi(\sigma(x_{n-1}, x_n)), \tag{44}$$

where

$$\begin{aligned} P_H(x_{n-1}, x_n) &= \min\{\sigma(Hx_{n-1}, Hx_n), \sigma(x_{n-1}, Hx_{n-1}), \sigma(x_n, Hx_n)\} \\ &\quad - \min\{\sigma(x_{n-1}, Hx_n), \sigma(Hx_{n-1}, x_n)\} \end{aligned}$$

After an elementary calculation, we find that

$$\begin{aligned} &\min\{\sigma(Hx_{n-1}, Hx_n), \sigma(x_{n-1}, Hx_{n-1}), \sigma(x_n, Hx_n)\} \\ &\quad - \min\{\sigma(x_{n-1}, Hx_n), \sigma(Hx_{n-1}, x_n)\} \\ &\leq \psi(\sigma(x_{n-1}, x_n)). \end{aligned} \tag{45}$$

It implies that

$$\min\{\sigma(x_n, x_{n+1}), \sigma(x_n, x_{n-1})\} \leq \psi(\sigma(x_{n-1}, x_n)). \tag{46}$$

Due to property of $\psi(t) < t$ for all $t > 0$, we find that the case $\sigma(x_n, x_{n-1}) \leq \psi(\sigma(x_{n-1}, x_n))$ is not possible. Accordingly, we get

$$\sigma(x_n, x_{n+1}) \leq \psi(\sigma(x_{n-1}, x_n)) < \sigma(x_{n-1}, x_n). \tag{47}$$

Iteratively, we find that

$$\sigma(x_n, x_{n+1}) \leq \psi(\sigma(x_{n-1}, x_n)) \leq \psi^2(\sigma(x_{n-2}, x_{n-1})) \leq \cdots \leq \psi^n(\sigma(x_0, x_1)). \tag{48}$$

Taking (47) into account, we find that the sequence $\{\sigma(x_n, x_{n+1})\}$ is non-increasing.

Since, for any $t \in [0, \infty)$, $\lim_{n \to \infty} \psi^n(t) = 0$, and $\psi(t) < t$ for $t > 0$, the Archimedean property implies that there exist a $q \in [0, 1)$ and a $M \in \mathbb{N}$ such that

$$\psi^k(t) \leq q^k \cdot t \text{ and } s \cdot q^k < 1 \text{ for each } n > M. \tag{49}$$

In what follows we prove that the sequence $\{x_n\}$ has no periodic point, i.e.,

$$x_n \neq x_{n+k} \text{ for all } (k, n) \in \mathbb{N} \times \mathbb{N}_0. \tag{50}$$

Actually, if $x_n = x_{n+k}$ for some $n \in \mathbb{N}_0$ and $k \in \mathbb{N}$, we find

$$x_{n+1} = Hx_n = Hx_{n+k} = x_{n+k+1}.$$

Regarding (47) and (55), we find that

$$\begin{aligned}
\sigma(x_n, x_{n+1}) &= \min\{\sigma(Hx_{n-1}, Hx_n), \sigma(x_{n-1}, Hx_{n-1}), \sigma(x_n, Hx_n)\} \\
&\quad - \min\{\sigma(x_{n-1}, Hx_n), \sigma(Hx_{n-1}, x_n)\} \\
&= \min\{\sigma(Hx_{n+k-1}, Hx_{n+k}), \sigma(x_{n+k-1}, Hx_{n+k-1}), \sigma(x_n, Hx_{n+k})\} \\
&\quad - \min\{\sigma(x_{n+k-1}, Hx_{n+k}), \sigma(Hx_{n+k-1}, x_{n+k})\} \tag{51} \\
&\leq \psi(\sigma(x_{n+k-1}, x_{n+k})) \\
&\leq \psi^{k-1}(\sigma(x_n, x_{n+1})) < \sigma(x_n, x_{n+1}),
\end{aligned}$$

a contradiction. Based on the discussion above, we presume that

$$x_n \neq x_m \text{ for all distinct } n, m \in \mathbb{N}. \tag{52}$$

Observe that $x_{n+k} \neq x_{m+k}$ for all distinct $n, m \in \mathbb{N}$ and $x_{n+k}, x_{m+k} \in \mathcal{S} \setminus \{x_n, x_m\}$.

Now, we assert that the sequence $\{x_n\}$ is fundamental. The modified quadrilateral inequality together with (48) and (49) yields that

$$\begin{aligned}
\sigma(x_m, x_n) &\leq s[\sigma(x_m, x_{m+k}) + \sigma(x_{m+k}, x_{n+k}) + \sigma(x_{n+k}, x_n)] \\
&\leq s\psi^m(\sigma(x_0, x_k)) + s\psi^k(\sigma(x_m, x_n)) + s\psi^n(\sigma(x_k, x_0)) \tag{53} \\
&\leq s\psi^m(\sigma(x_0, x_k)) + sq^k \cdot \sigma(x_m, x_n) + s\psi^n(\sigma(x_k, x_0)).
\end{aligned}$$

After a routine calculation, we get that

$$\sigma(x_m, x_n) \leq \frac{s}{1 - sq^k}[\psi^m(\sigma(x_0, x_k)) + \psi^n(\sigma(x_k, x_0))]. \tag{54}$$

Since $\lim_{n \to \infty} \psi^n(t) = 0$, for any $t \in [0, \infty)$, (54) implies that $\sigma(x_m, x_n) \to 0$ as $n, m \to \infty$. As a result, $\{x_n\}$ is a fundamental sequence in b-Branciari distance space $(\mathcal{S}, \sigma)$.

Here, H-orbitally completeness implies that there is $x^* \in \mathcal{S}$ such that $x_n \to x^*$. On account of the orbital continuity of H, we find that $x_n \to Fx^*$. On the other hand, Lemma 6 leads to $x^* = Fx^*$ which terminates the proof. □

Regarding Example 1 (i), we conclude the following result from Theorem 12.

Theorem 13 ([8]). *If there is $\psi \in \Psi$ such that*

$$\min\{\sigma(Hx, Hy), \sigma(x, Hx), \sigma(y, Hy)\} - \min\{\sigma(x, Hy), \sigma(Hx, y)\} \leq \psi(\sigma(x, y)), \tag{55}$$

for all $x, y \in \mathcal{S}$, then for each $x_0 \in \mathcal{S}$ the sequence $\{H^n x_0\}_{n \in \mathbb{N}}$ converges to a fixed point of H.

Corollary 3. *If there is $q \in [0, 1)$ such that*

$$\min\{\sigma(Hx, Hy), \sigma(x, Hx), \sigma(y, Hy)\} - \min\{\sigma(x, Hy), \sigma(Hx, y)\} \leq q\sigma(x, y), \tag{56}$$

for all $x, y \in \mathcal{S}$, then for each $x_0 \in \mathcal{S}$ the sequence $\{H^n x_0\}_{n \in \mathbb{N}}$ converges to a fixed point of H.

Proof. Employing Theorem 13 for $\psi(t) = qt$, where $q \in [0, 1)$, yields the desired result. □

Example 9 ([8]). Let $\mathcal{S} = A \cup B$ where $A = \{a_1, a_2, a_3, a_4\}$ and $B = [1, 2]$ with $A \cap B = \varnothing$ and each a_i distinct from a_j, whenever $i \neq j$. Define $\delta : \mathcal{S} \times \mathcal{S} \to [0, \infty)$ such that $\sigma(x, y) = \sigma(y, x)$ for all $x \in \mathcal{S}$,

$$\sigma(a_1, a_3) = 1, \quad \sigma(a_1, a_2) = \sigma(a_2, a_3) = \frac{1}{4},$$

$$\sigma(a_1, a_4) = \sigma(a_2, a_4) = \sigma(a_3, a_4) = \frac{1}{8},$$

$$\sigma(a, b) = \frac{1}{16}, \text{ for all } a \in A, b \in B, \text{ and,}$$

$$\sigma(x, y) = |x - y|^2 \text{ for any other case.}$$

Here, $(\mathcal{S}, \sigma)$ forms a complete b-Branciari distance space $(\mathcal{S}, \sigma)$ with $s = 2$. However, σ is not a Branciari distance. In addition, σ is neither a metric, nor b-metric. Define a mapping $H : X \to X$ as

$$f(a_1) = f(a_2) = a_1 \text{ and } f(a_3) = f(a_4) = a_4 \text{ and } f(b) = a_1 \text{ for all } b \in B.$$

Thus H fulfills all hypotheses of Theorem 13 for any choice of ψ. Please note that H has two distinct fixed points, namely, a_1 and a_3.

4.2. Ćirić-Jotić Type Non-Unique Fixed Point Results

Definition 14. *A self-mapping $H : \mathcal{S} \to \mathcal{S}$ is called ψ-Ćirić-Jotić type simulated if there exist $\zeta \in \mathcal{Z}$ and $\psi \in \Psi$ such that*

$$\zeta(P_H(x, y) - aQ_H(x, y), \psi(R_H(x, y))) \geq 0, \tag{57}$$

for all $x, y \in \mathcal{S}$, where

$$P_H(x, y) = \min \left\{ \begin{array}{c} \sigma(Hx, Hy), \sigma(x, y), \sigma(x, Hx), \sigma(y, Hy), \frac{\sigma(x, Hx)[1 + \sigma(y, Hy)]}{1 + \sigma(x, y)}, \\ \frac{\sigma(y, Hy)[1 + \sigma(x, Hx)]}{1 + \sigma(x, y)}, \frac{\min\{\sigma^2(Hx, Hy), \sigma^2(x, Hx), \sigma^2(y, Hy)\}}{\psi(\sigma(x, y))} \end{array} \right\},$$

$$Q_H(x, y) = \min\{\sigma(x, Hy), \sigma(y, Hx)\},$$

$$R(x, y) = \max\{\sigma(x, y), \sigma(x, Hx)\}.$$

Theorem 14. *If a mappings H is ψ-Ćirić-Jotić type simulated, then for each $x_0 \in \mathcal{S}$ the sequence $\{H^n x_0\}_{n \in \mathbb{N}}$ converges to a fixed point of H.*

Proof. By verbatim of the proof of Theorem 12, we shall built an recursive sequence $\{x_n = Hx_{n-1}\}_{n \in \mathbb{N}}$ by starting from an arbitrary initial value $x_0 := x \in S$. Recalling the discussion in the proof of Theorem 12, we presume that any adjacent terms are distinct from each other, i.e.,

$$x_n \neq x_{n-1} \text{ for all } n \in \mathbb{N}.$$

Letting $x = x_{n-1}$ and $y = Hx_{n-1} = x_n$ in the inequality (57), we derive that

$$0 \leq \zeta(P(x_{n-1}, x_n) - aQ(x_{n-1}, x_n), \psi(R(x_{n-1}, x_n))) \\ < \psi(R(x_{n-1}, x_n)) - [P(x_{n-1}, x_n) - aQ(x_{n-1}, x_n)],$$

which yields that

$$P(x_{n-1}, x_n) - aQ(x_{n-1}, x_n) \leq \psi(R(x_{n-1}, x_n)), \tag{58}$$

where

$$Q(x_{n-1}, x_n) = \min\{\sigma(x_{n-1}, x_{n+1}), \sigma(x_n, x_n)\} = 0,$$

$$R(x_{n-1}, x_n) = \max\{\sigma(x_{n-1}, x_n), \sigma(x_{n-1}, x_n)\} = \sigma(x_{n-1}, x_n).$$

and

$$P(x_{n-1}, x_n) = \min \left\{ \begin{array}{l} \sigma(x_n, x_{n+1}), \sigma(x_{n-1}, x_n), \sigma(x_{n-1}, x_n), \sigma(x_n, x_{n+1}), \\[6pt] \dfrac{\sigma(x_{n-1}, x_n)[1 + \sigma(x_n, x_{n+1})]}{1 + \sigma(x_{n-1}, x_n)}, \\[10pt] \dfrac{\sigma(x_n, x_{n+1})[1 + \sigma(x_{n-1}, x_n)]}{1 + \sigma(x_{n-1}, x_n)}, \\[10pt] \dfrac{\min\{\sigma^2(x_n, x_{n+1}), \sigma^2(x_{n-1}, x_n), \sigma^2(x_n, x_{n+1})\}}{\psi(\sigma(x_{n-1}, x_n))} \end{array} \right\}$$

$$= \min \left\{ \begin{array}{l} \sigma(x_n, x_{n+1}), \sigma(x_{n-1}, x_n), \\[6pt] \dfrac{\sigma(x_{n-1}, x_n)[1 + \sigma(x_n, x_{n+1})]}{1 + \sigma(x_{n-1}, x_n)}, \\[10pt] \dfrac{\sigma^2(x_n, x_{n+1})}{\psi(\sigma(x_{n-1}, x_n))} \end{array} \right\}$$

We examine the inequality (58) regarding the possible cases in $P(x_{n-1}, x_n)$. On the other hand, the case $P(x_{n-1}, x_n) = \sigma(x_{n-1}, x_n)$ is impossible. Indeed, if it would be the case the inequality (58) turns into

$$\sigma(x_{n-1}, x_n) \leq \psi(\sigma(x_{n-1}, x_n)) < \sigma(x_{n-1}, x_n),$$

since $\psi(t) < t$ for all $t > 0$. Thus, we observe that

$$\sigma(x_n, x_{n+1}) \leq \sigma(x_{n-1}, x_n).$$

Consequently, the inequality (58) yields the following three cases:

If $P(x_{n-1}, x_n) = \sigma(x_n, x_{n+1})$ or $P(x_{n-1}, x_n) = \dfrac{\sigma^2(x_n, x_{n+1})}{\psi(\sigma(x_{n-1}, x_n))}$, then the inequality (58) turns into

$$\sigma(x_n, x_{n+1}) \leq \psi(\sigma(x_{n-1}, x_n)) \tag{59}$$

If $P(x_{n-1}, x_n) = \dfrac{\sigma(x_{n-1}, x_n)[1+\sigma(x_n, x_{n+1})]}{1+\sigma(x_{n-1}, x_n)}$, then the inequality (58) becomes

$$\begin{aligned}\sigma(x_{n-1}, x_n)[1+\sigma(x_n, x_{n+1})] &\leq \psi(\sigma(x_{n-1}, x_n))[1+\sigma(x_{n-1}, x_n)] \\ &= \psi(\sigma(x_{n-1}, x_n)) + \psi(\sigma(x_{n-1}, x_n))\sigma(x_{n-1}, x_n) \\ &< \sigma(x_{n-1}, x_n) + \psi(\sigma(x_{n-1}, x_n))\sigma(x_{n-1}, x_n)\end{aligned}$$

The required simplification implies the (59). Consequently, for any choice of $P(x_{n-1}, x_n)$, the inequality (58) yields (59). Iteratively, we find that

$$\sigma(x_{n+1}, x_n) \leq \psi(\sigma(x_n, x_{n-1})) < \sigma(x_n, x_{n-1}),$$

and hence

$$\sigma(x_{n+1}, x_n) < \psi^n(\sigma(x_1, x_0)),$$

for all $n \in \mathbb{N}$.

Thus, the sequence $\{\sigma(x_n, x_{n+1})\}$ is non-increasing. As a next step, we claim that the sequence $\{x_n\}$ has no periodic point, i.e.,

$$x_n \neq x_{n+k} \text{ for all } (k, n) \in \mathbb{N} \times \mathbb{N}_0. \tag{60}$$

Indeed, if $x_n = x_{n+k}$ for some $n \in \mathbb{N}_0$ and $k \in \mathbb{N}$, we find

$$x_{n+1} = Hx_n = Hx_{n+k} = x_{n+k+1}.$$

Based on the discussion above, we have $P(x_{n-1}, x_n) = \sigma(x_n, x_{n+1})$. Thus, by taking the inequality (47) and (55) into account, we find that

$$\begin{aligned}\sigma(x_n, x_{n+1}) &= P(x_{n-1}, x_n) - aQ(x_{n-1}, x_n) \leq \psi(R(x_{n-1}, x_n)), \\ &\leq \psi(R(x_{n+k-1}, x_{n+k})), \\ &\leq \psi(\sigma(x_{n+k-1}, x_{n+k})) \\ &\leq \psi^{k-1}(\sigma(x_n, x_{n+1})) < \sigma(x_n, x_{n+1}),\end{aligned} \tag{61}$$

a contradiction. Attendantly, we have

$$x_n \neq x_m \text{ for all distinct } n, m \in \mathbb{N}. \tag{62}$$

By following the related lines in the proof of Theorem 12, one can complete the proof. □

Regarding Example 1 (i), we conclude the following result from Theorem 14.

Theorem 15 ([8]). Assume that there exist $\psi \in \Psi$ and $a \geq 0$ such that

$$P(x, y) - aQ(x, y) \leq \psi(R(x, y)),$$

for all distinct $x, y \in S$ where

$$P(x,y) = \min \left\{ \begin{array}{c} \sigma(Hx, Hy), \sigma(x,y), \sigma(x, Hx), \sigma(y, Hy), \\ \dfrac{\sigma(x, Hx)[1+\sigma(y, Hy)]}{1+\sigma(x,y)}, \dfrac{\sigma(y, Hy)[1+\sigma(x, Hx)]}{1+\sigma(x,y)}, \\ \dfrac{\min\{\sigma^2(Hx, Hy), \sigma^2(x, Hx), \sigma^2(y, Hy)\}}{\psi(\sigma(x,y))} \end{array} \right\},$$

$$Q(x,y) = \min\{\sigma(x, Hy), \sigma(y, Hx)\},$$

$$R(x,y) = \max\{\sigma(x,y), \sigma(x, Hx)\}.$$

Then, for each $x_0 \in S$ the sequence $\{H^n x_0\}_{n \in \mathbb{N}}$ converges to a fixed point of H.

Corollary 4. *Assume that there exist $q \in [0,1)$ and $a \geq 0$ such that*

$$P(x,y) - aQ(x,y) \leq qR(x,y),$$

for all distinct $x, y \in S$ where $P(x,y), Q(x,y), R(x,y)$ are defined as in Theorem 15 Then, for each $x_0 \in S$ the sequence $\{H^n x_0\}_{n \in \mathbb{N}}$ converges to a fixed point of H.

Corollary 5. *Assume that there exist $q \in [0,1)$ and $a \geq 0$ such that*

$$\min\{\sigma(Hx, Hy), \sigma(x,y), \sigma(x, Hx), \sigma(y, Hy)\} - aQ(x,y) \leq qR(x,y),$$

for $x, y \in S$ where $Q(x,y), R(x,y)$ are defined as in Theorem 15 Then, for each $x_0 \in S$ the sequence $\{H^n x_0\}_{n \in \mathbb{N}}$ converges to a fixed point of H.

Corollary 6. *If there exists $k, p \in [0,1)$ with $k+p < 1$ and $a \geq 0$ such that*

$$\min\{\sigma(Hx, Hy), \sigma(x,y), \sigma(x, Hx), \sigma(y, Hy)\} - aQ(x,y) \leq k\sigma(x,y) + p\sigma(x, Hx)$$

for $x, y \in S$ where $Q(x,y), R(x,y)$ are defined as in Theorem 15, then, for each $x_0 \in S$, the sequence $\{H^n x_0\}_{n \in \mathbb{N}}$ converges to a fixed point of H.

Definition 15. *A self-mapping $H : S \to S$ is called weakly-ψ-Ćirić-Jotić type simulated if there exist $\zeta \in \mathcal{Z}$ and $\psi \in \Psi$ such that*

$$\zeta(P(x,y) - aQ(x,y), \psi(R(x,y))) \geq 0, \tag{63}$$

for all $x, y \in S$, where

$$P_H(x,y) = \min \left\{ \begin{array}{c} \sigma(Hx, Hy), \sigma(x,y), \sigma(x, Hx), \sigma(y, Hy), \\ \dfrac{\sigma(x, Hx)[1+\sigma(y, Hy)]}{1+\sigma(x,y)}, \dfrac{\sigma(y, Hy)[1+\sigma(x, Hx)]}{1+\sigma(x,y)}, \\ \dfrac{\min\{\sigma^2(Hx, Hy), \sigma^2(x, Hx), \sigma^2(y, Hy)\}}{\psi(\sigma(x,y))} \end{array} \right\},$$

$$Q_H(x,y) = \min\{\sigma(x, Hy), \sigma(y, Hx)\},$$

$$R(x,y) = \max\{\sigma(x,y), \sigma(x, Hx)\},$$

with $R(x,y) \neq 0$.

Theorem 16. *If a mappings H is weakly-ψ-Ćirić-Jotić type simulated, then for each $x_0 \in S$ the sequence $\{H^n x_0\}_{n \in \mathbb{N}}$ converges to a fixed point of H.*

Proof. We use the same construction as in Theorem 12 to get an iterative sequence $\{x_n = Hx_{n-1}\}_{n \in \mathbb{N}}$, with an arbitrary initial value $x_0 := x \in S$. Repeating the same arguments in the proof of Theorem 12, we derive that adjacent terms of the sequence $\{x_n\}$ are distinct, i.e.,

$$x_n \neq x_{n-1} \text{ for all } n \in \mathbb{N}.$$

For $x = x_{n-1}$ and $y = x_n$, the inequality (80) infer that

$$\begin{aligned} 0 &\leq \zeta(K(x_{n-1}, x_n)) - aQ(x_{n-1}, x_n), \psi(S(x_{n-1}, x_n))) \\ &< \psi(S(x_{n-1}, x_n)) - K(x_{n-1}, x_n) - aQ(x_{n-1}, x_n) \end{aligned} \quad (64)$$

It yields that

$$K(x_{n-1}, x_n)) - aQ(x_{n-1}, x_n) \leq \psi(S(x_{n-1}, x_n)), \quad (65)$$

where

$$\begin{aligned} K(x_{n-1}, x_n) &= \min\{\sigma(Hx_{n-1}, Hx_n), \sigma(x_n, Hx_n)\} = \sigma(x_n, x_{n+1}), \\ Q(x_{n-1}, x_n) &= \min\{\sigma(x_{n-1}, Hx_n)\sigma(x_n, Hx_{n-1})\} = 0, \\ S(x_{n-1}, x_n) &= \min\{\sigma(x_{n-1}, x_n), \sigma(x_{n-1}, Hx_{n-1}), \sigma(x_n, Hx_n)\} \\ &= \min\{\sigma(x_{n-1}, x_n), \sigma(x_n, x_{n+1})\}. \end{aligned}$$

Since $\psi(t) < t$ for all $t > 0$, the case $S(x_{n-1}, x_n) = \sigma(x_n, x_{n+1})$ is impossible. More precisely, it is the case, the inequality (65) turns into

$$\sigma(x_n, x_{n+1}) \leq \psi\sigma(x_n, x_{n+1}) < \sigma(x_n, x_{n+1}),$$

a contradiction. Hence, the inequality (65) yields that

$$\sigma(x_n, x_{n+1}) \leq \psi\sigma(x_{n-1}, x_n) < \sigma(x_{n-1}, x_n) \text{ and } \sigma(x_n, x_{n+1}) \leq \psi^n \sigma(x_0, x_1)$$

for all $n \in \mathbb{N}$.

Hence, we conclude that the sequence $\{\sigma(x_n, x_{n+1})\}$ is non-increasing. On what follows that we show that the iterative sequence $\{x_n\}$ has no periodic point, i.e.,

$$x_n \neq x_{n+k} \text{ for all } k \in \mathbb{N} \text{ and for all } n \in \mathbb{N}_0. \quad (66)$$

Indeed, if $x_n = x_{n+k}$ for some $n \in \mathbb{N}_0$ and $k \in \mathbb{N}$, we have $x_{n+1} = Hx_n = Hx_{n+k} = x_{n+k+1}$. Based on the observations above, we obtain that $K(x_{n-1}, x_n) = \sigma(x_n, x_{n+1})$. Consequently, the inequality (66) and (80) implied that

$$\begin{aligned} \sigma(x_n, x_{n+1}) &= K(x_{n-1}, x_n) - aQ(x_{n-1}, x_n) \leq \psi(S(x_{n-1}, x_n)), \\ &\leq \psi(S(x_{n+k-1}, x_{n+k})), \\ &\leq \psi(\sigma(x_{n+k-1}, x_{n+k})) \\ &\leq \psi^{k-1}(\sigma(x_n, x_{n+1})) < \sigma(x_n, x_{n+1}), \end{aligned} \quad (67)$$

which is a contradiction. Hence, we assume that

$$x_n \neq x_m \text{ for all distinct } n, m \in \mathbb{N}. \tag{68}$$

A verbatim repetition of the related lines in the proof of Theorem 12 completes the proof. □

On account of Example 1 (i), we conclude the following result from Theorem 16.

Theorem 17 ([8]). *Suppose that there exists $\psi \in \Psi$ and $a \geq 0$ such that*

$$K(x,y) - aQ(x,y) \leq \psi(S(x,y)), \tag{69}$$

for all distinct $x, y \in S$ where

$$K(x,y) = \min\{\sigma(Hx, Hy), \sigma(y, Hy)\},$$

$$Q(x,y) = \min\{\sigma(x, Hy), \sigma(y, Hx)\},$$

$$S(x,y) = \max\{\sigma(x,y), \sigma(x, Hx), \sigma(y, Hy)\}.$$

Then, for each $x_0 \in S$ the sequence $\{H^n x_0\}_{n \in \mathbb{N}}$ converges to a fixed point of H.

Corollary 7. *If there exists $q \in [0,1)$ and $a \geq 0$ such that*

$$K(x,y) - aQ(x,y) \leq qS(x,y),$$

for all distinct $x, y \in S$ where $K(x,y), Q(x,y), S(x,y)$ are defined as in Theorem 17, then, for each $x_0 \in S$ the sequence $\{H^n x_0\}_{n \in \mathbb{N}}$ converges to a fixed point of H.

Corollary 8. *Suppose that there exists $k, p, r \in [0,1)$ with $k + p + r < 1$ and $a \geq 0$ such that*

$$K(x,y) - aQ(x,y) \leq k\sigma(x,y) + p\sigma(x, Hx) + r\sigma(x, Hx)$$

for $x, y \in S$ where $K(x,y), Q(x,y)$ are defined as in Theorem 17 Then, for each $x_0 \in S$ the sequence $\{H^n x_0\}_{n \in \mathbb{N}}$ converges to a fixed point of H.

4.3. Achari Type Non-Unique Fixed Point Results

Definition 16. *A self-mapping $H : S \to S$ is called ψ-Achari type simulated if there exists $\zeta \in \mathcal{Z}$ and $\psi \in \Psi$ such that*

$$\zeta\left(\frac{A(x,y) - B(x,y)}{C(x,y)}, \psi(\sigma(x,y))\right) \geq 0, \tag{70}$$

for all $x, y \in S$, where

$$A(x,y) = \min\{\sigma(Hx, Hy)\sigma(x,y), \sigma(x, Hx)\sigma(y, Hy)\},$$
$$B(x,y) = \min\{\sigma(x, Hx)\sigma(x, Hy), \sigma(y, Hy)\sigma(Hx, y)\},$$
$$C(x,y) = \min\{\sigma(x, Hx), \sigma(y, Hy)\},$$

with $C(x,y) \neq 0$.

Theorem 18. *If a mappings H is ψ-Achari type simulated, then for each $x_0 \in S$ the sequence $\{H^n x_0\}_{n \in \mathbb{N}}$ converges to a fixed point of H.*

Proof. By following line by line the proof of Theorem 12, we construct an iterative sequence $\{x_n = Hx_{n-1}\}_{n \in \mathbb{N}}$, starting from an arbitrary initial value $x_0 := x \in \mathcal{S}$. Regarding the discussion in the proof of Theorem 12, we know that the terms of the sequence $\{x_n\}$ are distinct, i.e.,

$$x_n \neq x_{n-1} \text{ for all } n \in \mathbb{N}.$$

Taking the inequality (79) into account, by letting $x = x_{n-1}$ and $y = x_n$ in, we attain that

$$\begin{aligned} 0 &\leq \zeta\left(\frac{A(x_{n-1}, x_n) - B(x_{n-1}, x_n)}{C(x_{n-1}, x_n)}, \psi(\sigma(x_{n-1}, x_n))\right) \\ &< \psi(\sigma(x_{n-1}, x_n)) - \frac{A(x_{n-1}, x_n) - B(x_{n-1}, x_n)}{C(x_{n-1}, x_n)}, \end{aligned} \tag{71}$$

which implies that

$$\frac{A(x_{n-1}, x_n) - B(x_{n-1}, x_n)}{C(x_{n-1}, x_n)} \leq \psi(\sigma(x_{n-1}, x_n)),$$

where

$$\begin{aligned} A(x_{n-1}, x_n) &= \min\{\sigma(Hx_{n-1}, Hx_n)\sigma(x_{n-1}, x_n), \sigma(x_{n-1}, Hx_{n-1})\sigma(x_n, Hx_n)\}, \\ B(x_{n-1}, x_n) &= \min\{\sigma(x_{n-1}, Hx_{n-1})\sigma(x_{n-1}, Hx_n), \sigma(x_n, Hx_n)\sigma(Hx_{n-1}, x_n)\}, \\ C(x_{n-1}, x_n) &= \min\{\sigma(x_{n-1}, Hx_{n-1}), \sigma(x_n, Hx_n)\}. \end{aligned}$$

On account of b-BDS, we simplify the above the inequality as

$$\frac{\sigma(x_n, x_{n+1})\sigma(x_{n-1}, x_n)}{\min\{\sigma(x_{n-1}, x_n), \sigma(x_n, x_{n+1})\}} \leq \psi(\sigma(x_{n-1}, x_n)). \tag{72}$$

Notice that for the case $\min\{\sigma(x_{n-1}, x_n), \sigma(x_n, x_{n+1})\} = \sigma(x_n, x_{n+1})$, the inequality (72) turns into

$$\sigma(x_{n-1}, x_n) \leq \psi(\sigma(x_{n-1}, x_n)) < \sigma(x_{n-1}, x_n),$$

a contraction (since $\psi(t) < t$ for all $t > 0$). Accordingly, we conclude that

$$\sigma(x_n, x_{n+1}) \leq \psi(\sigma(x_{n-1}, x_n)).$$

Recursively, we get

$$\sigma(x_n, x_{n+1}) \leq \psi(\sigma(x_{n-1}, x_n)) \leq \psi^2(\sigma(x_{n-2}, x_{n-1})) \leq \cdots \leq \psi^n(\sigma(x_0, x_1)). \tag{73}$$

Due to definition of comparison function, we have

$$\lim_{n \to \infty} \sigma(x_{n+1}, x_n) = 0.$$

Furthermore, one can easily show that the sequence $\{x_n\}$ has no periodic point, i.e.,

$$x_n \neq x_{n+k} \text{ for all } k \in \mathbb{N} \text{ and for all } n \in \mathbb{N}_0. \tag{74}$$

Indeed, if $x_n = x_{n+k}$ for some $n \in \mathbb{N}_0$ and $k \in \mathbb{N}$, we get $x_{n+1} = Hx_n = Hx_{n+k} = x_{n+k+1}$. On account of (73), we derive that

$$\sigma(x_n, x_{n+1}) = \sigma(x_{n+k}, x_{n+k+1}) \leq \psi^k(\sigma(x_n, x_{n+1})) < \sigma(x_n, x_{n+1}), \tag{75}$$

a contradiction. Accordingly, we suppose that

$$x_n \neq x_m \text{ for all distinct } n, m \in \mathbb{N}. \tag{76}$$

A verbatim repetition of the related lines in the proof of Theorem 12 completes the proof. □

On account of Example 1 (i), we conclude the following result from Theorem 18.

Theorem 19 ([8]). *Suppose that there exists $\psi \in \Psi$ such that*

$$\frac{A(x,y) - B(x,y)}{C(x,y)} \leq \psi(\sigma(x,y)), \tag{77}$$

for all $x, y \in S$, where

$$\begin{aligned} A(x,y) &= \min\{\sigma(Hx, Hy)\sigma(x,y), \sigma(x, Hx)\sigma(y, Hy)\}, \\ B(x,y) &= \min\{\sigma(x, Hx)\sigma(x, Hy), \sigma(y, Hy)\sigma(Hx, y)\}, \\ C(x,y) &= \min\{\sigma(x, Hx), \sigma(y, Hy)\}. \end{aligned}$$

with $C(x,y) \neq 0$. Then, for each $x_0 \in S$ the sequence $\{H^n x_0\}_{n \in \mathbb{N}}$ converges to a fixed point of H.

Corollary 9. *Suppose that there exists $\psi \in \Psi$ such that*

$$\frac{A(x,y) - B(x,y)}{C(x,y)} \leq \psi(\sigma(x,y)), \tag{78}$$

for all $x, y \in S$, where $A(x,y), B(x,y), C(x,y)$ are defined as in Theorem 19. Then, for each $x_0 \in S$ the sequence $\{H^n x_0\}_{n \in \mathbb{N}}$ converges to a fixed point of H.

The following is an immediate consequence of Theorem 19 by letting $\psi(t) = qt$, where $q \in [0,1)$.

Corollary 10. *Suppose that there exists $q \in [0,1)$ such that*

$$\frac{A(x,y) - B(x,y)}{C(x,y)} \leq q\sigma(x,y), \tag{79}$$

for all $x, y \in S$, where $A(x,y), B(x,y), C(x,y)$ are defined as in Theorem 19. Then, for each $x_0 \in S$ the sequence $\{H^n x_0\}_{n \in \mathbb{N}}$ converges to a fixed point of H.

4.4. Pachpatte Type Non-Unique Fixed Point Results

Definition 17. *A self-mapping $H : S \to S$ is called ψ-Pachpatte type simulated if there exists $\zeta \in \mathcal{Z}$ and $\psi \in \Psi$ such that*

$$\zeta(m(x,y) - n(x,y), \psi(\sigma(x,y))) \geq 0, \tag{80}$$

for all $x, y \in S$, where

$$\begin{aligned} m(x,y) &= \min\{[d(Tx, Ty)]^2, d(x,y)d(Tx, Ty), [d(y, Ty)]^2\}, \\ n(x,y) &= \min\{d(x, Tx)d(y, Ty), d(x, Ty)d(y, Tx)\} \end{aligned}$$

Theorem 20. *If a mappings H is ψ-Pachpatte type simulated, then for each $x_0 \in S$ the sequence $\{H^n x_0\}_{n \in \mathbb{N}}$ converges to a fixed point of H.*

Proof. Again by following line by line the proof of Theorem 12, we construct an iterative sequence $\{x_n = Hx_{n-1}\}_{n \in \mathbb{N}}$ whose terms are distinct from each other, by starting from an arbitrary initial value $x_0 := x \in S$.

Taking the inequality (87) into consideration by letting $x = x_{n-1}$ and $y = x_n$, we find that

$$0 \leq \zeta(m(x_{n-1}, x_n) - n(x_{n-1}, x_n), \psi(\sigma(x_{n-1}, Hx_{n-1})\sigma(x_n, Hx_n)))$$
$$< \psi(\sigma(x_{n-1}, Hx_{n-1})\sigma(x_n, Hx_n)) - m(x_{n-1}, x_n) - n(x_{n-1}, x_n),$$

which yields that

$$m(x_{n-1}, x_n) - n(x_{n-1}, x_n) \leq \psi(\sigma(x_{n-1}, Hx_{n-1})\sigma(x_n, Hx_n)), \quad (81)$$

where

$$m(x_{n-1}, x_n) = \min\{[\sigma(Hx_{n-1}, Hx_n)]^2, \sigma(x_{n-1}, x_n)\sigma(Hx_{n-1}, Hx_n), [\sigma(x_n, Hx_n)]^2\},$$
$$n(x_{n-1}, x_n) = \min\{\sigma(x_{n-1}, Hx_{n-1})\sigma(x_n, Hx_n), \sigma(x_{n-1}, Hx_n)\sigma(x_n, Hx_{n-1})\}.$$

By simplifying the inequality above inequality, we find that

$$m(x_{n-1}, x_n) \leq \psi(\sigma(x_{n-1}, x_n)\sigma(x_n, x_{n+1})), \quad (82)$$

where

$$m(x_{n-1}, x_n) = \min\{[\sigma(x_n, x_{n+1})]^2, \sigma(x_{n-1}, x_n)\sigma(x_n, x_{n+1})\}.$$

It is clear that the case

$$m(x_{n-1}, x_n) = \sigma(x_{n-1}, x_n)\sigma(x_n, x_{n+1})$$

is not possible. If it would be the case, the inequality (83) turns into

$$\sigma(x_{n-1}, x_n)\sigma(x_n, x_{n+1}) \leq \psi(\sigma(x_{n-1}, x_n)\sigma(x_n, x_{n+1})) < \sigma(x_{n-1}, x_n)\sigma(x_n, x_{n+1}), \quad (83)$$

a contraction (since $\psi(t) < t$ for all $t > 0$). Consequently, we derive

$$[\sigma(x_n, x_{n+1})]^2 \leq \psi(\sigma(x_{n-1}, x_n)\sigma(x_n, x_{n+1})) < \sigma(x_{n-1}, x_n)\sigma(x_n, x_{n+1}), \quad (84)$$

which yields

$$\sigma(x_n, x_{n+1}) < \sigma(x_{n-1}, x_n). \quad (85)$$

Regarding the fact that ψ is nondecreasing, and combining the inequalities (84) and (85), we obtain that

$$[\sigma(x_n, x_{n+1})]^2 \leq \psi(\sigma(x_{n-1}, x_n)\sigma(x_n, x_{n+1})) < \psi([\sigma(x_{n-1}, x_n)]^2), \quad (86)$$

Iteratively, we get that

$$[\sigma(x_n, x_{n+1})]^2 \leq \psi([\sigma(x_{n-1}, x_n)]^2) \leq \psi^2([\sigma(x_{n-2}, x_{n-1})]^2) \leq \cdots \leq \psi^n([\sigma(x_0, x_1)]^2).$$

Hence, we have

$$\lim_{n \to \infty} [\sigma(x_{n+1}, x_n)]^2 = 0 \iff \lim_{n \to \infty} \sigma(x_{n+1}, x_n) = 0.$$

The rest of the proof is a verbatim repetition of the related lines in the proof of Theorem 12. □

Due to Example 1 (*i*), Theorem 22 yields the next result.

Theorem 21 ([8]). *Suppose that there exists $\psi \in \Psi$ such that*

$$m(x, y) - n(x, y) \leq \psi(\sigma(x, Hx)\sigma(y, Hy)), \quad (87)$$

for all $x, y \in \mathcal{S}$, where

$$m(x,y) = \min\{[\sigma(Hx, Hy)]^2, \sigma(x,y)\sigma(Hx, Hy), [\sigma(y, Hy)]^2\},$$
$$n(x,y) = \min\{\sigma(x, Hx)\sigma(y, Hy), \sigma(x, Hy)\sigma(y, Hx)\}.$$

Then, for each $x_0 \in \mathcal{S}$ the sequence $\{H^n x_0\}_{n \in \mathbb{N}}$ converges to a fixed point of H.

If we take $\psi(t) = qt$, then Theorem 21 implies the following result.

Corollary 11. *If there exists $q \in [0, 1)$ such that*

$$m(x,y) - n(x,y) \leq q\sigma(x, Hx)\sigma(y, Hy), \tag{88}$$

for all $x, y \in \mathcal{S}$, where $m(x,y)$ and $n(x,y)$ are defined as in Theorem 21, then, for each $x_0 \in \mathcal{S}$ the sequence $\{H^n x_0\}_{n \in \mathbb{N}}$ converges to a fixed point of H.

4.5. Karapınar Type Non-Unique Fixed Point Results

Definition 18. *A self-mapping $H : \mathcal{S} \to \mathcal{S}$ is called ψ-Karapınar type simulated if there exist $\zeta \in \mathcal{Z}$ and $\psi \in \Psi$ such that*

$$0 \leq \frac{a_4 - a_2}{a_1 + a_2} < 1, \ a_1 + a_2 \neq 0, \ a_1 + a_2 + a_3 > 0 \text{ and } 0 \leq a_3 - a_5 \tag{89}$$

$$\zeta(L(x,y), R(x,y)) \tag{90}$$

for all $x, y \in \mathcal{S}$, where

$$L(x,y) := a_1 \sigma(Hx, Hy) + a_2[\sigma(x, Hx) + \sigma(y, Hy)] + a_3[\sigma(y, Hx) + \sigma(x, Hy)],$$
$$R(x,y) := a_4 \sigma(x,y) + a_5 \sigma(x, F^2 x).$$

Theorem 22. *If a mappings H is ψ-Karapınar type simulated, then for each $x_0 \in \mathcal{S}$ the sequence $\{H^n x_0\}_{n \in \mathbb{N}}$ converges to a fixed point of H.*

Proof. For an arbitrary $x_0 \in \mathcal{S}$, we shall built a construct a sequence $\{x_n\}$ as follows:

$$x_{n+1} := Hx_n \quad n = 0, 1, 2, \dots \tag{91}$$

Utilizing the inequality by taking $x = x_n$ and $y = x_{n+1}$ we find that

$$0 \leq \zeta(L(x,y), R(x,y)) < R(x,y) - L(x,y),$$

which infer to

$$a_1 \sigma(Hx_n, Hx_{n+1}) + a_2[\sigma(x_n, Hx_n) + \sigma(x_{n+1}, Hx_{n+1})] + a_3[\sigma(x_{n+1}, Hx_n) + \sigma(x_n, Hx_{n+1})]$$
$$\leq a_4 \sigma(x_n, x_{n+1}) + a_5 \sigma(x_n, F^2 x_n) \tag{92}$$

for all a_1, a_2, a_3, a_4, a_5 which fulfils (89). On account of (91), the statement (92) becomes

$$a_1 \sigma(x_{n+1}, x_{n+2}) + a_2[\sigma(x_n, x_{n+1}) + \sigma(x_{n+1}, x_{n+2})] + a_3[\sigma(x_{n+1}, x_{n+1}) + \sigma(x_n, x_{n+2})]$$
$$\leq a_4 \sigma(x_n, x_{n+1}) + a_5 \sigma(x_n, x_{n+2}). \tag{93}$$

By a simple computation, we derive

$$(a_1 + a_2)\sigma(x_{n+1}, x_{n+2}) + (a_3 - a_5)\sigma(x_n, x_{n+2}) \leq (a_4 - a_2)\sigma(x_n, x_{n+1}). \tag{94}$$

So, the inequality above yields that

$$\sigma(x_{n+1}, x_{n+2}) \leq q\sigma(x_n, x_{n+1}) \tag{95}$$

where $q = \frac{a_4 - a_2}{a_1 + a_2}$. Due to (89), we have $0 \leq q < 1$. Regarding (95), we recursively obtain

$$\sigma(x_n, x_{n+1}) \leq q\sigma(x_{n-1}, x_n) \leq q^2 \sigma(x_{n-2}, x_{n-1}) \leq \cdots \leq q^n \sigma(x_0, x_1). \tag{96}$$

Thus, the sequence $\{\sigma(x_n, x_{n+1})\}$ is non-increasing.

On what follows that we shall prove that the sequence $\{x_n\}$ has no periodic point, i.e.,

$$x_n \neq x_{n+k} \text{ for all } k \in \mathbb{N} \text{ and for all } n \in \mathbb{N}_0. \tag{97}$$

Actually, if $x_n = x_{n+k}$ for some $n \in \mathbb{N}_0$ and $k \in \mathbb{N}$, we find $x_{n+1} = Hx_n = Hx_{n+k} = x_{n+k+1}$. Keeping the inequality (95) in the mind, we derive that

$$\sigma(x_n, x_{n+1}) = \sigma(x_{n+k}, x_{n+k+1}) \leq q^k \sigma(x_n, x_{n+1}), \tag{98}$$

which is a contradiction. Consequently, we suppose that

$$x_n \neq x_m \text{ for all distinct } n, m \in \mathbb{N}. \tag{99}$$

One can easily discover that $x_{n+k} \neq x_{m+k}$ for all distinct $n, m \in \mathbb{N}$ and $x_{n+k}, x_{m+k} \in \mathcal{S} \setminus \{x_n, x_m\}$. There exists a natural number M such that

$$0 < q^k s < 1 \text{ for all } k \geq M,$$

since $k \in [0, 1)$ and hence $\lim_{n \to \infty} k^n = 0$.

As a next step, we shall indicate that $\{x_n\}$ is a Cauchy sequence. By regarding the modified quadrilateral inequality, we find

$$\sigma(x_m, x_n) \leq s[\sigma(x_m, x_{m+k}) + \sigma(x_{m+k}, x_{n+k}) + \sigma(x_{n+k}, x_n)] \\ \leq s\, q^m \sigma(x_0, x_k) + sq^k \sigma(x_m, x_n) + sq^n \sigma(x_k, x_0) \tag{100}$$

By rearranging the term in the inequality above, we attain that

$$\sigma(x_m, x_n) \leq \frac{s(q^m + q^n)}{1 - q^k s} \sigma(x_k, x_0) \tag{101}$$

Consequently, we derive that $\{x_n\}_{n \in \mathbb{N}}$ is a Cauchy sequence.

The rest of the proof is deduced by following the corresponding lines in the proof of Theorem 12. □

We deduce the following results, by employing Example 1 (i) on Theorem 22.

Theorem 23 ([8]). *Let H be an orbitally continuous self-map on the H-orbitally complete b-Branciari distance space $(\mathcal{S}, \sigma)$. Suppose there exist real numbers a_1, a_2, a_3, a_4, a_5 and a self mapping $H : \mathcal{S} \to \mathcal{S}$ satisfies the conditions*

$$0 \leq \frac{a_4 - a_2}{a_1 + a_2} < 1, \quad a_1 + a_2 \neq 0, \quad a_1 + a_2 + a_3 > 0 \text{ and } 0 \leq a_3 - a_5 \tag{102}$$

$$L(x, y) \leq R(x, y) \tag{103}$$

for all $x, y \in \mathcal{S}$, where

$$L(x,y) := a_1\sigma(Hx, Hy) + a_2[\sigma(x, Hx) + \sigma(y, Hy)] + a_3[\sigma(y, Hx) + \sigma(x, Hy)],$$
$$R(x,y) := a_4\sigma(x,y) + a_5\sigma(x, F^2x).$$

Then, H has at least one fixed point.

It is clear that all results in these section can be stated in the context of Branciari distance space by letting $s = 1$. For avoiding the repetition, we skip to list these immediate consequences of Chapter 4. In addition, one can also get several more consequences by modifying the contraction inequality.

5. Conclusions

One of the most attractive research topic of nonlinear functional analysis is metric fixed point theory [1–129]. In this paper, we aim to underline the importance of the existence of a fixed point rather than uniqueness. Such non-unique fixed point theorems can be more applicable not only in nonlinear analysis, but also, in several qualitative sciences. It seems that the analog of the presented results can be derived in some other abstract spaces, such as in the setting of modular metric spaces.

Funding: This research received no external funding.

Conflicts of Interest: The author declares no conflict of interest.

References

1. Ćirić, L.B. On some maps with a non-unique fixed point. *Publ. Inst. Math.* **1974**, *17*, 52–58.
2. Achari, J. On Ćirić's non-unique fixed points. *Mat. Vesnik.* **1976**, *13*, 255–257.
3. Alqahtani, B.; Fulga, A.; Karapınar, E. Non-Unique Fixed Point Results in Extended b-Metric Space. *Mathematics* **2018**, *6*, 68. [CrossRef]
4. Aydi, H.; Karapınar, E.; Rakočević, V. Nonunique Fixed Point Theorems on b-Metric Spaces via Simulation Functions. *Jordan J. Math. Stat.* **2019**, in press.
5. Alsulami, H.H.; Karapınar, E.; Rakočević, V. Ciric Type Nonunique Fixed Point Theorems on b-Metric Spaces. *Filomat* **2017**, *31*, 3147–3156. [CrossRef]
6. Karapınar, E.; Romaguera, S. Nonunique fixed point theorems in partial metric spaces. *Filomat* **2013**, *27*, 1305–1314. [CrossRef]
7. Gupta, S.; Ram, B. Non-unique fixed point theorems of Ćirić type, (Hindi). *Vijnana Parishad Anusandhan Patrika* **1998**, *41*, 217–231.
8. Karapınar, E.; Agarwal, R.P. A note on Ćirić type non-unique fixed point theorems. *Fixed Point Theory Appl.* **2017**, *2017*, 20. [CrossRef]
9. Liu, Z.Q. On Ćirić type mappings with a non-unique coincidence points. *Mathematica (Cluj)* **1993**, *35*, 221–225.
10. Liu, Z.Q.; Guo, Z.; Kang, S.M.; Lee, S.K. On Ćirić type mappings with non-unique fixed and periodic points. *Int. J. Pure Appl. Math.* **2006**, *26*, 399–408.
11. Pachpatte, B.G. On Ćirić type maps with a non-unique fixed point. *Indian J. Pure Appl. Math.* **1979**, *10*, 1039–1043.
12. Zhang, F.; Kang, S.M.; Xie, L. Ćirić type mappings with a non-unique coincidence points. *Fixed Point Theory Appl.* **2007**, *6*, 187–190.
13. Karapınar, E. Ćirić types non-unique fixed point results: A Review. *Appl. Comput. Math.* **2019**, *1*, 3–21.
14. Ćirić, L.B.; Jotić, N. A further extension of maps with non-unique fixed points. *Mat. Vesnik.* **1998**, *50*, 1–4.
15. Karapınar, E. A new non-unique fixed point theorem. *J. Appl. Funct. Anal.* **2012**, *7*, 92–97.
16. Browder, F.E. On the convergence of successive approximations for nonlinear functional equations. *Nederl. Akad. Wetensch. Ser. A71 Indag. Math.* **1968**, *30*, 27–35. [CrossRef]
17. Rus, I.A. *Generalized Contractions and Applications*; Cluj University Press: Cluj-Napoca, Romania, 2001.
18. Khojasteh, F.; Shukla, S.; Radenović, S. A new approach to the study of fixed point theorems via simulation functions. *Filomat* **2015**, *29*, 1189–1194. [CrossRef]

19. Argoubi, H.; Samet, B.; Vetro, C. Nonlinear contractions involving simulation functions in a metric space with a partial order. *J. Nonlinear Sci. Appl.* **2015**, *8*, 1082–1094. [CrossRef]
20. Alsulami, H.H.; Karapınar, E.; Khojasteh, F.; Roldán-López-de-Hierro, A.F. A proposal to the study of contractions in quasi-metric spaces. *Discret. Dyn. Nat. Soc.* **2014**, *2014*, 269286. [CrossRef]
21. Roldán-López-de-Hierro, A.F.; Karapınar, E.; Roldán-López-de-Hierro, C.; Martínez-Moreno, J. Coincidence point theorems on metric spaces via simulation functions. *J. Comput. Appl. Math.* **2015**, *275*, 345–355. [CrossRef]
22. Matthews, S.G. *Partial Metric Topology*; Research Report 212; Department of Computer Science, University of Warwick: Coventry, UK, 1992.
23. Matthews, S.G. Partial metric topology, Proc. 8th Summer Conference on General Topology and Applications. *Ann. N. Y. Acad. Sci.* **1994**, *728*, 183–197. [CrossRef]
24. Karapınar, E.; Shobkolaei, N.; Sedghi, S.; Vaezpour, S.M. A common fixed point theorem for cyclic operators on partial metric spaces. *Filomat* **2012**, *26*, 407–414. [CrossRef]
25. Shobkolaei, N.; Vaezpour, S.M.; Sedghi, S. A common fixed point theorem on ordered partial metric spaces. *J. Basic Appl. Sci. Res.* **2011**, *1*, 3433–3439.
26. Hitzler, P.; Seda, A. *Mathematical Aspects of Logic Programming Semantics, Studies in Informatics Series*; CRC Press: Boca Raton, FL, USA, 2011.
27. Karapınar, E.; Erhan, I.M. Fixed point theorems for operators on partial metric spaces. *Appl. Math. Lett.* **2011**, *24*, 1900–1904. [CrossRef]
28. Karapınar, E. Ćirić types non-unique fixed point theorems on partial metric spaces. *J. Nonlinear Sci. Appl.* **2012**, *5*, 74–83. [CrossRef]
29. Branciari, A. A fixed point theorem of Banach-Caccioppoli type on a class of generalized metric spaces. *Publ. Math. Debrecen* **2000**, *57*, 31–37.
30. Branciari, A. A fixed point theorem for mappings satisfying a general contractive condition of integral type. *Int. J. Math. Math. Sci.* **2002**, *29*, 531–536. [CrossRef]
31. Aydi, H.; Karapınar, E.; Samet, B. Fixed points for generalized (α, ψ)-contractions on generalized metric spaces. *J. Inequal. Appl.* **2014**, *2014*, 229. [CrossRef]
32. Aydi, H.; Karapınar, E.; Lakzian, H. Fixed point results on the class of generalized metric spaces. *Math. Sci.* **2012**, *6*, 46. [CrossRef]
33. Azam, A.; Arshad, M. Kannan fixed point theorems on generalized metric spaces. *J. Nonlinear Sci. Appl.* **2008**, *1*, 45–48. [CrossRef]
34. Bilgili, N.; Karapınar, E. A note on "common fixed points for (ψ, α, β)-weakly contractive mappings in generalized metric spaces". *Fixed Point Theory Appl.* **2013**, *2013*, 287. [CrossRef]
35. Das, P.; Lahiri, B.K. Fixed point of a Ljubomir Ćirić's quasi-contraction mapping in a generalized metric space. *Publ. Math. Debrecen* **2002**, *61*, 589–594.
36. Jleli, M.; Samet, B. The Kannan's fixed point theorem in a cone rectangular metric space. *J. Nonlinear Sci. Appl.* **2009**, *2*, 161–167. [CrossRef]
37. Kadeburg, Z.; Radenoviç, S. On generalized metric spaces: A survey. *TWMS J. Pure Appl. Math.* **2014**, *5*, 3–13.
38. Karapınar, E. Discussion on (α, ψ) contractions on generalized metric spaces. *Abstr. Appl. Anal.* **2014**, *2014*, 962784. [CrossRef]
39. Karapınar, E. Fixed points results for α-admissible mapping of integral type on generalized metric spaces. *Abstr. Appl. Anal.* **2014**, *2014*, 141409. [CrossRef]
40. Karapınar, E. On (α, ψ) contractions of integral type on generalized metric spaces. In Proceedings of the 9th ISAAC Congress, Krakow, Poland, 5–9 August 2013; Mityushevand, V., Ruzhansky, M., Eds.; Springer: Krakow, Poland, 2013.
41. Kikina, L.; Kikina, K. A fixed point theorem in generalized metric space. *Demonstr. Math.* **2013**, *XLVI*, 181–190. [CrossRef]
42. Mihet, D. On Kannan fixed point principle in generalized metric spaces. *J. Nonlinear Sci. Appl.* **2009**, *2*, 92–96. [CrossRef]
43. Samet, B. Discussion on: A fixed point theorem of Banach-Caccioppoli type on a class of generalized metric spaces by A. Branciari. *Publ. Math. Debrecen* **2010**, *76*, 493–494.

44. Suzuki, T. Generalized metric space do not have the compatible topology. *Abstr. Appl. Anal.* **2014**, *2014*, 458098. [CrossRef]
45. Sarma, I.R.; Rao, J.M.; Rao, S.S. Contractions over generalized metric spaces. *J. Nonlinear Sci. Appl.* **2009**, *2*, 180–182. [CrossRef]
46. Czerwik, S. Contraction mappings in b-metric spaces. *Acta Math. Inf. Univ. Ostrav.* **1993**, *1*, 5–11.
47. George, R.; Radenovic, S.; Reshma, K.P.; Shukla, S. Rectangular b-metric space and contraction principles. *J. Nonlinear Sci. Appl.* **2015**, *8*, 1005–1013. [CrossRef]
48. Almezel, S.; Chen, C.M.; Karapınar, E.; Rakočević, V. Fixed point results for various α-admissible contractive mappings on metric-like spaces. *Abstr. Appl. Anal.* **2014**, *2014*, 379358.
49. Liouville, J. Second mémoire sur le développement des fonctions ou parties de fonctions en séries dont divers termes sont assujettis á satisfaire a une m eme équation différentielle du second ordre contenant un paramétre variable. *J. Math. Pure Appl.* **1837**, *2*, 16–35.
50. Picard, E. Memoire sur la theorie des equations aux derivees partielles et la methode des approximations successives. *J. Math. Pures Appl.* **1890**, *6*, 145–210.
51. Banach, S. Sur les opérations dans les ensembles abstraits et leur application aux équations intégrales. *Fund. Math.* **1922**, *3*, 133–181. [CrossRef]
52. Brouwer, L.E.J. Uber Abbildung von Mannigfaltigkeiten. *Math Ann.* **1912**, *71*, 97–115. [CrossRef]
53. Schauder, J. Der Fixpunktsatz in Funktionalraumen. *Stud. Math.* **1930**, *2*, 171–180. [CrossRef]
54. Poincaré, H. Surless courbes define barles equations differentiate less. *J. Math.* **1886**, *2*, 54–65.
55. Bohl, P. über die Bewegung eines mechanischen Systems in der Nähe einer Gleichgewichtslage. *J. Reine Angew. Math.* **1904**, *127*, 179–276.
56. Hadamard, J. *Note sur Quelques Applications de L'indice de Kronecker in Jules Tannery: Introduction á la Théorie des Fonctions D'une Variable*, 2nd ed.; A. Hermann & Fils: Paris, France, 1910; Volume 2, pp. 437–477.
57. Tarski, A. A lattice theoretical fixpoint theorem and its applications. *Pac. J. Math.* **1955**, *5*, 285–309. [CrossRef]
58. Abedelljawad, T.; Karapınar, E.; Taş, K. Existence and uniqueness of common fixed point on partial metric spaces. *Appl. Math. Lett.* **2011**, *24*, 1894–1899. [CrossRef]
59. Abedelljawad, T.; Karapınar, E.; Taş, K. A generalized contraction principle with control functions on partial metric spaces. *Comput. Math. Appl.* **2012**, *63*, 716–719. [CrossRef]
60. Afshari, H.; Aydi, H.; Karapınar, E. Existence of Fixed Points of Set-Valued Mappings in b-Metric Spaces. *East Asian Math. J.* **2016**, *32*, 319–332. [CrossRef]
61. Aksoy, U.; Karapınar, E.; Erhan, I.M. Fixed points of generalized α-admissible contractions on b-metric spaces with an application to boundary value problems. *J. Nonlinear Convex A* **2016**, *17*, 1095–1108.
62. Alharbi, A.S.; Alsulami, H.H.; Karapınar, E. On the Power of Simulation and Admissible Functions in Metric Fixed Point Theory. *J. Funct. Spaces* **2017**, *2017*, 2068163. [CrossRef]
63. Ali, M.U.; Kamram, T.; Karapınar, E. An approach to existence of fixed points of generalized contractive multivalued mappings of integral type via admissible mapping. *Abstr. Appl. Anal.* **2014**, *2014*, 141489. [CrossRef]
64. Ali, M.U.; Kamran, T.; Nar, E.K. On (α, ψ, η)-contractive multivalued mappings. *Fixed Point Theory Appl.* **2014**, *2014*, 7. [CrossRef]
65. Alsulami, H.; Gulyaz, S.; Karapınar, E.; Erhan, I.M. Fixed point theorems for a class of α-admissible contractions and applications to boundary value problem. *Abstr. Appl. Anal.* **2014**, *2014*, 187031. [CrossRef]
66. Arshad, M.; Ameer, E.; Karapınar, E. Generalized contractions with triangular α-orbital admissible mapping on Branciari metric spaces. *J. Inequal. Appl.* **2016**, *2016*, 63. [CrossRef]
67. Aydi, H.; Karapınar, E.; Yazidi, H. Modified F-Contractions via α-Admissible Mappings and Application to Integral Equations. *Filomat* **2017**, *31*, 1141–1148. [CrossRef]
68. Aydi, H.; Karapınar, E.; Zhang, D. A note on generalized admissible-Meir-Keeler-contractions in the context of generalized metric spaces. *Results Math.* **2017**, *71*, 73–92. [CrossRef]
69. Aydi, H.; Jellali, M.; Karapınar, E. On fixed point results for α-implicit contractions in quasi-metric spaces and consequences. *Nonlinear Anal. Model. Control.* **2016**, *21*, 40–56. [CrossRef]
70. Aydi, H.; Karapınar, E.; Shatanawi, W. Coupled fixed point results for (ψ, φ)-weakly contractive condition in ordered partial metric spaces. *Comput. Math. Appl.* **2011**, *62*, 4449–4460. [CrossRef]
71. Aydi, H.; Karapınar, M.B.E.; Mitrović, S. A fixed point theorem for set-valued quasi-contractions in b-metric spaces. *Fixed Point Theory Appl.* **2012**, *2012*, 88. [CrossRef]

72. Aydi, H.; Karapınar, M.B.E.; Moradi, S. A common fixed point for weak ϕ-contractions in b-metric spaces. *Fixed Point Theory* **2012**, *13*, 337–346.
73. Bakhtin, I.A. The contraction mapping principle in quasimetric spaces. *Funct. Anal.* **1989**, *30*, 26–37.
74. Berinde, V. Generalized contractions in quasi-metric spaces, Seminar on Fixed Point Theory, Babeş-Bolyai University. *Res. Sem.* **1993**, *3*, 3–9.
75. Berinde, V. Sequences of operators and fixed points in quasimetric spaces. *Mathematica* **1996**, *41*, 23–27.
76. Berinde, V. *Contracţii Generalizate şi Aplicaţii*; Editura Cub Press: Baie Mare, Romania, 1997; Volume 2.
77. Boriceanu, M. Strict fixed point theorems for multivalued operators in b-metric spaces. *Int. J. Mod. Math.* **2009**, *4*, 285–301.
78. Boriceanu, M. Fixed point theory for multivalued generalized contraction on a set with two b-metrics. *Mathematica* **2009**, *54*, 3–14.
79. Boriceanu, M.; Sel, A.P.; Rus, I.A. Fixed point theorems for some multivalued generalized contractions in b-metric spaces. *Int. J. Math. Stat.* **2010**, *6*, 65–76.
80. Bota, M. *Dynamical Aspects in the Theory of Multivalued Operators*; Cluj University Press: Cluj-Napoka, Romania, 2010.
81. Bota, M.; Molnár, A.; Varga, C. On Ekeland's variational principle in b-metric spaces. *Fixed Point Theory* **2011**, *12*, 21–28.
82. Bota, M.; Karapınar, E. A note on "Some results on multi-valued weakly Jungck mappings in b-metric space". *Cent. Eur. J. Math.* **2013**, *11*, 1711–1712. [CrossRef]
83. Karapınar, M.B.E.; Te, O.M.S. Ulam-Hyers stability for fixed point problems via $\alpha - \phi$-contractive mapping in b-metric spaces. *Abstr. Appl. Anal.* **2013**, *2013*, 855293.
84. Bota, M.; Chifu, C.; Karapınar, E. Fixed point theorems for generalized (alpha-psi)-Ciric-type contractive multivalued operators in b-metric spaces. *J. Nonlinear Sci. Appl.* **2016**, *9*, 1165–1177. [CrossRef]
85. Bourbaki, N. *Topologie Générale*; Herman: Paris, France, 1974.
86. Caristi, J. Fixed point theorems for mapping satisfying inwardness conditions. *Trans. Am. Math. Soc.* **1976**, *215*, 241–251. [CrossRef]
87. Chen, C.M.; Abkar, A.; Ghods, S.; Karapınar, E. Fixed Point Theory for the α-Admissible Meir-Keeler Type Set Contractions Having KKM* Property on Almost Convex Sets. *Appl. Math. Inf. Sci.* **2017**, *11*, 171–176. [CrossRef]
88. Ding, H.S.; Li, L. Coupled fixed point theorems in partially ordered cone metric spaces. *Filomat* **2011**, *25*, 137–149. [CrossRef]
89. Gulyaz, S.; Karapınar, E.; Erhan, I.M. Generalized α-Meir-Keeler Contraction Mappings on Branciari b-metric Spaces. *Filomat* **2017**, *31*, 5445–5456. [CrossRef]
90. Gulyaz, S.; Karapınar, E. Coupled fixed point result in partially ordered partial metric spaces through implicit function. *Hacet. J. Math. Stat.* **2013**, *42*, 347–357.
91. Gulyaz, S.; Karapınar, E.; Rakocevic, V.; Salimi, P. Existence of a solution of integral equations via fixed point theorem. *J. Inequal. Appl.* **2013**, *2013*, 529. [CrossRef]
92. Gulyaz, S.; Karapınar, E.; Yuce, I.S. A coupled coincidence point theorem in partially ordered metric spaces with an implicit relation. *Fixed Point Theory Appl.* **2013**, *2013*, 38. [CrossRef]
93. Hadžić, O.; Pap, E. A fixed point theorem for multivalued mappings in probabilistic metric spaces and an application in fuzzy metric spaces. *Fuzzy Sets Syst.* **2002**, *127*, 333–344. [CrossRef]
94. Hammache, K.; Karapınar, E.; Ould-Hammouda, A. On Admissible weak contractions in b-metric-like space. *J. Math. Anal.* **2017**, *8*, 167–180.
95. Hicks, T.L. Fixed point theory in probabilistic metric spaces. *Univ. Novom Sadu Zb. Rad. Prirod.-Mat. Fak. Ser. Mat.* **1983**, *13*, 63–72.
96. Hicks, T.L. Fixed point theorems for quasi-metric spaces. *Math. Japonica* **1988**, *33*, 231–236.
97. Ilić, D.; Pavlović, V.; Rakocecić, V. Some new extensions of Banach's contraction principle to partial metric space. *Appl. Math. Lett.* **2011**, *24*, 1326–1330. [CrossRef]
98. Janković, S.; Kadelburg, Z.; Radenović, S. On cone metric spaces: A survey. *Nonlinear Anal.* **2011**, *74*, 2591–2601. [CrossRef]
99. Jleli, M.; Karapınar, E.; Samet, B. Best proximity points for generalized $\alpha - \psi$-proximal contractive type mappings. *J. Appl. Math.* **2013**, *2013*, 534127. [CrossRef]

100. Jleli, M.; Karapınar, E.; Samet, B. Fixed point results for $\alpha - \psi_\lambda$ contractions on gauge spaces and applications. *Abstr. Appl. Anal.* **2013**, *2013*, 730825. [CrossRef]
101. Karapınar, E.; Piri, H.; AlSulami, H. Fixed Points of Generalized F-Suzuki Type Contraction in Complete b-Metric Spaces. *Discret. Dyn. Nat. Soc.* **2015**, *2015*, 969726.
102. Karapınar, E.; Samet, B. Generalized α-ψ-contractive type mappings and related fixed point theorems with applications. *Abstr. Appl. Anal.* **2012**, *2012*, 793486. [CrossRef]
103. Karapınar, E. Fixed point theorems in cone Banach spaces. *Fixed Point Theory Appl.* **2009**, *2009*, 609281. [CrossRef]
104. Karapınar, E. A note on common fixed point theorems in partial metric spaces. *Miskolc Math. Notes* **2011**, *12*, 185–191. [CrossRef]
105. Karapınar, E.; Yuksel, U. Some common fixed point theorems in partial metric spaces. *J. Appl. Math.* **2011**, *2011*, 263621. [CrossRef]
106. Karapınar, E. Some fixed point theorems on the class of comparable partial metric spaces on comparable partial metric spaces. *Appl. Gen. Topol.* **2011**, *12*, 187–192.
107. Karapınar, E. Weak ϕ-contraction on partial metric spaces. *J. Comput. Anal. Appl.* **2012**, *14*, 206–210.
108. Karapınar, E.; Erhan, I.M. Cyclic Contractions and Fixed Point Theorems. *Filomat* **2012**, *26*, 777–782. [CrossRef]
109. Karapınar, E. Some non-unique fixed point theorems of Ćirić type on cone metric spaces. *Abstr. Appl. Anal.* **2010**, *2010*, 123094. [CrossRef]
110. Karapınar, E.; Kumam, P.; Salimi, P. On $\alpha - \psi$-Meir-Keeler contractive mappings. *Fixed Point Theory Appl.* **2013**, *2013*, 94. [CrossRef]
111. Karapinar, E.K.; Czerwik, S.; Aydi, H. (α, ψ)-Meir-Keeler contraction mappings in generalized b-metric spaces. *J. Funct. Spaces* **2018**, *2018*, 3264620. [CrossRef]
112. Kopperman, R.; Matthews, S.G.; Pajoohesh, H. *What Do Partial Metrics Represent?, Spatial Representation: Discrete vs. Continuous Computational Models, Dagstuhl Seminar Proceedings*; No. 04351; Internationales Begegnungs- und Forschungszentrum für Informatik (IBFI): Schloss Dagstuhl, Germany, 2005.
113. Kutbi, M.A.; Karapınar, E.; Ahmed, J.; Azam, A. Some fixed point results for multi-valued mappings in b-metric spaces. *J. Inequal. Appl.* **2014**, *2014*, 126. [CrossRef]
114. Künzi, H.P.A.; Pajoohesh, H.; Schellekens, M.P. Partial quasi-metrics. *Theoret. Comput. Sci.* **2006**, *365*, 237–246. [CrossRef]
115. Oltra, S.; Valero, O. Banach's fixed point theorem for partial metric spaces. *Rend. Ist. Mat. Univ. Trieste* **2004**, *36*, 17–26.
116. O'Neill, S.J. *Two Topologies Are Better Than One*; Tech. Report; University of Warwick: Coventry, UK, 1995.
117. Popescu, O. Some new fixed point theorems for α−Geraghty-contraction type maps in metric spaces. *Fixed Point Theory Appl.* **2014**, *2014*, 190. [CrossRef]
118. Romaguera, S. Fixed point theorems for generalized contractions on partial metric spaces. *Topol. Appl.* **2012**, *159*, 194–199. [CrossRef]
119. Romaguera, S. Matkowski's type theorems for generalized contractions on (ordered) partial metric spaces. *Appl. Gen. Topol.* **2011**, *12*, 213–220. [CrossRef]
120. Romaguera, S.; Schellekens, M. Partial metric monoids and semivaluation spaces. *Topol. Appl.* **2005**, *153*, 948–962. [CrossRef]
121. Romaguera, S.; Valero, O. A quantitative computational model for complete partial metric spaces via formal balls. *Math. Struct. Comput. Sci.* **2009**, *19*, 541–563. [CrossRef]
122. Samet, B. A fixed point theorem in a generalized metric space for mappings satisfying a contractive condition of integral type. *Int. J. Math. Anal.* **2009**, *26*, 1265–1271.
123. Suzuki, T. Some results on recent generalization of Banach contraction principle. In Proceedings of the 8th International Conference of Fixed Point Theory and its Applications, Chiang Mai, Thailand, 16–22 July 2007; pp. 751–761.
124. Suzuki, T. A generalized Banach contraction principle that characterizes metric completeness. *Proc. Am. Math. Soc.* **2008**, *163*, 1861–1869. [CrossRef]
125. Suzuki, T. Fixed point theorems and convergence theorems for some generalized nonexpansive mappings. *J. Math. Anal. Appl.* **2008**, *340*, 1088–1095. [CrossRef]

126. Suzuki, T. A new type of fixed point theorem on metric spaces. *Nonlinear Anal.* **2009**, *71*, 5313–5317. [CrossRef]
127. Turinici, M. *Topics in Mathematical Analysis and Applications*; Themistocles, M.R., László, T., Eds.; Springer: Berlin/Heidelberg, Germany, 2014; Volume 94, p. 715746
128. Valero, O. On Banach fixed point theorems for partial metric spaces. *Appl. Gen. Topol.* **2005**, *6*, 229–240. [CrossRef]
129. Wilson, W.A. On semimetric spaces. *Am. J. Math.* **1931**, *53*, 361–373. [CrossRef]

© 2019 by the authors. Licensee MDPI, Basel, Switzerland. This article is an open access article distributed under the terms and conditions of the Creative Commons Attribution (CC BY) license (http://creativecommons.org/licenses/by/4.0/).

Article

Fixed Point Theorems via α-ϱ-Fuzzy Contraction

Badshah-e-Rome [1], Muhammad Sarwar [1,*] and Poom Kumam [2,3,*]

[1] Department of Mathematics, University of Malakand, Chakdara Dir(L), Pakistan; baadeshah1@gmail.com
[2] KMUTTFixed Point Research Laboratory, Room SCL 802 Fixed Point Laboratory, Science Laboratory Building, Department of Mathematics, Faculty of Science, King Mongkut's University of Technology Thonburi (KMUTT), 126 Pracha-Uthit Road, Bang Mod, Thrung Khru, Bangkok 10140, Thailand
[3] Center of Excellence in Theoretical and Computational Science (TaCS-CoE), Science Laboratory Building, Faculty of Science, King Mongkut's University of Technology Thonburi (KMUTT), 126 Pracha-Uthit Road, Bang Mod, Thrung Khru, Bangkok 10140, Thailand
* Correspondence: sarwar@uom.edu.pk (M.S.); poom.kum@kmut.ac.th (P.K.)

Received: 12 March 2019 ; Accepted: 9 May 2019; Published: 31 May 2019

Abstract: Some well known results from the existing literature are extended and generalized via new contractive type mappings in fuzzy metric spaces. A non trivial supporting example is also provided to demonstrate the validity of the obtained results.

Keywords: fuzzy metric space; α-ϱ-fuzzy contraction; M-cauchy sequence; G-cauchy sequence

1. Introduction

The Banach contraction principle [1] plays an important role in the study of nonlinear equations and is one of the most useful mathematical tools for establishing the existence and uniqueness of a solution of an operator equation $Tx = x$. Many researchers have extended and generalized this principle in different spaces such as b-metric spaces, vector valued metric spaces, G-metric spaces, partially ordered complete metric spaces, cone metric spaces etc. Zadeh [2] introduced the notions of fuzzy logic and fuzzy sets. With this introduction, fuzzy mathematics began to evolve. Kramosil and Michalek [3] initiated the concept of fuzzy metric space as a generalization of the probabilistic metric space.

Fixed point theory in fuzzy metric space has been an attractive area for researchers. Heilpern [4] introduced fuzzy mappings and proved the fixed point theorem for such mappings. Grabiec [5] defined complete fuzzy metric space (G-complete fuzzy metric space) and extended the Banach fixed point theorem to fuzzy metric space (in the sense of Kramosil and Michalek). Besides the extension of the illustrious Banach contraction principle, several results concerning fixed point were established in G-complete fuzzy metric spaces (see, e.g, [6]). Gregori and Sapena [6] defined fuzzy contraction and established a fixed point result in fuzzy metric space in the sense of George and Veeramani. Afterwards many fixed point results were established for complete fuzzy metric spaces introduced by George and Veeramani [7], called M-complete fuzzy metric.

Gopal et al. [8] proposed the notion of α-ϕ-fuzzy contractive mapping and proved some fixed point results in G-complete fuzzy metric spaces in the sense of Grabiec. In this paper, we propose the notion of α-ϱ-fuzzy contractive mapping and establish some fixed point results for such mappings. Our work generalizes several corresponding results given in the literature, in particular, the Grabiec fixed point theorem is extended. A supporting example is also given.

2. Preliminaries

In this section we recall some basic definitions which will be needed in the sequel.

Definition 1 ([9]). *A binary operation* $* : [0,1] \times [0,1] \to [0,1]$ *satisfying conditions (1)–(4) is called continuous t-norm:*

1. $*$ is associative and commutative,
2. $*$ is continuous,
3. $1 * r = r$ for all $r \in [0,1]$,
4. if $r \leq s$ and $w \leq z$ then $r * w \leq s * z$ for all $r, s, w, z \in [0,1]$.

$\alpha *_L \beta = max\{\alpha + \beta - 1, 0\}$, called Lukasievicz t-norm,
$\alpha *_P \beta = \alpha\beta$, called product t-norm, and
$\alpha *_M \beta = min\{\alpha, \beta\}$, minimum t-norm are examples of continuous t-norms.
Michalek and Kramosil [3] defined fuzzy metric space in the following way.

Definition 2. *Having a nonempty set S, let ς be a fuzzy set on $S^2 \times [0, \infty)$ and * be a continuous t-norm. Then the triplet $(S, ς, *)$ is said to be fuzzy metric space if the following conditions are satisfied:*

(K1) $ς(r, s, 0) = 0$;
(K2) $ς(r, s, \ell) = 1$ iff $r = s$ for all $r, s \in S$ and $\ell > 0$;
(K3) $ς(r, s, \ell) = ς(s, r, \ell)$ for all $\ell > 0$;
(K4) $ς(r, s, \ell) * ς(s, w, t) \leq ς(r, w, \ell + t)$ for all $r, s, w \in S$ and $\ell, t > 0$;
(K5) $ς(r, s, \ell) : [0, \infty) \to [0, 1]$ is left continuous and non-decreasing function of ℓ;
(K6) $\lim_{\ell \to \infty} ς(r, s, \ell) = 1$, for all $r, s, w \in S$.

The value of $ς(r, s, \ell)$ represents the degree of closeness between r and s with respect to $\ell \geq 0$.
Veeramani and George modified Kramosil's definition of fuzzy metric space in the following way.

Definition 3 ([10]). *The triplet $(S, ς, *)$ is called fuzzy metric space, if S is a nonempty set, * is a continuous t-norm and ς is a fuzzy set on $S^2 \times [0, \infty)$ such that for all $r, s, w \in S$ and $\ell, t > 0$ the following assertions are satisfied.*

(G1) $ς(r, s, \ell) > 0$,
(G2) $ς(r, s, \ell) = 1$ iff $r = s$,
(G3) $ς(r, s, \ell) = ς(s, r, \ell)$,
(G4) $ς(r, s, \ell) * ς(s, w, t) \leq ς(r, w, \ell + t)$,
(G5) $ς(r, s, .) : (0, \infty) \to [0, 1]$ is continuous.

Remark 1 ([11]). *It should be noted that $0 < ς(r, s, \ell) < 1$ if $r \neq s$ and $\ell > 0$.*

Lemma 1 ([6]). $ς(r, s, .)$ *is nondecreasing for all $r, s \in S$.*

Example 1 ([10]). *For a metric space (S, d), let $M : S^2 \times (0, \infty) \to [0, 1]$ be defined as*

$$ς(r, s, \ell) = \frac{k\ell^n}{k\ell^n + md(r,s)}; \; \forall \; r, s \in S \text{ and } \ell > 0. \text{ where } k, m, n \in \mathbb{R}^+,$$

*where * is product t-norm (also true for minimum t-norm). Then ς is a fuzzy metric on S and is referred to as a fuzzy metric induced by the metric d.*

If we take $k = m = n = 1$, then the above fuzzy metric reduces to the well known *standard fuzzy metric*. For further examples of fuzzy metrics see [12].

Definition 4 ([7]). *In a fuzzy metric space $(S, ς, *)$:*

1. *A sequence $\{r_n\}$ will converge to $r \in S$ if $\lim_{n \to \infty} ς(r_n, r, \ell) = 1, \; \forall \; \ell > 0$.*

2. $\{r_n\}_{n\in\mathbb{N}}$ is said to be an M-cauchy sequence if for every positive real number $\epsilon \in (0,1)$ and $\ell > 0$ there exists $n_\epsilon \in \mathbb{N}$. such that $\varsigma(r_n, r_m, \ell) > 1 - \epsilon$, $\forall\ m, n \geq n_\epsilon$.
3. $\{r_n\}_{n\in\mathbb{N}}$ is called G-cauchy sequence if $\lim_{n\to\infty} \varsigma(r_{n+k}, r_n, \ell) = 1$, for all $\ell > 0$ and each $k \in \mathbb{N}$.

If every M-Cauchy sequence converges to some point of a fuzzy metric space $(S, \varsigma, *)$, then $(S, \varsigma, *)$ is called M-complete. Similarly $(S, \varsigma, *)$ will be G-complete if every G-Cauchy sequence converges in it. It is worth mentioning that G-completeness implies M-completeness.

3. Main Results

Definition 5. *Let $(S, \varsigma, *)$ be a fuzzy metric space and Ω be the class of all mappings $\varrho : [0,1] \to [1,\infty)$ such that for any sequence $\{r_n\} \subset [0,1]$, of positive real numbers $r_n \to 1 \Rightarrow \varrho(r_n) \to 1$. Then a self mapping $F : S \to S$ is said to be α-ϱ-fuzzy contraction if there exists two functions $\alpha : S^2 \times (0,\infty) \to [0,\infty)$ and $\varrho \in \Omega$ such that*

$$(\varsigma(Fr, Fs, \kappa\ell))^{\alpha(r, Fr, \ell)\alpha(s, Fs, \ell)} \geq \varrho(\varsigma(r, s, \ell))\varsigma(r, s, \ell), \tag{1}$$

for all $r, s \in S$, $\ell > 0$ and $\kappa \in (0, 1)$.

Now we have proved our first result.

Theorem 1. *Let $(S, \varsigma, *)$ be a G-complete fuzzy metric space, $F : S \to S$ be α-ϱ-fuzzy contraction where $\alpha : S^2 \times (0, \infty) \to [0, \infty)$ is such that $\alpha(r, Fr, \ell) \geq 1$, for all $r \in S$ $\ell > 0$.*
Then F has a unique fixed point.

Proof. Define sequence $\{r_n\}$ by $r_{n+1} = Fr_n$, $n \in \mathbb{N} \cup \{0\}$ where r_0 is an arbitrary but fixed element in S. Then by the hypothesis it follows that $\alpha(r_n, Fr_n, \ell) \geq 1$, for $n \in \mathbb{N} \cup \{0\}$. If $r_{n+1} = r_n$ for any $n \in \mathbb{N}$, then r_n is a fixed point of F. Therefore we assume that $r_{n+1} \neq r_n$ for all $n \in \mathbb{N}$, i.e., that no consecutive terms of the sequence $\{r_n\}$ are equal.

Further, if $r_n = r_m$ for some $n < m$, then as no consecutive terms of the sequence $\{r_n\}$ are equal from (1), we have

$$\begin{aligned}\varsigma(r_{n+1}, r_{n+2}, \ell) &= \varsigma(Fr_n, Fr_{n+1}, \ell) \\ &> (\varsigma(Fr_n, Fr_{n+1}, \kappa\ell))^{\alpha(r_n, Fr_n, \ell)\alpha(r_{n+1}, Fr_{n+1}, \ell)} \\ &\geq \varrho(\varsigma(r_n, r_{n+1}, \ell))\varsigma(r_n, r_{n+1}, \ell) \geq \varsigma(r_n, r_{n+1}, \ell),\end{aligned}$$

i.e., $\varsigma(r_n, r_{n+1}, \ell) < \varsigma(r_{n+1}, r_{n+2}, \ell)$. Similarly one can show that

$$\varsigma(r_n, r_{n+1}, \ell) < \varsigma(r_{n+1}, r_{n+2}, \ell) < \cdots < \varsigma(r_m, r_{m+1}, \ell).$$

Now $r_n = r_m$ implies that $r_{n+1} = Fr_n = Fr_m = r_{m+1}$, and so, the above inequality yields a contradiction. Thus we can suppose $r_n \neq r_m$ for all distinct $m, n \in \mathbb{N}$. Using (1), we get

$$\begin{aligned}\varsigma(r_n, r_{n+1}, \kappa\ell) &\geq (\varsigma(Fr_{n-1}, Fr_n, \kappa\ell))^{\alpha(r_{n-1}, Fr_{n-1}, \ell)\alpha(r_n, Fr_n, \ell)} \\ &\geq \varrho(\varsigma(r_{n-1}, r_n, \ell))\varsigma(r_{n-1}, r_n, \ell) \geq \varsigma(r_{n-1}, r_n, \ell).\end{aligned}$$

Therefore

$$\varsigma(r_n, r_{n+1}, \kappa\ell) \geq \varsigma(r_{n-1}, r_n, \ell). \tag{2}$$

Continuing in this manner, one can conclude by simple induction that

$$\varsigma(r_n, r_{n+1}, \kappa\ell) \geq \varsigma(r_0, r_1, \frac{\ell}{\kappa^{n-1}}). \tag{3}$$

Let q be a positive integer, then using $(K4)$, we have

$$\varsigma(r_n, r_{n+q}, \ell) \geq \varsigma(r_n, r_{n+1}, \frac{\ell}{q}) * \varsigma(r_{n+1}, r_{n+2}, \frac{\ell}{q}) * \overbrace{\cdots\cdots}^{q} * \varsigma(r_{n+q-1}, r_{n+q}, \frac{\ell}{q}).$$

Using (3), we have

$$\varsigma(r_n, r_{n+q}, \ell) \geq \varsigma(r_0, r_1, \frac{\ell}{q\kappa^n}) * \varsigma(r_0, r_1, \frac{\ell}{q\kappa^{n+1}}) * \overbrace{\cdots\cdots}^{q} * \varsigma(r_0, r_1, \frac{\ell}{q\kappa^{n+q-1}}).$$

For $n \to \infty$, the above inequality becomes

$$\lim_{n \to \infty} \varsigma(r_n, r_{n+q}, \ell) = 1.$$

Hence $\{r_n\}$ is G-cauchy. Therefore there will be some $w \in S$ such that $r_n \to w$ as $n \to \infty$, that is $\lim_{n \to \infty} \varsigma(r_n, w, \ell) = 1$ for each $\ell > 0$.
Now using $(K4)$ and (1) we have

$$\begin{aligned}
\varsigma(Fw, w, \ell) &\geq \varsigma(Fw, Fr_n, \frac{\ell}{2}) * \varsigma(r_{n+1}, w, \frac{\ell}{2}) \\
&\geq \varsigma(Fw, Fr_n, \frac{\ell}{2})^{\alpha(w, Fw, \ell)\alpha(r_n, Fr_n, \ell)} * \varsigma(r_{n+1}, w, \frac{\ell}{2}) \\
&\geq \varrho(\varsigma(w, r_n, \frac{\ell}{2}))\varsigma(w, r_n, \frac{\ell}{2}) * \varsigma(r_{n+1}, w, \frac{\ell}{2}) \\
&\geq \varsigma(w, r_n, \frac{\ell}{2}) * \varsigma(r_{n+1}, w, \frac{\ell}{2}) \to 1 * 1 = 1.
\end{aligned}$$

Thus $Fw = w$. To show uniqueness, let w and z be two distinct fixed points of F. That is $w = Fw \neq Fz = z$. Then for all $\ell > 0$, $0 < \varsigma(w, z, \ell) = \varsigma(Fw, Fz, \ell) < 1$. Therefore using (1), we have

$$\begin{aligned}
1 > \varsigma(w, z, \ell) &= \varsigma(Fw, Fz, \ell) \geq (\varsigma(Fw, Fz, \ell))^{\alpha(w, Fw, \ell)\alpha(z, Fz, \ell)} \\
&\geq \varrho(\varsigma(w, z, \frac{\ell}{\kappa}))\varsigma(w, z, \frac{\ell}{\kappa}) \geq \varsigma(w, z, \frac{\ell}{\kappa}).
\end{aligned}$$

Applying (1) repeatedly, we have $1 > \varsigma(w, z, \ell) \geq \varsigma(w, z, \frac{\ell}{\kappa}) \geq \varsigma(w, z, \frac{\ell}{\kappa^2}) \geq \cdots \geq \varsigma(w, z, \frac{\ell}{\kappa^n})$.
Letting $n \to \infty$, we have $1 \leq \varsigma(w, z, \ell) < 1$. Which is a contradiction. Hence $w = z$. □

Theorem 2. *Let $(S, \varsigma, *)$ be a G-complete fuzzy metric space, $F : S \to S$ be a mapping. If there exists two mappings $\alpha : S^2 \times (0, \infty) \to [0, \infty)$ and $\varrho \in \Omega$ such that $\alpha(r, Fr, \ell) \geq 1$, for all $r \in S, \ell > 0$ and*

$$2\varsigma(Fr, Fs, \kappa\ell) \geq (\alpha(r, Fr, \ell)\alpha(s, Fs, \ell) + 1)^{\varrho(\varsigma(r,s,\ell))\varsigma(r,s,\ell)} \tag{4}$$

for all $r, s \in S, 0 < \kappa < 1$ and $\ell > 0$, then F has a unique fixed point.

Proof. Let r_0 be an arbitrary element in S. Set $r_{n+1} = Fr_n$, $n \in \mathbb{N}$. Then by the hypothesis of the theorem it follows that $\alpha(r_n, Fr_n, \ell) \geq 1$, where $n \in \mathbb{N} \cup \{0\}$. If $r_{n+1} = r_n$ for any $n \in \mathbb{N}$, then r_n is a fixed point of F. Therefore we assume that $r_{n+1} \neq r_n$ for all $n \in \mathbb{N}$, i.e., that no consecutive terms of the sequence $\{r_n\}$ are equal.

Further, if $r_n = r_m$ for some $n < m$, then as no consecutive terms of the sequence $\{r_n\}$ are equal from (4), we have

$$2^{\varsigma(r_{n+1},r_{n+2},\ell)} = 2^{\varsigma(Fr_n,Fr_{n+1},\ell)}$$
$$> 2^{\varsigma(Fr_n,Fr_{n+1},\kappa\ell)}$$
$$\geq (\alpha(r_n,r_{n+1},\ell)\alpha(r_{n+1},r_{n+2},\ell)+1)^{\varrho((\varsigma(r_n,r_{n+1},\ell))\varsigma(r_n,r_{n+1},\ell)}$$
$$> 2^{\varsigma(r_n,r_{n+1},\ell)},$$

i.e., $\varsigma(r_n,r_{n+1},\ell) < \varsigma(r_{n+1},r_{n+2},\ell)$. Similarly one can show that

$$\varsigma(r_n,r_{n+1},\ell) < \varsigma(r_{n+1},r_{n+2},\ell) < \cdots < \varsigma(r_m,r_{m+1},\ell).$$

Now $r_n = r_m$ implies that $r_{n+1} = Fr_n = Fr_m = r_{m+1}$, and so, the above inequality yields a contradiction. Thus we can suppose $r_n \neq r_m$ for all distinct $m, n \in \mathbb{N}$. Using (4), we get

$$2^{\varsigma(r_n,r_{n+1},\kappa\ell)} = 2^{(\varsigma(Fr_{n-1},Fr_n,\kappa\ell))}$$
$$\geq (\alpha(r_{n-1},r_n,\ell)\alpha(r_n,r_{n+1},\ell)+1)^{\varrho((\varsigma(r_{n-1},r_n,\ell))\varsigma(r_{n-1},r_n,\ell)}$$
$$\geq 2^{\varrho(\varsigma(r_{n-1},r_n,\ell))\varsigma(r_{n-1},r_n,\ell)}.$$

Therefore

$$\varsigma(r_n,r_{n+1},\kappa\ell) \geq \varrho(\varsigma(r_{n-1},r_n,\ell))(\varsigma(r_{n-1},r_n,\ell)) \tag{5}$$
$$\Rightarrow \varsigma(r_n,r_{n+1},\kappa\ell) \geq \varsigma(r_{n-1},r_n,\ell).$$

Continuing in this manner one can conclude, by simple induction, that

$$\varsigma(r_n,r_{n+1},\kappa\ell) \geq \varsigma(r_0,r_1,\frac{\ell}{\kappa^{n-1}}). \tag{6}$$

Using (K4), we have for any positive integer q,

$$\varsigma(r_n,r_{n+q},\ell) \geq \varsigma(r_n,r_{n+1},\frac{\ell}{q}) * \varsigma(r_{n+1},r_{n+2},\frac{\ell}{q}) * \overbrace{\cdots\cdots}^{q} * \varsigma(r_{n+q-1},r_{n+q},\frac{\ell}{q}).$$

Using (6), we have

$$\varsigma(r_n,r_{n+q},\ell) \geq \varsigma(r_0,r_1,\frac{\ell}{q\kappa^n}) * \varsigma(r_0,r_1,\frac{\ell}{q\kappa^{n+1}}) * \overbrace{\cdots\cdots}^{q} * \varsigma(r_0,r_1,\frac{\ell}{q\kappa^{n+q-1}}).$$

For $n \to \infty$ the above inequality gives

$$\lim_{n\to\infty} \varsigma(r_n,r_{n+q},\ell) = 1.$$

Hence $\{r_n\}$ is G-cauchy. As S is complete, there will be $w \in S$ such that $r_n \to w$ as $n \to \infty$, that is $\lim_{n\to\infty} \varsigma(r_n,w,\ell) = 1$ for each $\ell > 0$.
Using (4) we have

$$2^{\varsigma(Fw,r_{n+1},\kappa\ell)} = 2^{(\varsigma(Fw,Fr_n,\kappa\ell))} \geq (\alpha(w,Fw,\ell)\alpha(r_n,Fr_n,\ell)+1)^{\varrho((\varsigma(w,r_n,\ell))(\varsigma(w,r_n,\ell)}$$
$$\geq 2^{\varrho((\varsigma(w,r_n,\ell))(\varsigma(w,r_n,\ell)}.$$

This implies
$$\varsigma(Fw, r_{n+1}, \kappa\ell) \geq \varrho((\varsigma(w, r_n, \ell))(\varsigma(w, r_n, \ell)). \tag{7}$$

Using (K4) and (7) we get
$$\begin{aligned}
\varsigma(Fw, w, \kappa\ell) &\geq \varsigma(Fw, r_{n+1}, \kappa\frac{\ell}{2}) * \varsigma(w, r_{n+1}, \kappa\frac{\ell}{2}) \\
&\geq \varrho(\varsigma(w, r_n, \frac{\ell}{2}))\varsigma(w, r_n, \frac{\ell}{2}) * \varsigma(w, r_{n+1}, \kappa\frac{\ell}{2}) \\
&\geq \varsigma(w, r_n, \frac{\ell}{2}) * \varsigma(w, r_{n+1}, \kappa\frac{\ell}{2}).
\end{aligned}$$

For $n \to \infty$ the above inequality gives
$$\lim_{n\to\infty} \varsigma(Fw, w, \kappa\ell) = 1 \Rightarrow Fw = w.$$

To prove uniqueness of the fixed point, assume w and z be two distinct fixed points of F. That is $w = Fw \neq Fz = z$. Then for all $\ell > 0$, $0 < \varsigma(w, z, \ell) = \varsigma(Fw, Fz, \ell) < 1$. Therefore using (4), we have
$$\begin{aligned}
2 > 2^{\varsigma(w,z,\ell)} &= 2^{\varsigma(Fw,Fz,\ell)} \\
&\geq (\alpha(w, Fw, \frac{\ell}{\kappa})\alpha(z, Fz, \frac{\ell}{\kappa}) + 1)^{\varrho(\varsigma(w,z,\frac{\ell}{\kappa}))\varsigma(w,z,\frac{\ell}{\kappa})} \\
&\geq 2^{\varrho(\varsigma(w,z,\frac{\ell}{\kappa}))\varsigma(w,z,\frac{\ell}{\kappa})} \\
&\geq 2^{\varsigma(w,z,\frac{\ell}{\kappa})}.
\end{aligned}$$

which implies $1 > \varsigma(w, z, \ell) \geq \varsigma(w, z, \frac{\ell}{\kappa})$. With repeated use of (4), it turns out that
$$1 > \varsigma(w, z, \ell) \geq \varsigma(w, z, \frac{\ell}{\kappa}) \geq \varsigma(w, z, \frac{\ell}{\kappa^2}) \geq \cdots \geq \varsigma(w, z, \frac{\ell}{\kappa^n}).$$

For $n \to \infty$, we get $1 \leq \varsigma(w, z, \ell) < 1$. Which is a contradiction. Therefore $w = z$. □

Theorem 3. *Let $(S, \varsigma, *)$ be a G-complete fuzzy metric space, $F : S \to S$ be a mapping. If there exist two mappings $\alpha : S^2 \times (0, \infty) \to [0, \infty)$ and $\varrho \in \Omega$ such that $\alpha(r, Fr, \ell) \geq 1$, for all $r \in S$, $\ell > 0$ and*
$$\frac{\varsigma(Fr, Fs, \kappa\ell)}{\alpha(r, Fr, \ell)\alpha(s, Fs, \ell)} \geq \varrho(\varsigma(r, s, \ell))\varsigma(r, s, \ell) \tag{8}$$

for all $r, s \in S$, $0 < \kappa < 1$ and $\ell > 0$, then F has a unique fixed point.

Proof. Set $r_{n+1} = Fr_n$, $n = 0, 1, \cdots$, for a fixed element $r_0 \in S$. By hypothesis of the theorem we have $\alpha(r_n, Fr_n, \ell) = \alpha(r_n, r_{n+1}, \ell) \geq 1$ where $n \in \mathbb{N} \cup \{0\}$. Let $r_{n+1} \neq r_n$, for $n \geq 0$. Otherwise r_n is fixed point of F and hence the result is proved. Further, if $r_n = r_m$ for some $n < m$, then as no consecutive terms of the sequence $\{r_n\}$ are equal from (8), we have
$$\begin{aligned}
\varsigma(r_{n+1}, r_{n+2}, \ell) &= \varsigma(Fr_n, Fr_{n+1}, \ell) \\
&> \varsigma(Fr_n, Fr_{n+1}, \kappa\ell) \geq \frac{\varsigma(Fr_n, Fr_{n+1}, \kappa\ell)}{\alpha((r_n, r_{n+1})\alpha(r_{n+1}, r_{n+2}, \ell)} \\
&\geq \varrho(\varsigma(r_n, r_{n+1}, \ell))\varsigma(r_n, r_{n+1}, \ell) \\
&> \varsigma(r_n, r_{n+1}, \ell),
\end{aligned}$$

i.e., $\varsigma(r_n, r_{n+1}, \ell) < \varsigma(r_{n+1}, r_{n+2}, \ell)$. Similarly it can be proved that

$$\varsigma(r_n, r_{n+1}, \ell) < \varsigma(r_{n+1}, r_{n+2}, \ell) < \cdots < \varsigma(r_m, r_{m+1}, \ell).$$

Now $r_n = r_m$ implies that $r_{n+1} = Fr_n = Fr_m = r_{m+1}$, and so, the above inequality yields a contradiction. Thus we can suppose $r_n \neq r_m$ for all distinct $m, n \in \mathbb{N}$. Using (8), we have

$$\varsigma(r_n, r_{n+1}, \kappa\ell) = \varsigma(Fr_{n-1}, Fr_n, \kappa\ell) \geq \frac{\varsigma(Fr_{n-1}, Fr_n, \kappa\ell)}{\alpha(r_{n-1}, r_n)\alpha(r_n, r_{n+1}, \ell)}$$
$$\geq \varrho(\varsigma(r_{n-1}, r_n, \ell))\varsigma(r_{n-1}, r_n, \ell).$$

Therefore

$$\varsigma(r_n, r_{n+1}, \kappa\ell) \geq \varrho(\varsigma(r_{n-1}, r_n, \ell))(\varsigma(r_{n-1}, r_n, \ell)) \quad (9)$$
$$\Rightarrow \varsigma(r_n, r_{n+1}, \kappa\ell) \geq \varsigma(r_{n-1}, r_n, \ell).$$

Following the related arguments in the proof of Theorem (1), we conclude that $\{r_n\}$ is a G-cauchy sequence. Due to the completeness of S, there will be $w \in S$ such that $r_n \to w$ as $n \to \infty$, that is $\lim_{n \to \infty} \varsigma(r_n, w, \ell) = 1$ for each $\ell > 0$.

Then using (K4) and (8) we have

$$\varsigma(Fw, w, \kappa\ell) \geq \varsigma(Fw, r_{n+1}, \kappa\frac{\ell}{2}) * \varsigma(w, r_{n+1}, \kappa\frac{\ell}{2})$$
$$= \varsigma(Fw, Fr_n, \kappa\frac{\ell}{2}) * \varsigma(w, r_{n+1}, \kappa\frac{\ell}{2})$$
$$\geq \frac{\varsigma(Fw, Fr_n, \kappa\frac{\ell}{2})}{\alpha(w, Fw, \ell)\alpha(r_n, r_{n+1}, \ell)} * \varsigma(w, r_{n+1}, \kappa\frac{\ell}{2})$$
$$\geq \varrho((\varsigma(w, r_n, \frac{\ell}{2}))(\varsigma(w, r_n, \frac{\ell}{2}) * \varsigma(w, r_{n+1}, \kappa\frac{\ell}{2}))$$
$$\geq \varsigma(w, r_n, \frac{\ell}{2}) * \varsigma(w, r_{n+1}, \kappa\frac{\ell}{2}).$$

For $n \to \infty$ the above inequality gives

$$\lim_{n \to \infty} \varsigma(Fw, w, \kappa\ell) = 1 \Rightarrow Fw = w.$$

For uniqueness, assume w and z be two distinct fixed points of F. That is $w = Fw \neq Fz = z$. Then for all $\ell > 0, 0 < \varsigma(w, z, \ell) = \varsigma(Fw, Fz, \ell) < 1$. Therefore using (8), we have

$$1 > \varsigma(w, z, \ell) = \varsigma(Fw, Fz, \ell)$$
$$\geq \frac{\varsigma(Fw, Fz, \ell)}{\alpha(w, Fw, \ell)\alpha(z, Fz, \ell)}$$
$$\geq \varrho(\varsigma(w, z, \frac{\ell}{\kappa}))\varsigma(w, z, \frac{\ell}{\kappa})) \geq \varsigma(w, z, \frac{\ell}{\kappa}).$$

Using (8), it can be shown that $1 > \varsigma(w, z, \ell) \geq \varsigma(w, z, \frac{\ell}{\kappa}) \geq \varsigma(w, z, \frac{\ell}{\kappa^2}) \geq \cdots \geq \varsigma(w, z, \frac{\ell}{\kappa^n})$. Letting $n \to \infty$, we get $1 \leq \varsigma(w, z, \ell) < 1$, a contradiction. Hence $w = z$. □

By taking $\alpha(r, s, \ell) = 1$ and $\varrho(t) = 1$ in Theorems (1), (2) and (3), we have the following corollary which is actually the fixed point result established by Grabiec [5].

157

Corollary 1. Let $(S, \varsigma, *)$ be a G-complete fuzzy metric space and $F : S \to S$ be be a self mapping such that

$$\varsigma(Fr, Fs, \kappa\ell) \geq \varsigma(r, s, \ell) \tag{10}$$

for all $r, s \in S$, $\ell > 0$ and $\kappa \in (0, 1)$.
Then F has a unique fixed point.

4. Example

In this section we present a supporting example to demonstrate the validity of our results.

Example 2. Let $S = [0, \infty)$, $r * s = rs$ for all $r, s \in [0, 1]$ and $\varsigma(r, s, \ell) = e^{\frac{-|u-v|}{\ell}}$ for all $r, s \in S$ and $t > 0$. Then $(S, \varsigma, *)$ is a complete fuzzy metric space. Let $F : S \to S$ be defined as

$$Fu = \begin{cases} \frac{u}{9}, & \text{if } r \in [0, 1], \\ \sqrt{u} & \text{if } r \in (1, \infty). \end{cases}$$

Further, define $\alpha : S^2 \times (0, \infty) \to [0, \infty)$ as

$$\alpha(r, s, \ell) = \begin{cases} \sqrt{2} & \text{if } r, s \in [0, 1], \\ (\frac{3}{2})^{0.25} & \text{if } r, s \in (1, \infty), \\ 0 & \text{otherwise.} \end{cases}$$

Also for all $r, s \in S$ and $\ell > 0$, we have $\alpha(r, Fr, \ell) \geq 1$, and

$$(\varsigma(Fr, Fv, \ell))^{\alpha(r, Fr, \ell)\alpha(s, Fs, \ell)} \geq e^{\frac{-|u-v|}{4\ell}}$$
$$= (\varsigma(r, s, \ell))^{-\frac{3}{4}} \varsigma(r, s, \ell).$$

That is F is α-ϱ-fuzzy contraction with $\varrho(t) = t^{-\frac{3}{4}}$, where $t \in [0, 1]$.
Thus all conditions of Theorem (1) are fulfilled. Obviously 0 is a unique fixed point of F.

Similarly supporting examples for other results do exist and can be constructed easily.

5. Conclusions

We proposed the concept of the α-ϱ-Fuzzy Contraction and some new types of fuzzy contractive mappings. We proved three theorems which ensure the existence and uniqueness of fixed points of these new types of contractive mappings. The new concepts may lead to further investigation and applications. For example, using the recent ideas in the literature, it is possible to extend our results to the case of coupled fixed points in fuzzy metric spaces.

Author Contributions: All authors contribute equally to the writing of this manuscript. All authors reads and improve the final version.

Funding: This project was supported by the Theoretical and Computational Science (TaCS) Center under the Computational and Applied Science for Smart Innovation Cluster (CLASSIC), Faculty of Science, KMUTT.

Acknowledgments: The authors wish to thank the editor and anonymous referees for their comments and suggestions, which helped to improve this paper.

Conflicts of Interest: The authors declare no conflict of interest.

References

1. Banach, S. Sur les opérations dans les ensembles abstraits et leur application aux equations integrales. *Fund. Math.* **1922**, *3*, 133–181. [CrossRef]

2. Zadeh, A. Fuzzy sets. *Inf. Control* **1965**, *8*, 338–353. [CrossRef]
3. Kramosil, I.; Michalek, J. Fuzzy metric and statistical metric spaces. *Kybernetika* **1975**, *11*, 336–344.
4. Heilpern, S. Fuzzy mappings and fixed point theorems. *J. Math. Anal. Appl.* **1981**, *83*, 566–569. [CrossRef]
5. Grabiec, M. Fixed points in fuzzy metric spaces. *Fuzzy Sets Syst.* **1988**, *27*, 385–389. [CrossRef]
6. Gregori, V.; Sapena, A. On fixed point theorems in fuzzy metric spaces. *Fuzzy Sets Syst.* **2002**, *125*, 245–252. [CrossRef]
7. Vasuki, R.; Veeramani, P. Fixed point theorems and Cauchy sequences in fuzzy metric spaces. *Fuzzy Sets Syst.* **2003**, *135*, 415–417. [CrossRef]
8. Gopal, D.; Vetro, C. Some new fixed point theorems in fuzzy metric spaces. *Iran. J. Fuzzy Syst.* submitted. [CrossRef]
9. Schweizer, B.; Sklar, A. Statistical metric spaces. *Pac. J. Math.* **1960**, *10*, 313–334. [CrossRef]
10. George, A.; Veeramani, P. On some results in fuzzy metric spaces. *Fuzzy Sets Syst.* **1994**, *64*, 395–399. [CrossRef]
11. Mihet, D. Fixed point theorems in fuzzy metric spaces using property (E.A). *Nonlinear Anal.* **2010**, *73*, 2184–2188. [CrossRef]
12. Gregori, V.; Morillas, S.; Sapena, A. Examples of fuzzy metrics and applications. *Fuzzy Sets Syst.* **2011**, *170*, 95–111. [CrossRef]

© 2019 by the authors. Licensee MDPI, Basel, Switzerland. This article is an open access article distributed under the terms and conditions of the Creative Commons Attribution (CC BY) license (http://creativecommons.org/licenses/by/4.0/).

Article
Best Approximation Results in Various Frameworks

Taoufik Sabar, Abdelhafid Bassou * and Mohamed Aamri

Laboratory of Algebra, Analysis and Applications (L3A), Departement of Mathematics and Computer Science, Faculty of Sciences Ben M'sik, Hassan II University of Casablanca, P.B 7955 Sidi Othmane, Casablanca, Morocco; sabarsaw@gmail.com (T.S.); aamrimohamed82@gmail.com (M.A.)
* Correspondence: hbassou@gmail.com

Received: 20 April 2019; Accepted: 16 May 2019; Published: 27 May 2019

Abstract: We first provide a best proximity point result for quasi-noncyclic relatively nonexpansive mappings in the setting of dualistic partial metric spaces. Then, those spaces will be endowed with convexity and a result for a cyclic mapping will be obtained. Afterwards, we prove a best proximity point result for tricyclic mappings in the framework of the newly introduced extended partial S_b-metric spaces. In this way, we obtain extensions of some results in the literature.

Keywords: best proximity point; dualistic partial metric space; tricyclic mappings; extended partial S_b-metric space

1. Introduction

Whether a self mapping has fixed points or not is a problem that has been exhaustively studied ever since Banach stated his contraction principle. In the beginning of the current century, an issue of equivalent importance to that of the fixed point problem appeared: Let T be a cyclic (resp. noncyclic) mapping on $A \cup B$ where A and B are nonempty subsets of a metric space (X, d), that is, $T(A) \subseteq B$ and $T(B) \subseteq A$ resp. $T(A) \subseteq A$ and $T(B) \subseteq B$). The equation $Tx = x$ may not possess a soltution, in this case, we wish to determine an element (resp. a pair) which is as close to its image as possible, i.e., an element $x \in A \cup B$ such that $d(x, Tx) = dist(A, B)$ (resp. a pair $(x, y) \in A \times B$ of fixed points such that $d(x, y) = dist(A, B)$). Such a point (resp. pair) is called a best proximity point (resp. pair). The problem of best approximation for cyclic and noncyclic mappings attracted a good many authors and many pertinent results were obtained in different frameworks [1–7].

In 2011, the notion of P-property was introduced in [8] and best proximity point results for weakly contractive non-self-mappings were obtained. Two years later, using the aforementioned property, Abkar and Gabaleh [9] proved that some existence and uniqueness results in best proximity point theory can be acquired from existing results in the fixed point theory. In the same year, Almeida, Karapinar and Sadarangani [10] showed that best proximity point results can be obtained from fixed point results using only the weaker condition of weak P-property. In 2016, Ref. [11] presented a new approach to best proximity point results by means of the so-called simulation functions.

In 2017, Sabar, Aamri and Bassou [12] introduced the class of tricyclic mappings and best proximity points thereof. Let A, B and C be nonempty subsets of a metric space (X, d). A mapping $T : A \cup B \cup C \longrightarrow A \cup B \cup C$ is said to be tricyclic if $T(A) \subseteq B$, $T(B) \subseteq C$ and $T(C) \subseteq A$, and a best proximity point for T is an element $x \in A \cup B \cup C$ such that $D(x, Tx, T^2x) = dist(A, B, C)$ where $D(x, y, z) = d(x, y) + d(y, z) + d(z, x)$ and

$$dist(A, B, C) = \inf \{D(x, y, z) : x \in A, y \in B \text{ and } z \in C\}.$$

This paper aims to establish best proximity point results for subclasses of cyclic, noncyclic and tricylic mappings in the framework of partial dualistic metric spaces and the lately introduced extended partial S_b-metric spaces [13].

2. Best Proximity Point Results in Dualistic Partial Metric Spaces

This section deals with cyclic and noncyclic mappings in dualistic partial metric spaces; these spaces were first introduced as follows.

Definition 1 ([14]). *Let X be a nonempty set. A function $\mathcal{D} : X \times X \longrightarrow \mathbb{R}$ is called a dualistic partial metric if*

(D_1) $x = y$ if and only if $\mathcal{D}(x,x) = \mathcal{D}(y,y) = \mathcal{D}(x,y)$,
(D_2) $\mathcal{D}(x,x) \leq \mathcal{D}(x,y)$,
(D_3) $\mathcal{D}(x,y) = \mathcal{D}(y,x)$,
(D_4) $\mathcal{D}(x,y) \leq \mathcal{D}(x,z) + \mathcal{D}(z,y) - \mathcal{D}(z,z)$,
for all $x, y, z \in X$.

Complying with [14], $\mathcal{D}$ generates a T_0 topology on X, denoted by $\tau(\mathcal{D})$ in which the open balls are

$$\{B_\mathcal{D}(x,\varepsilon) : x \in X, \ \varepsilon > 0\} \quad \text{where} \quad B_\mathcal{D}(x,\varepsilon) = \{y \in X : \mathcal{D}(x,y) < \varepsilon + \mathcal{D}(x,x)\}.$$

Now, we are able to introduce the notions of convergence and Cauchy sequences in the setting of dualistic partial metric spaces.

Definition 2 ([15]). *A sequence (x_n) in $(X, \mathcal{D})$ converges to a point x if and only if $\mathcal{D}(x,x) = \lim_{n \to \infty} \mathcal{D}(x_n, x)$ and it is a Cauchy sequence if $\lim_{n \to \infty} \mathcal{D}(x_n, x_m)$ exists and it is finite.*

To present our results, we need to mention some basic concepts related to noncyclic mappings. In this section, unless stated otherwise, A and B are nonempty subsets of a dualistic partial metric space $(X, \mathcal{D})$ and $T : A \cup B \longrightarrow A \cup B$ is a noncyclic mapping:

$$\begin{aligned} F_A(T) &= \{x \in A : Tx = x\} \text{ and } F_B(T) = \{y \in B : Ty = y\}, \\ \text{dist}(A, B) &= \inf\{\mathcal{D}(x,y) : x \in A, \ y \in B\}, \\ A_0 &= \{x \in A : \mathcal{D}(x,y) = \text{dist}(A,B) \text{ for some } y \in B\}, \\ B_0 &= \{y \in B : \mathcal{D}(x,y) = \text{dist}(A,B) \text{ for some } x \in A\}. \end{aligned}$$

Definition 3. *The mapping T is said to be relatively nonexpansive if*

$$\mathcal{D}(Tx, Ty) \leq \mathcal{D}(x, y) \text{ for all } x \in A \text{ and } y \in B.$$

In addition, a pair $(x, y) \in A \times B$ is said to be a best proximity pair if

$$x \in F_A(T), \ y \in F_B(T) \text{ and } \mathcal{D}(x,y) = \text{dist}(A, B).$$

In [16], Gabeleh and Otafudu introduced the class of quasi-noncyclic relatively nonexpansive mappings as follows.

Definition 4. *Suppose $A_0 \neq \emptyset$. The mapping T is said to be quasi-noncyclic relatively nonexpansive mapping provided that $(F_{A_0}(T), F_{B_0}(T)) \neq \emptyset$ and, for all $(a, b) \in F_{A_0}(T) \times F_{B_0}(T)$, we have*

$$\begin{cases} \mathcal{D}(Tx, b) \leq \mathcal{D}(x, b) \text{ for all } x \in A, \\ \mathcal{D}(a, Ty) \leq \mathcal{D}(a, y) \text{ for all } y \in B. \end{cases}$$

The class of quasi-noncyclic relatively nonexpansive mappings is not a subclass of noncyclic relatively nonexpansive mappings. To check that out and for more constructions on quasi-noncyclic relatively nonexpansive mappings, we refer the reader to [17,18].

Definition 5. *A is said to be approximatively compact with respect to B if and only if every sequence (x_n) in A such that $\mathcal{D}(y, x_n) \longrightarrow \mathcal{D}(y, A)$ for some $y \in B$ has a convergent subsequence.*

Remark 1.

- If A is a compact set, then it is approximatively compact with respect to B.
- If $A \cap B \neq \emptyset$, then A is approximatively compact with respect to $A \cap B$. Indeed, let (x_n) in A such that $\mathcal{D}(y, x_n) \longrightarrow \mathcal{D}(y, A)$ for some $y \in A \cap B$. Since $\mathcal{D}(y, y) \leq \mathcal{D}(y, x)$ for all $x \in X$, $\mathcal{D}(y, A) = \mathcal{D}(y, y)$ and that means (x_n) converges to y.

Definition 6 ([19]). *The pair (A, B) is called sharp (resp. semi-sharp) proximal if and only if, for each x in A and y in B, there exist a unique (resp. at most one) element x' in B and a unique element y' in A such that*

$$\mathcal{D}(x, x') = \mathcal{D}(y', y) = \text{dist}(A, B).$$

Now, we're entitled to state our first main result.

Theorem 1. *Let $(X, \mathcal{D})$ be a dualistic partial metric space such that $\mathcal{D}$ is continuous and let A, B be nonempty subsets of X such that $A_0 \neq \emptyset$, B is approximatively compact with respect to A and the pair (A, B) is semi-sharp proximal. Then,, each quasi-noncyclic relatively nonexpansive mapping defined on $A \cup B$ possesses a best proximity pair.*

Proof. Let (x_n) be a sequence of elements of A_0 which converges to some $x \in F_{A_0}(T)$. (The fact that $F_{A_0}(T)$ is nonempty guarantees the existence of such a sequence). Choose a point y_n in B_0 such that

$$\mathcal{D}(x_n, y_n) = \text{dist}(A, B) \quad \text{for all } n \in \mathbb{N}.$$

Now, we get

$$\begin{aligned}
\mathcal{D}(x, y_n) &\leq \mathcal{D}(x, x_n) + \mathcal{D}(x_n, y_n) - \mathcal{D}(x_n, x_n) \\
&= \mathcal{D}(x, x_n) + \text{dist}(A, B) - \mathcal{D}(x_n, x_n) \\
&\leq \mathcal{D}(x, x_n) + \text{dist}(x, B) - \mathcal{D}(x_n, x_n).
\end{aligned}$$

Taking into account that $\mathcal{D}$ is a continuous mapping on $X \times X$, we get

$$\mathcal{D}(x_n, x_n) \longrightarrow \mathcal{D}(x, x) \text{ as } n \longrightarrow \infty.$$

Therefore, letting $n \longrightarrow \infty$, we obtain $\mathcal{D}(x, y_n) \longrightarrow \text{dist}(x, B)$. The hypothesis that B is approximatively compact with respect to A implies the existence of a subsequence (y_{n_k}) of (y_n) and a $y \in B$ such that $y_{n_k} \longrightarrow y$ as $k \longrightarrow \infty$. Hence, $\text{dist}(A, B) = \mathcal{D}(x_{n_k}, y_{n_k}) \longrightarrow \mathcal{D}(x, y)$, which means

$$\mathcal{D}(x, y) = \text{dist}(A, B).$$

Since T is quasi-noncyclic relatively nonexpansive,

$$\mathcal{D}(x, Ty) \leq \mathcal{D}(x, y) = \text{dist}(A, B).$$

Now, we use the assumption that the pair (A, B) is semi-sharp proximal to conclude that y is a fixed point and therefore (x, y) is a best proximity pair. □

Example 1. Let $X = \mathbb{R}^2$ with the dualistic partial metric $\mathcal{D}\left((x,y),(x',y')\right) = \max\{x,x'\} + \max\{y,y'\}$. Let $A = \{0\} \times [0,\infty)$ and $B = \{1\} \times [0,\infty)$. Then, $A_0 = \{(0,0)\}$ and $\text{dist}(A,B) = 1$. Moreover, the pair (A,B) is semi-sharp proximal. Let $T : A \cup B \longrightarrow A \cup B$ be a noncyclic mapping such that $T(0,x) = (0,x/2)$ and $T(1,x) = (1,x/2)$ for all $x \in [0,\infty)$. Clearly, T is a quasi-noncyclic relatively nonexpansive and its best proximity pair is $((0,0),(1,0))$.

As a special case of the previous theorem, we obtain the following result which was proven in [20].

Corollary 1. *(Theorem 1 of [20]) Let (X,d) be a complete metric space and A, B be nonempty subsets of X such that A is closed and $A_0 \neq \emptyset$. Suppose that B is approximatively compact with respect to A and that $T : A \cup B \longrightarrow A \cup B$ is a quasi-noncyclic mapping such that $T|A$ is a contraction in the sense of Banach, $T(A_0) \subseteq A_0$ and the pair (A,B) is semi-sharp proximal. Then, T has a best proximity pair.*

The notion of convexity in metric spaces was firstly introduced in [21] and the exact same notion can be given in dualistic partial metric spaces.

Definition 7. *A mapping $W : X \times X \times [0,1] \longrightarrow X$ is said to be a convex structure on X if, for each $(x,y) \in X \times X$ and $\lambda \in [0,1]$,*

$$\mathcal{D}(u, W(x,y,\lambda)) \leq \lambda \mathcal{D}(u,x) + (1-\lambda)\mathcal{D}(u,y) \text{ for all } u \in X.$$

In addition, $(X, \mathcal{D}, W)$ is said to be a convex dualistic partial metric space.

Definition 8. *A subset K of a convex dualistic partial metric space $(X, \mathcal{D}, W)$ is said to be convex if $W(x,y,\lambda) \in K$ for all $x,y \in K$ and $\lambda \in [0,1]$.*

The following propositions are immediate.

Proposition 1 ([21])**.** *Let $\{K_\alpha\}_{\alpha \in A}$ be a family of convex subsets of the convex dualistic partial metric space X; then, $\cap_{\alpha \in A} K_\alpha$ is also a convex subset of X.*

Proposition 2. *The closed ball centered at $a \in X$ with radius $r \in \mathbb{R}$ is a convex subset of X.*

Proof. Let $x, y \in B(a,r)$ and $\lambda \in [0,1]$,

$$\begin{aligned}\mathcal{D}(a, W(x,y,\lambda)) &\leq \lambda \mathcal{D}(a,x) + (1-\lambda)\mathcal{D}(a,y) \\ &\leq \lambda(r + \mathcal{D}(a,a)) + (1-\lambda)(r + \mathcal{D}(a,a)) \\ &\leq r + \mathcal{D}(a,a).\end{aligned}$$

In addition, this means that the closed ball is convex. □

Definition 9. *A convex dualistic partial metric space $(X, \mathcal{D}, W)$ is said to verify property (C) if every bounded increasing net of nonempty, closed and convex subsets of X is of nonempty intersection.*

A weakly compact convex subset of a Banach space has property (C) for instance. For more examples, we allude to [22].

Let A and B be nonempty subsets of a convex dualistic partial metric space $(X, \mathcal{D}, W)$. We set

$$\begin{aligned}\delta(A,B) &= \sup\{\mathcal{D}(x,y) : x \in A \text{ and } y \in B\}, \\ \delta_{(x)}(B) &= \sup\{\mathcal{D}(x,y) : y \in B\} \text{ for all } x \in A.\end{aligned}$$

By $\overline{con}(A)$, we denote the closed and convex hull of A and it is defined by

$$\overline{con}(A) = \cap\{C : C \text{ is a closed and convex subset of } X \text{ such that } C \supseteq A\}.$$

The following lemma is used in the proof of our second main result of this section.

Lemma 1. *Let (A, B) be a nonempty, bounded, closed, and convex pair in a convex dualistic partial metric space $(X, \mathcal{D}, W)$. Suppose that $T : A \cup B \to A \cup B$ is a cyclic mapping. If X has the property (C), then there exists a pair $(K_1, K_2) \subseteq (A, B)$ which is maximal with respect to being nonempty, closed and convex such that T is cyclic on $K_1 \cup K_2$. Furthermore,*

$$\overline{co}(T(K_1)) = K_2 \text{ and } \overline{co}(T(K_2)) = K_1.$$

Proof. The set of all nonempty, closed, and convex pairs $(C, D) \subseteq (A, B)$ such that T is cyclic on $C \cup D$ is partially ordered by reverse inclusion, i.e.,

$$(C_1, D_1) \leq (C_2, D_2) \iff (C_2, D_2) \subseteq (C_1, D_1).$$

For each increasing chain $\{(C_\alpha, D_\alpha)\}_\alpha$, we set $C := \cap C_\alpha$ and $D := \cap D_\alpha$. Since X has the property (C) and from the fact that every intersection of convex subsets is a convex subset, (C, D) is a nonempty, closed and convex pair. In addition,

$$T(C) \subseteq T(\cap C_\alpha) \subseteq \cap T(C_\alpha) \subseteq \cap D_\alpha = D.$$

Similarly, $T(D) \subseteq C$, which means that T is cyclic on $C \cup D$. Therefore, every increasing chain is bounded above and Zorn's Lemma assures the existence of the maximal pair (K_1, K_2). Now, we note that the pair $(\overline{co}(T(K_2)), \overline{co}(T(K_1))) \subseteq (K_1, K_2)$ is nonempty, closed and convex. We also have

$$T(\overline{co}(T(K_2))) \subseteq T(K_1) \subseteq \overline{co}(T(K_1)).$$

Similarly, $T(\overline{co}(T(K_1))) \subseteq \overline{co}(T(K_2))$, that is, T is cyclic on $\overline{co}(T(K_2)) \cup \overline{co}(T(K_1))$. The maximality of (K_1, K_2) finishes the proof. □

Theorem 2. *Let (A, B) be a nonempty, bounded, closed, and convex pair in a convex dualistic partial metric space $(X, \mathcal{D}, W)$ such that $\mathcal{D}$ is continuous and $\mathcal{D}(x, x) \leq 0$ for all $x \in A \cup B$. Let $(K_1, K_2) \subseteq (A, B)$ be a maximal pair with respect to being nonempty, closed and convex such that T is cyclic on $K_1 \cup K_2$. Suppose that $T : A \cup B \to A \cup B$ is a cyclic. Suppose that, for all $x \in K_1$ and $y \in K_2$,*

$$\mathcal{D}(Tx, Ty) \leq \Lambda := \{k\delta(K_1, K_2) + (1 - k) \operatorname{dist}(A, B)\} + \min\{\mathcal{D}(Tx, Tx), \mathcal{D}(Ty, Ty)\}.$$

If X has the property (C), then T has a best proximity pair.

Proof. Let $x \in K_1$ and $y \in K_2$; from the inequality fulfilled by the mapping T, we get $Ty \in B(Tx, \Lambda)$ and then

$$T(K_2) \subseteq B(Tx, \Lambda);$$

thus,

$$K_1 = \overline{co}(T(K_2)) \subseteq B(Tx, \Lambda),$$

which means,

$$\mathcal{D}(Tx, z) \leq \Lambda + \mathcal{D}(Tx, Tx), \text{ for all } z \in K_1,$$

that is, $\delta_{Tx}(K_1) \leq \Lambda + \mathcal{D}(Tx, Tx)$ and similarly we get $\delta_{Ty}(K_2) \leq \Lambda + \mathcal{D}(Ty, Ty)$. Put

$$L_1 := \{x \in K_1 : \delta_x(K_2) \leq \Lambda + \mathcal{D}(x,x)\} \text{ and } L_2 := \{y \in K_2 : \delta_y(K_1) \leq \Lambda + \mathcal{D}(y,y)\}.$$

Clearly, (L_1, L_2) is a pair of nonempty, closed and convex subsets such that T is cyclic on $L_1 \cup L_2$. We take account of the maximilaty of (K_1, K_2) to conclude that $L_1 = K_1$ and $L_2 = K_2$—from which we get

$$\delta_x(K_2) \leq r\delta(K_1, K_2) + (1-r)\, \text{dist}(A, B) + \mathcal{D}(x,x) \text{ for all } x \in K_1.$$

Hence,

$$\delta(K_1, K_2) = \text{dist}(A, B).$$

Consequently,

$$\text{dist}(A, B) \leq \mathcal{D}(p, Tp), \mathcal{D}(Tq, q) \leq \delta(K_1, K_2) = \text{dist}(A, B), \text{ for all } (p, q) \in K_1 \times K_2.$$

In addition, that is the desired result. □

The next corollary follows immediately.

Corollary 2 ([1]). *Let (A, B) be a nonempty, bounded, closed, and convex pair in a convex metric space (X, d, W). Suppose that $T : A \cup B \to A \cup B$ is a generalized cyclic contraction. If X has the (C) property, then T has a best proximity pair.*

3. Tricyclic Mappings in Convex Extended Partial S_b Metric Spaces

Lately, extended partial S_b-metric spaces were introduced as comes

Definition 10 ([7]). *Let X be a nonempty subset and let $\theta : X^3 \longrightarrow [1, \infty)$. If a mapping $S_\theta : X^3 \longrightarrow [0, \infty)$ satisfies*
1. *$x = y = z$ if and only if $S_\theta(x, y, z) = S_\theta(x, x, x) = S_\theta(y, y, y) = S_\theta(z, z, z)$,*
2. *$S_\theta(x, x, x) \leq S_\theta(x, y, z)$,*
3. *$S_\theta(x, y, z) \leq \theta(x, y, z)[S_\theta(x, x, t) + S_\theta(y, y, t) + S_\theta(z, z, t)]$,*

for all $x, y, z, t \in X$. Then, (X, S_θ) is called an extended partial S_b-metric space.

Next, we introduce the notion of convexity in extended partial S_b-metric spaces.

Definition 11. *Let (X, S_θ) be an extended partial S_b-metric space. A mapping $W : X \times X \times [0, 1] \longrightarrow X$ is said to be a convex structure on X if, for each $(x, y) \in X \times X$ and $\lambda \in [0, 1]$,*

$$S_\theta(u, v, W(x, y, \lambda)) \leq \lambda S_\theta(u, v, x) + (1 - \lambda) S_\theta(u, v, y) \text{ for all } u, v \in X.$$

In addition, (X, S_θ, W) is said to be a convex extended partial S_b-metric space.

It is easy to see that every convex metric space in the sense of [15] is a convex extended partial S_b-metric space. Now, we present a yet stronger version of convexity.

Definition 12. *Retaining the same notations as in the previous definition, W is said to be a double convex structure on X if it is a convex structure and if, for each $(x_1, y_1), (x_2, y_2) \in X \times X$, $\lambda \in [0, 1]$ and $u \in X$,*

$$S_\theta(u, W(x_1, y_1, \lambda), W(x_2, y_2, \lambda)) \leq \lambda S_\theta(u, x_1, x_2) + (1 - \lambda) S_\theta(u, y_1, y_2).$$

Example 2. *Let $(X, \|.\|)$ be a normed linear space and $S_\theta : X^3 \longrightarrow [0, \infty)$ be defined as $S_\theta(x, y, z) = \|x - y\| + \|y - z\| + \|z - x\|$. Then, (X, S_θ) is an extended partial S_b-metric space and the mapping W :*

$X \times X \times [0,1] \longrightarrow X$ defined by $W(x,y,\lambda) = \lambda x + (1-\lambda) y$ is a convex structure on X. Moreover, W is a double convex structure. Indeed, fix $(x_1, y_1), (x_2, y_2) \in X \times X, \lambda \in [0,1]$ and $u \in X$, we have

$$
\begin{aligned}
S_\theta(u, W(x_1, y_1, \lambda), W(x_2, y_2, \lambda)) &= \|u - \lambda x_1 - (1-\lambda) y_1\| \\
&\quad + \|u - \lambda x_2 - (1-\lambda) y_2\| \\
&\quad + \|\lambda x_1 + (1-\lambda) y_1 - \lambda x_2 - (1-\lambda) y_2\| \\
&\leq \lambda \|u - x_1\| + (1-\lambda) \|u - y_1\| \\
&\quad + \lambda \|u - x_2\| + (1-\lambda) \|u - y_2\| \\
&\quad + \lambda \|x_1 - x_2\| + (1-\lambda) \|y_1 - y_2\| \\
&= \lambda S_\theta(u, x_1, x_2) + (1-\lambda) S_\theta(u, y_1, y_2).
\end{aligned}
$$

From now on, (X, S_θ, W) will denote a convex extended partial S_b-metric space.

Definition 13. *A subset K of X is said to be convex if $W(x,y,\lambda) \in K$ for all $x, y \in X$ and $\lambda \in [0,1]$.*

Definition 14. *For all $x, y \in X$ and $\varepsilon > 0$, the ball of foci x and y, and of ray ε is given by*

$$B(x,y,\varepsilon) = \{z \in X : S_\theta(x,y,z) \leq \varepsilon\}.$$

The following propositions follow from the aforementioned definitions immediately.

Proposition 3 ([21]). *Let $\{K_\alpha\}_\alpha$ be a family of convex subsets of the convex extended partial S_b-metric space X, then $\cap K_\alpha$ is a convex subset of X as well.*

Proposition 4. *The balls $B(x,y,\varepsilon)$ are convex subsets of X. Moreover, they are closed subsets whenever S_θ is a continuous mapping.*

Proof. Let $a, b \in B(x,y,\varepsilon)$ and $\lambda \in [0,1]$.

$$
\begin{aligned}
S_\theta(x, y, W(a, b, \lambda)) &\leq \lambda S_\theta(x, y, a) + (1-\lambda) S_\theta(x, y, b) \\
&\leq \lambda \varepsilon + (1-\lambda) \varepsilon = \varepsilon.
\end{aligned}
$$

Furthermore, $B(x, y, \varepsilon) = T^{-1}([0, \varepsilon])$ where $T(z) = S_\theta(x, y, z)$ for all $z \in X$. The balls $B(x, y, \varepsilon)$ are closed subsets if S_θ is continuous. □

Before getting to our main result of this section, we fix some notations. Let A, B and C be nonempty subsets of (X, S_θ, W):

$$
\begin{aligned}
\text{dist}(A, B, C) &= \inf\{S_\theta(x, y, z) : x \in A, y \in B \text{ and } z \in C\}, \\
\delta(A, B, C) &= \sup\{S_\theta(x, y, z) : x \in A, y \in B \text{ and } z \in C\}, \\
\delta_{(x,y)}(C) &= \sup\{S_\theta(x, y, z) : z \in C\} \text{ for all } x \in A \text{ and } y \in B.
\end{aligned}
$$

Take note that extended partial S_b-metric spaces are, sort of, three-dimensional metric spaces and, since a tricyclic mapping is defined on the union of three subsets, the definition of a best proximity point for a tricylic mapping is naturally given by:

Definition 15. *Let $T : A \cup B \cup C \longrightarrow A \cup B \cup C$ be a tricyclic mapping where A, B and C are nonempty subsets of (X, S_θ). A point $x \in A \cup B \cup C$ is said to be a best proximity point for T provided that*

$$S_\theta\left(x, Tx, T^2 x\right) = \text{dist}(A, B, C).$$

Lemma 2. Let (A, B, C) be a nonempty, bounded, closed, and convex triad in X. Suppose that $T : A \cup B \cup C \longrightarrow A \cup B \cup C$ is a tricyclic mapping. If X has the property (C), then there exists a triad $(K_1, K_2, K_3) \subseteq (A, B)$ which is maximal with respect to being nonempty, closed and convex such that T is tricyclic on $K_1 \cup K_2$. Furthermore,

$$\overline{co}\,(T(K_1)) = K_2,\ \overline{co}\,(T(K_2)) = K_3\ \text{and}\ \overline{co}\,(T(K_3)) = K_1.$$

Proof. Let Γ denote the set of all nonempty, closed, and convex triads $(I, J, H) \subseteq (A, B, C)$ such that T is tricyclic on $I \cup J \cup H$. Note that Γ is partially ordered by reverse inclusion, that is,

$$(I_1, J_1, H_1) \leq (I_2, J_2, H_2) \iff (I_2, J_2, H_2) \subseteq (I_1, J_1, H_1).$$

Let $\{(I_\alpha, J_\alpha, H_\alpha)\}_\alpha$ be an increasing chain of Γ. Since X has the property (C) and from the fact that every intersection of convex subsets is a convex subset, $(\cap I_\alpha, \cap J_\alpha, \cap H_\alpha)$ is a nonempty, closed and convex triad. In addition, the maximal triad (K_1, K_2, K_3) is obtained as Zorn's Lemma states. Now, the triad $(\overline{co}\,(T(K_3)), \overline{co}\,(T(K_1)), \overline{co}\,(T(K_2))) \subseteq (K_1, K_2, K_3)$ is nonempty, closed and convex. We also have

$$T(\overline{co}\,(T(K_3))) \subseteq T(K_1) \subseteq \overline{co}\,(T(K_1)).$$

Similarly, we see that T is tricyclic on $\overline{co}\,(T(K_3)) \cup \overline{co}\,(T(K_1)) \cup \overline{co}\,(T(K_3))$. The desired result follows from the maximality of (K_1, K_2, K_3). $\square$

Theorem 3. Let (A, B, C) be a nonempty, bounded, closed, and convex triad in X such that S_θ is continuous and W is a double convex strusture. Let $(K_1, K_2, K_3) \subseteq (A, B, C)$ be a maximal triad with respect to being nonempty, closed and convex such that T is tricyclic on $K_1 \cup K_2 \cup K_3$. Suppose that $T : A \cup B \cup C \longrightarrow A \cup B \cup C$ is a tricyclic mapping such that

$$S_\theta(Tx, Ty, Tz) \leq \Lambda := k\delta(K_1, K_2, K_3) + (1-k)\,\text{dist}(A, B, C)$$

for all $(x, y, z) \in K_1 \times K_2 \times K_3$. If X has the property (C) then T has a best proximity triad.

Proof. Let $x \in K_1, y \in K_2$; the inequality satisfied by the mapping T implies that $Tz \in B(Tx, Ty, \Lambda)$ for all $z \in K_3$ and that means

$$T(K_3) \subseteq B(Tx, Ty, \Lambda).$$

Since S_θ is continuous, $B(Tx, Ty, \Lambda)$ is closed. Thus,

$$K_1 = \overline{co}\,(T(K_3)) \subseteq B(Tx, Ty, \Lambda).$$

Thus,

$$\delta_{(Tx,Ty)}(K_1) \leq \Lambda.$$

Put

$$L_1 := \left\{(x,y) \in K_1 \times K_2 : \delta_{(x,y)}(K_3) \leq \Lambda\right\},$$
$$L_2 := \left\{(y,z) \in K_2 \times K_3 : \delta_{(y,z)}(K_1) \leq \Lambda\right\},$$
$$L_3 := \left\{(z,x) \in K_3 \times K_1 : \delta_{(z,x)}(K_2) \leq \Lambda\right\}.$$

Clearly, (L_1, L_2, L_3) is a triad of nonempty, closed and convex subsets. Define

$$\widetilde{T} : (A \times B) \cup (B \times C) \cup (C \times A) \longrightarrow (A \times B) \cup (B \times C) \cup (C \times A)$$

$$(x, y) \longmapsto \widetilde{T}(x, y) = (Tx, Ty).$$

Since T is tricyclic on $A \cup B \cup C$, $\widetilde{T}$ is tricyclic on $(A \times B) \cup (B \times C) \cup (C \times A)$. For all $(x,y) \in K_1 \times K_2$, $\widetilde{T}(x,y) = (Tx, Ty) \in L_2$, then $\widetilde{T}(K_1 \times K_2) \subseteq L_2$. Thus, $\widetilde{T}$ is tricyclic on $L_1 \cup L_2 \cup L_3$. Furthermore, $(K_1 \times K_2, K_2 \times K_3, K_3 \times K_1)$ is maximal in

$$\widetilde{\Gamma} = \left\{ \begin{array}{l} ((I \times J), (J \times H), (H \times I)) \subseteq ((A \times B), (B \times C), (C \times A)) / \\ (I \times J), (J \times H) \text{ and } (H \times I) \text{ are non-empty, bounded, closed} \\ \text{and convex with } \widetilde{T} \text{ is tricyclic on } (I \times J) \cup (J \times H) \cup (H \times I) \end{array} \right\},$$

which is partially ordered by

$$((I_1 \times J_1), (J_1 \times H_1), (H_1 \times I_1)) \widetilde{\leq} ((I_2 \times J_2), (J_2 \times H_2), (H_2 \times I_2)) \iff$$

$$((I_2 \times J_2), (J_2 \times H_2), (H_2 \times I_2)) \subseteq ((I_1 \times J_1), (J_1 \times H_1), (H_1 \times I_1)).$$

Therefore,
$$L_1 = K_1 \times K_2, \; L_2 = K_2 \times K_3 \text{ and } L_3 = K_3 \times K_1.$$

Consequently, for all $(x,y) \in K_1 \times K_2$,

$$\delta_{(x,y)}(K_3) - k\delta(K_1, K_2, K_3) \leq (1-k) \, dist(A, B, C).$$

That is,
$$\delta(K_1, K_2, K_3) \leq dist(A, B, C).$$

Now, for all $(p, q, r) \in K_1 \times K_2 \times K_3$, we get

$$\begin{aligned} dist(A, B, C) &\leq S_\theta\left(p, Tp, T^2p\right), S_\theta\left(q, Tq, T^2q\right), S_\theta\left(r, Tr, T^2r\right) \\ &\leq \delta(K_1, K_2, K_3) \leq dist(A, B, C). \end{aligned}$$

In addition, this is a best proximity triad. $\square$

As a particular case of the previous theorem, we get the following result.

Corollary 3 ([12]). *Let A, B and C be nonempty, closed, bounded and convex subsets of reflexive Banach space X, let $T : A \cup B \cup C \longrightarrow A \cup B \cup C$ be a tricyclic contraction map i.e.,*

$$D(Tx, Ty, Tz) \leq kD(x, y, z) + (1-k) \, dist(A, B, C) \text{ for all } (x, y, z) \in A \times B \times C,$$

where $D(x, y, z) = \|x - y\| + \|y - z\| + \|z - x\|$. Then, T has a best proximity triad.

4. Conclusions

In this work, we have provided two best approximation result for cyclic mappings in thesetting of dualistic partial and convex, metric spaces. Next, we have provided best proximity point existence result for a new class of tricyclic mappings. Our three results extend and improve some results in the literature.

Author Contributions: Conceptualization, T.S.; Supervision, A.B. and M.A.; Validation, M.A.; Writing—original draft, T.S. and A.B.

Funding: This research received no external funding.

Acknowledgments: Research was supported by a National Centre of Scientific and Technological Research grant. The authors would like to express their gratitude to the editor and the anonymous referees for their constructive comments and suggestions, which have improved the quality of the manuscript.

Conflicts of Interest: The authors declare no conflict of interest.

References

1. Gabeleh, M.; Shahzad, N. Some new results on cyclic relatively nonexpansive mappings in convex metric spaces. *J. Inequal. Appl.* **2014**, *2014*, 350. [CrossRef]
2. Aydi, H.; Karapınar, E.; Erhan, I.M. Best proximity points of generalized almost Ψ-Geraghty contractive non-self-mappings. *Fixed Point Theory Appl.* **2014**, *2014*, 32. [CrossRef]
3. Kumam, P.; Aydi, H.; Karapınar, E.; Sintunavarat, W. Best proximity points and extension of Mizoguchi-Takahashi's fixed point theorems. *Fixed Point Theory Appl.* **2013**, *2013*, 242. [CrossRef]
4. Pitea, A. Best proximity results on dualistic partial metric spaces. *Symmetry* **2019**, *11*, 306. [CrossRef]
5. Eldered, A.A; Veeramani, P. Convergence and existence for best proximity points. *J. Math. Anal. Appl.* **2006**, *323*, 1001–1006. [CrossRef]
6. Sankar Raj, V.; Veeramani, P. Best proximity pair theorems for relatively nonexpansive mappings. *Appl. Gen. Topol.* **2009**, *10*, 21–28. [CrossRef]
7. Eldred, A.; Kirk, W.A.; Veeramani, P. Proximal normal structure and relatively nonexpansive mappings. *Stud. Math.* **2005**, *171*, 283–293. [CrossRef]
8. Sankar Raj, V. A best proximity point theorem for weakly contractive non-self-mappings. *Nonlinear Anal. Theory Methods Appl.* **2011**, *74*, 4804–4808. [CrossRef]
9. Abkar, A.; Gabaleh, M. A note on some best proximity point theorems proved under P-property. *Abstr. Appl. Anal.* **2013**, *2013*, 189567. [CrossRef]
10. Almeida, A.; Karapinar, E.; Sadrangani, K. A note on best proximity point theorems under weak P-property. *Abstr. Appl. Anal.* **2014**, *2014*, 716825. [CrossRef]
11. Karapinar, E.; Khojasteh, F. An approach to best proximity points results via simulation functions. *J. Fixed Point Theory Appl.* **2017**, *19*, 1983–1995. [CrossRef]
12. Sabar, T.; Aamri, M.; Bassou, A. Best proximity point of tricyclic contractions. *Adv. Fixed Point Theory* **2017**, *7*, 512–523.
13. Mukheimer, A. Extended partial Sb-metric spaces. *Axioms* **2018**, *7*, 87. [CrossRef]
14. O'Neill, S.J. Partial metric, valuations and domain theory. *Ann. N. Y. Acad. Sci.* **1996**, *806*, 304–315. [CrossRef]
15. Arshad, M.; Nazam, M.; Beg, I. Fixed point theorems in ordered dualistic partial metric spaces. *Korean J. Math.* **2016**, *24*, 169–179. [CrossRef]
16. Gabeleh, M.; Otafudu, O.O. Generalized pointwise noncyclic relatively nonexpansive mappings in strictly convex Banach spaces. *J. Nonlinear Convex Anal.* **2016**, *17*, 1117–1128.
17. Gabeleh, M.; Otafudu, O.O. Markov-Kakutani's theorem for best proximity pairs in Hadamard spaces. *Indag. Math.* **2017**, *28*, 680–693. [CrossRef]
18. Diaz, J.B.; Metcalf, F.T. On the structure of the set of subsequential limit points of successive approximations. *Bull. Am. Math. Soc.* **1967**, *73*, 51–59. [CrossRef]
19. Espínola, R.; Kosuru, G.S.R.; Veeramani, P. Pythagorean Property and Best-Proximity Point Theorems. *J. Optim. Theory Appl.* **2015**, *164*, 534–550. [CrossRef]
20. Kumam, P.; Mongkolkeha, C. Global optimization for quasi-noncyclic relatively nonexpansive mappings with application to analytic complex functions. *Mathematics* **2019**, *7*, 46. [CrossRef]
21. Takahashi, W. A convexity in metric space and nonexpansive mappings. *Kodai Math. Semin. Rep.* **1970**, *22*, 142–149. [CrossRef]
22. Shimizu, T.; Takahashi, W. Fixed points of multivalued mappings in certain convex metric spaces. *Topol. Methods Nonlinear Anal.* **1996**, *8*, 197–203. [CrossRef]

© 2019 by the authors. Licensee MDPI, Basel, Switzerland. This article is an open access article distributed under the terms and conditions of the Creative Commons Attribution (CC BY) license (http://creativecommons.org/licenses/by/4.0/).

Article

Relation Theoretic Common Fixed Point Results for Generalized Weak Nonlinear Contractions with an Application

Atiya Perveen [1],*, Idrees A. Khan [2] and Mohammad Imdad [1]

1. Department of Mathematics, Aligarh Muslim University, Aligarh 202002, India; mhimdad@yahoo.co.in
2. Department of Mathematics, Integral University, Lucknow 226026, India; idrees_maths@yahoo.com
* Correspondence: atiya.rs@amu.ac.in

Received: 11 March 2019; Accepted: 18 April 2019; Published: 24 April 2019

Abstract: In this paper, by introducing the concept of generalized Ćirić-type weak $(\phi_g, \mathcal{R})$-contraction, we prove some common fixed point results in partial metric spaces endowed with binary relation $\mathcal{R}$. We also deduce some useful consequences showing the usability of our results. Finally, we present an application to establish the solution of a system of integral equations.

Keywords: common fixed point; binary relation; preserving mapping; $(\phi_g, \mathcal{R})$-contraction; partial ordering

MSC: 54H25; 47H10

---

1. Introduction

With a view to enhance the domain of applicability, Matthews [1] initiated the idea of a partial metric space by weakening the metric conditions and also proved an analogue of Banach contraction principle in such spaces. Thereafter, many well-known results of metric fixed point theory were extended to partial metric spaces (see [2–16] and references therein).

On the other hand, Turinici [17] initiated the idea of order theoretic metric fixed point results, which was put in more natural and systematic forms by Ran and Reurings [18], Nieto and Rodríguez-López [19,20], and some others. Very recently, Alam and Imdad [21] extended the Banach contraction principle to complete metric space endowed with an arbitrary binary relation. This idea has inspired intense activity in this theme, and by now, there exists considerable literature around this result (e.g., [6,21–25]).

Proving new results in metric fixed point theory by replacing contraction conditions with a generalized one continues to be the natural approach. In recent years, several well-known contraction conditions such as Kannan type, Chatterjee type, Ciric type, phi-contractions, and some others were introduced in this direction.

In this paper, we introduce some useful notions, namely, $\mathcal{R}$-precompleteness, $\mathcal{R}$-g-continuity and $\mathcal{R}$-compatibility, and utilize the same to establish common fixed point results for generalized weak ϕ-contraction mappings in partial metric spaces endowed with an arbitrary binary relation $\mathcal{R}$. We also derive several useful corollaries which are either new results in their own right or sharpened versions of some known results. Finally, an application is provided to validate the utility of our result.

2. Preliminaries

Matthews [1] defined partial metric space as follows:

Definition 1. [1] Let M be a non-empty set. A mapping $\rho : M \times M \to [0, \infty)$ is said to be a partial metric if (for all $z_1, z_2, z_3 \in M$):

(a) $z_1 = z_2 \iff \rho(z_1, z_1) = \rho(z_1, z_2) = \rho(z_2, z_2)$;
(b) $\rho(z_1, z_1) \leq \rho(z_1, z_2)$;
(c) $\rho(z_1, z_2) = \rho(z_2, z_1)$;
(d) $\rho(z_1, z_2) \leq \rho(z_1, z_3) + \rho(z_3, z_2) - \rho(z_3, z_3)$.

The pair (M, ρ) is called a partial metric space.

Notice that in partial metric, the self-distance of any point need not be zero. A metric on a non-empty set M is a partial metric with the condition that for all $z \in M$, $\rho(z, z) = 0$.

A partial metric ρ generates a T_0-topology, say τ_ρ on M, with base the family of open balls $\mathcal{B}_\rho(z, \epsilon)$ ($z \in M$ and $\epsilon > 0$) defined as:

$$\mathcal{B}_\rho(z, \epsilon) = \{w \in M : \rho(z, w) \leq \rho(z, z) + \epsilon\}.$$

If ρ is a partial metric on M, then the function $d_\rho : M \times M \to [0, \infty)$ defined by:

$$d_\rho(z_1, z_2) = 2\rho(z_1, z_2) - \rho(z_1, z_1) - \rho(z_2, z_2),$$

is a metric on M.

Definition 2. [1] Let (M, ρ) be a partial metric space. Then:

(a) A sequence $\{z_n\}$ is said to be convergent to a point $z \in M$ if $\lim_{n \to \infty} \rho(z_n, z) = \rho(z, z)$.
(b) A sequence $\{z_n\}$ is said to be Cauchy if $\lim_{m,n \to \infty} \rho(z_n, z_m)$ exists and is finite.
(c) (M, ρ) is said to be complete if every Cauchy sequence $\{z_n\}$ in M converges (with respect to τ_ρ) to a point a $z \in M$ and $\rho(z, z) = \lim_{n \to \infty} \rho(z_n, z_m)$.

Remark 1. In a complete partial metric space, every closed subset is complete.

The following lemmas are needed in the sequel.

Lemma 1. [1] Let (M, ρ) be a partial metric space. Then:

(a) A sequence $\{z_n\}$ is Cauchy in (M, ρ) if and only if it is Cauchy in (M, d_ρ).
(b) (M, ρ) is complete if and only if the metric space (M, d_ρ) is complete. In addition:

$$\lim_{n \to \infty} d_\rho(z_n, z) = 0 \iff \rho(z, z) = \lim_{n \to \infty} \rho(z_n, z) = \lim_{m,n \to \infty} \rho(z_n, z_m).$$

Lemma 2. [2] Let (M, ρ) be a partial metric space and $\{z_n\}$ a sequence in M such that $\{z_n\} \to w$, for some $w \in M$ with $\rho(w, w) = 0$. Then, for any $z \in M$, we have $\lim_{n \to \infty} \rho(z_n, z) = \rho(w, z)$.

Definition 3. Let S and g be two self-mappings on a non-empty set M.

(a) An element $z \in M$ is said to be a coincidence point of S and g if $Sz = gz$.
(b) An element $z^* \in M$ is said to be a point of coincidence if $z^* = Sz = gz$, for some $z \in M$.
(c) If $z \in M$ is a point of coincidence of S and g such that $z = Sz = gz$, then z is called a common fixed point.

3. Relation Theoretic Notions and Auxiliary Results

Let M be a non-empty set. A binary relation $\mathcal{R}$ on M is a subset of $M \times M$. For $z_1, z_2 \in M$, we write $(z_1, z_2) \in \mathcal{R}$ if z_1 is related to z_2 under $\mathcal{R}$. Sometimes, we denote it as $z_1 \mathcal{R} z_2$ instead of $(z_1, z_2) \in \mathcal{R}$.

Further, if $(z_1,z_2) \in \mathcal{R}$ such that z_1 and z_2 are distinct, then we write $(z_1,z_2) \in \mathcal{R}^{\neq}$ (sometimes as $z_1\mathcal{R}^{\neq}z_2$). It is observed that $\mathcal{R}^{\neq} \subseteq \mathcal{R}$ is also a binary relation on M. $M \times M$ and $\emptyset$ are trivial binary relations on M, specifically called a universal relation and empty relation, respectively. The inverse, transpose or dual relation of $\mathcal{R}$ is denoted by $\mathcal{R}^{-1}$ and is defined as $\mathcal{R}^{-1} = \{(z_1,z_2) \in M \times M : (z_2,z_1) \in \mathcal{R}\}$. We denote by $\mathcal{R}^s$ the symmetric closure of $\mathcal{R}$, which is defined as $\mathcal{R}^s = \mathcal{R} \cup \mathcal{R}^{-1}$.

Throughout this manuscript, M is a non-empty set, $\mathcal{R}$ stands for a binary relation on M and I_M denotes an identity mapping, and S and g are self-mappings on M.

Definition 4. *[26] For a binary relation $\mathcal{R}$:*

(a) *Two elements $z_1, z_2 \in M$ are said to be $\mathcal{R}$-comparative if $(z_1,z_2) \in \mathcal{R}$ or $(z_2,z_1) \in \mathcal{R}$. We denote it by $[z_1,z_2] \in \mathcal{R}$.*
(b) *$\mathcal{R}$ is said to be complete if $[z_1,z_2] \in \mathcal{R}$, for all $z_1, z_2 \in M$.*

Proposition 1. *[21] For a binary relation $\mathcal{R}$ on M, we have (for all $z_1, z_2 \in M$):*

$$(z_1,z_2) \in \mathcal{R}^s \iff [z_1,z_2] \in \mathcal{R}.$$

Definition 5. *[21] A sequence $\{z_n\} \subseteq M$ is said to be $\mathcal{R}$-preserving if $(z_n, z_{n+1}) \in \mathcal{R}$, for all $n \in \mathbb{N}_0$.*

Here, we follow the notion (of $\mathcal{R}$-preserving) as used by Alam and Imdad [21]. Notice that Roldán and Shahzad [27] and Shahzad et al. [28] used the term "$\mathcal{R}$-nondecreasing" instead of "$\mathcal{R}$-preserving".

Definition 6. *[29] Let $N \subseteq M$. If for each $z_1, z_2 \in N$, there exists a point $z_3 \in M$ such that $(z_1,z_3) \in \mathcal{R}$ and $(z_2,z_3) \in \mathcal{R}$, then N is said to be $\mathcal{R}$-directed.*

Definition 7. *[30] For $z_1, z_2 \in M$, a path of length $l \in \mathbb{N}$ in $\mathcal{R}$ from z_1 to z_2 is a finite sequence $\{p_0, p_1, ..., p_l\} \subseteq M$ such that $p_0 = z_1$, $p_l = z_2$ and $(p_i, p_{i+1}) \in \mathcal{R}$, for each $0 \leq i \leq l-1$.*

Definition 8. *[31] Let $N \subseteq M$. If for each $z_1, z_2 \in N$, there exists a path in $\mathcal{R}$ from z_1 to z_2, then N is said to be $\mathcal{R}$-connected.*

Definition 9. *[21] $\mathcal{R}$ is said to be S-closed if $(z_1,z_2) \in \mathcal{R}$ implies that $(Sz_1, Sz_2) \in \mathcal{R}$, for all $z_1, z_2 \in M$.*

Definition 10. *[31] $\mathcal{R}$ is said to be (S,g)-closed if $(gz_1, gz_2) \in \mathcal{R}$ implies that $(Sz_1, Sz_2) \in \mathcal{R}$, for all $z_1, z_2 \in M$.*

Observe that on setting $g = I_M$, Definition 10 reduces to Definition 9.

Proposition 2. *[31] If $\mathcal{R}$ is (S,g)-closed, then $\mathcal{R}^s$ is also (S,g)-closed.*

Definition 11. *[23] $\mathcal{R}$ is said to be locally S-transitive if for each $\mathcal{R}$-preserving sequence $\{z_n\} \subseteq S(M)$ with range $E = \{z_n : n \in \mathbb{N}_0\}$, the binary relation $\mathcal{R}|_E$ is transitive.*

Motivated by Alam and Imdad [31], we introduce the notion of $\mathcal{R}$-continuity and $\mathcal{R}$-g-continuity in the context of partial metric space as follows:

Definition 12. *Let (M,ρ) be a partial metric space endowed with a binary relation $\mathcal{R}$. A self-mapping S on M is said to be $\mathcal{R}$-continuous at a point $z \in M$ if for any $\mathcal{R}$-preserving sequence $\{z_n\} \subseteq M$ such that $\{z_n\} \to z$, we have $\{Sz_n\} \to Sz$. S is $\mathcal{R}$-continuous if it is $\mathcal{R}$-continuous at each point of M.*

Definition 13. Let (M, ρ) be a partial metric space endowed with a binary relation $\mathcal{R}$. A self mapping S is said to be $(g, \mathcal{R})$-continuous at a point $z \in M$ if for any sequence $\{z_n\} \subseteq M$ with $\{gz_n\}$ $\mathcal{R}$-preserving and $\{gz_n\} \to gz$, we have $\{Sz_n\} \to Sz$. S is $\mathcal{R}$-g-continuous if it is $\mathcal{R}$-g-continuous at each point of M.

Remark 2. Notice that for $g = I_M$, Definition 13 reduces to Definition 12.

In the next definition, we introduce $\mathcal{R}$-compatibility.

Definition 14. Let (M, ρ) be a partial metric space endowed with binary relation $\mathcal{R}$ and $S, g : M \to M$. S and g are said to be $\mathcal{R}$-compatible if for any sequence $\{z_n\}$ such that $\{Sz_n\}$ and $\{gz_n\}$ are $\mathcal{R}$-preserving and $\lim_{n\to\infty} Sz_n = \lim_{n\to\infty} gz_n$, we have:
$$\lim_{n\to\infty} d_\rho(g(Sz_n), S(gz_n)) = 0.$$

Inspired by Imdad et al. [24], we introduce the following notions in the setting of partial metric spaces in the similar way.

Definition 15. Let (M, ρ) be a partial metric space endowed with a binary relation $\mathcal{R}$. A subset $N \subseteq M$ is said to be $\mathcal{R}$-precomplete if each $\mathcal{R}$-preserving Cauchy sequence $\{z_n\} \subseteq N$ converges to some $z \in M$.

Remark 3. Every $\mathcal{R}$-complete subset of M is $\mathcal{R}$-precomplete.

Proposition 3. Every $\mathcal{R}$-closed subspace of an $\mathcal{R}$-complete partial metric space is $\mathcal{R}$-complete.

Proposition 4. An $\mathcal{R}$-complete subspace of a partial metric space is $\mathcal{R}$-closed.

Next, we introduce the notion of ρ-self closedness in the setting of partial metric spaces.

Definition 16. Let (M, ρ) be a partial metric space endowed with binary relation $\mathcal{R}$. Then $\mathcal{R}$ is said to be ρ-self closed if for each $\mathcal{R}$-preserving sequence $\{z_n\} \subseteq M$ with $\{z_n\} \to z$, there exists a subsequence $\{z_{n_k}\}$ of $\{z_n\}$ such that $[z_{n_k}, z] \in \mathcal{R}$, for all $k \in \mathbb{N}_0$.

We now state the following lemma needed in our subsequent discussion.

Lemma 3. Let M be a non-empty set and $g : M \to M$. Then there exists a subset $N \subseteq M$ with $g(N) = g(M)$ and $g : N \to M$ is one–one.

We use the following notations in our subsequent discussions:
$Coin(S, g)$: Set of all coincidence points of S and g;
$M(g, S, \mathcal{R})$: The collection of all points $z \in M$ such that $[gz, Sz] \in \mathcal{R}$.

4. Main Results

Let Φ denote the set of all mappings $\phi : [0, \infty) \to [0, \infty)$ satisfying the following:

(Φ1) ϕ is non-decreasing;
(Φ2) $\phi(\delta) = 0$ iff $\delta = 0$ and $\liminf_{n\to\infty} \phi(\delta_n) > 0$ if $\lim_{n\to\infty} \delta_n > 0$.

Notice that Reference [32] used the condition that ϕ is continuous. Inspired by Reference [33], we replace their condition by a more weaker condition (Φ2). In fact, this condition is also weaker than that ϕ is lower semi-continuous. Indeed, if ϕ is a lower semi-continuous function, then for a sequence $\{\delta_n\}$ with $\lim_{n\to\infty} \delta_n = \delta > 0$, we have $\liminf_{n\to\infty} \phi(\delta_n) \geq \phi(\delta) > 0$.

Before presenting our main result, we define the following.

Definition 17. Let M be a non-empty set endowed with an arbitrary binary relation $\mathcal{R}$ and $N \subseteq M$. Then, N is said to be $(S, g, \mathcal{R})$-directed if for each $z_1, z_2 \in N$, there exists a point $z_3 \in M$ such that $(gz_i, gz_3) \in \mathcal{R}$, for $i = 1, 2$ and $(gz_3, Sz_3) \in \mathcal{R}$.

Definition 18. Let M be a non-empty set endowed with an arbitrary binary relation $\mathcal{R}$ and $N \subseteq M$. Then, N is said to be $(S, g, \mathcal{R})$-connected if for each $z_1, z_2 \in N$, there exists a path $\{gp_0, gp_1, ..., gp_l\} \subseteq g(M)$ between z_1 and z_2 such that $(gp_i, Sp_i) \in \mathcal{R}$, for $1 \leq i \leq l-1$.

Remark 4. For $g = I_M$, Definitions 17 and 18 reduce to $(S, \mathcal{R})$-directed and $(S, \mathcal{R})$-connected.

Now, we state and prove our first main result, which runs as follows:

Theorem 1. Let (M, ρ) be a partial metric space equipped with a binary relation $\mathcal{R}$, $N \subseteq M$, an $\mathcal{R}^{\neq}$-precomplete subspace in M and $S, g : M \to M$. Assume that the following conditions are satisfied:

(a) $M(g, S, \mathcal{R}) \neq \emptyset$;
(b) $\mathcal{R}$ is (S, g)-closed;
(c) $S(M) \subseteq g(M) \cap N$;
(d) $\mathcal{R}$ is locally S-transitive;
(e) S satisfies generalized Ćirić-type weak $(\phi_g, \mathcal{R})$-contraction, i.e.,

$$\rho(Sz, Sw) \leq \mathcal{M}_{\rho,g}(z,w) - \phi(\rho(Sz, Sw)), \quad (1)$$

for all $z, w \in M$ with $(gz, gw) \in \mathcal{R}^{\neq}$ and $\phi \in \Phi$, where:

$$\mathcal{M}_{\rho,g}(z,w) = \max\left\{\rho(gz, gw), \rho(gz, Sz), \rho(gw, Sw), \frac{\rho(gz, Sw) + \rho(gw, Sz)}{2}\right\};$$

(f) (f1) S and g are $\mathcal{R}^{\neq}$-compatible;
    (f2) S and g are $\mathcal{R}^{\neq}$-continuous;
    or alternatively:
(f*) (f*1) $N \subseteq g(M)$;
    (f*2) either S is $(g, \mathcal{R}^{\neq})$-continuous or S and g are continuous or $\mathcal{R}^{\neq}|_N$ is ρ-self closed.

Then, S and g have a coincidence point.

Proof. Choose $z_0 \in M$ as in (a) and construct a sequence $\{gz_n\}$ in M as follows:

$$gz_n = Sz_{n-1} = S^n z_0, \quad \forall n \in \mathbb{N}_0.$$

If there is some $m_0 \in \mathbb{N}_0$ such that $gz_{m_0} = gz_{m_0+1}$, then z_{m_0} is the coincidence point of the pair (S, g) and we are done. Henceforth, assume that $gz_n \neq gz_{n+1}$, for all $n \in \mathbb{N}_0$. In view of condition (b), we have $(gz_n, gz_{n+1}) \in \mathcal{R}$, for all $n \in \mathbb{N}_0$. Employing condition (e), we have:

$$\rho(Sz_{n-1}, Sz_n) \leq \mathcal{M}_{\rho,g}(z_{n-1}, z_n) - \phi(\rho(Sz_{n-1}, Sz_n)), \quad (2)$$

which implies:

$$\rho(gz_n, gz_{n+1}) = \rho(Sz_{n-1}, Sz_n) \leq \mathcal{M}_{\rho,g}(z_{n-1}, z_n), \quad (3)$$

where:

$$M_{\rho,g}(z_{n-1}, z_n) = \max\Big\{\rho(gz_{n-1}, gz_n), \rho(gz_{n-1}, Sz_{n-1}), \rho(gz_n, Sz_n),$$
$$\frac{\rho(gz_{n-1}, Sz_n) + \rho(gz_n, Sz_{n-1})}{2}\Big\}$$
$$= \max\Big\{\rho(gz_{n-1}, gz_n), \rho(gz_{n-1}, gz_n), \rho(gz_n, gz_{n+1}),$$
$$\frac{\rho(gz_{n-1}, gz_{n+1}) + \rho(gz_n, gz_n)}{2}\Big\}$$
$$\leq \max\Big\{\rho(gz_{n-1}, gz_n), \rho(gz_n, gz_{n+1}), \frac{\rho(gz_{n-1}, gz_n) + \rho(gz_n, gz_{n+1})}{2}\Big\}$$
$$= \max\{\rho(gz_{n-1}, gz_n), \rho(gz_n, gz_{n+1})\}.$$

Now, if $M_{\rho,g}(z_{n-1}, z_n) = \rho(gz_n, gz_{n+1})$, then Equation (2) becomes:

$$\rho(gz_n, gz_{n+1}) \leq \rho(gz_n, gz_{n+1}) - \phi(\rho(gz_n, gz_{n+1})),$$

a contradiction. Hence, we have $M_{\rho,g}(z_{n-1}, z_n) = \rho(gz_{n-1}, gz_n)$ and Equation (3) implies that $\{\rho(gz_n, gz_{n+1})\}$ is non-decreasing (also bounded below by 0). Thus, there exists $r \geq 0$ such that $\lim_{n\to\infty} \rho(gz_n, gz_{n+1}) = r$. Next, we show that $r = 0$. Suppose, by contrast, that it is not so, i.e., $r > 0$. Passing the limit $n \to \infty$ in Equation (2), we get:

$$r \leq r - \liminf_{n\to\infty} \phi(\rho(gz_n, gz_{n+1}))$$

which is a contradiction. Hence:

$$\lim_{n\to\infty} \rho(gz_n, gz_{n+1}) = 0. \qquad (4)$$

We also have:

$$d_\rho(gz_n, gz_{n+1}) = 2\rho(gz_n, gz_{n+1}) - \rho(gz_n, gz_n) - \rho(gz_{n+1}, gz_{n+1})$$
$$\leq 2\rho(gz_n, gz_{n+1}),$$

which, on letting $n \to \infty$ and applying Equation (4), yields that:

$$\lim_{n\to\infty} d_\rho(gz_n, gz_{n+1}) = 0.$$

Now, our claim is that $\{gz_n\}$ is a Cauchy sequence in (N, d_ρ). Otherwise, there exist two subsequences $\{gz_{m_k}\}$ and $\{gz_{n_k}\}$ of $\{gz_n\}$ such that n_k is the smallest integer for which:

$$n_k > m_k > k \text{ and } d_\rho(gz_{m_k}, gz_{n_k}) \geq \epsilon. \qquad (5)$$

Since $d_\rho(z, w) \leq 2\rho(z, w)$, for all $z, w \in M$, Equation (5) gives:

$$n_k > m_k > k, \ \rho(gz_{m_k}, gz_{n_k}) \geq \frac{\epsilon}{2} \text{ and } \rho(gz_{m_k}, gz_{n_k}) < \frac{\epsilon}{2}.$$

Now, using triangular inequality, we have:

$$\frac{\epsilon}{2} \leq \rho(gz_{m_k}, gz_{n_k}) \leq \rho(gz_{m_k}, gz_{n_k-1}) + \rho(gz_{n_k-1}, gz_{n_k})$$
$$< \frac{\epsilon}{2} + \rho(gz_{n_k-1}, gz_{n_k}).$$

Letting $k \to \infty$ in the above inequality, we obtain:

$$\lim_{k \to \infty} \rho(gz_{m_k}, gz_{n_k}) = \frac{\epsilon}{2}. \tag{6}$$

Again, the triangle inequality yields the following:

$$\rho(gz_{m_k}, gz_{n_k-1}) \leq \rho(gz_{m_k}, gz_{n_k}) + \rho(gz_{n_k}, gz_{n_k-1})$$

and:

$$\rho(gz_{m_k}, gz_{n_k}) \leq \rho(gz_{m_k}, gz_{n_k-1}) + \rho(gz_{n_k-1}, gz_{n_k})$$

which together give rise to:

$$|\rho(gz_{m_k}, gz_{n_k-1}) - \rho(gz_{m_k}, gz_{n_k})| \leq \rho(gz_{n_k-1}, gz_{n_k}).$$

Now, on taking $k \to \infty$, the above inequality gives:

$$\lim_{k \to \infty} \rho(gz_{m_k}, gz_{n_k-1}) = \frac{\epsilon}{2}.$$

In a similar manner, one can show that:

$$\lim_{k \to \infty} \rho(gz_{m_k-1}, gz_{n_k-1}) = \lim_{k \to \infty} \rho(gz_{m_k-1}, gz_{n_k}) = \frac{\epsilon}{2}.$$

Thus, we get:

$$\lim_{k \to \infty} M_{\rho,g}(z_{m_k-1}, z_{n_k-1}) = \frac{\epsilon}{2}. \tag{7}$$

Using (d), we have $(gz_{m_k-1}, gz_{n_k-1}) \in \mathcal{R}$ and hence, Equation (1) implies:

$$\rho(gz_{m_k}, gz_{n_k}) \leq M_{\rho,g}(z_{m_k-1}, z_{n_k-1}) - \phi(\rho(gz_{m_k}, gz_{n_k})).$$

Using Equations (6) and (7) and letting $k \to \infty$ in the above inequality, we get:

$$\frac{\epsilon}{2} \leq \frac{\epsilon}{2} - \liminf_{k \to \infty} \phi(\rho(gz_{m_k}, gz_{n_k})),$$

a contradiction. Hence, $\{gz_n\}$ is Cauchy in (N, d_ρ) (as $\{gz_n\} \subseteq S(M) \subseteq N$) which is also $\mathcal{R}^{\neq}$-preserving. Lemma 1 ensures that it is also Cauchy in (N, ρ). Thus, the $\mathcal{R}^{\neq}$-precompleteness of N in M ensures the existence of a point $\bar{z} \in M$ such that:

$$\lim_{n \to \infty} gz_n = \bar{z}. \tag{8}$$

Thus, we also have:

$$\lim_{n \to \infty} d_\rho(gz_n, \bar{z}) = 0. \tag{9}$$

Now, by Equation (9) and Lemma 1, we get:

$$\rho(\bar{z}, \bar{z}) = \lim_{m,n \to \infty} \rho(gz_n, \bar{z}) = \lim_{m,n \to \infty} \rho(gz_m, gz_n) = 0. \tag{10}$$

Further, by the definition of $\{gz_n\}$ and Equation (8), we have:

$$\lim_{n \to \infty} Sz_n = \bar{z}. \tag{11}$$

Finally, to prove the existence of coincidence point of S and g, we make use of conditions (f) and (f^*). Firstly, assume that (f) holds. Now, as $(gz_n, gz_{n+1}) \in \mathcal{R}^{\neq}$, so using assumption $(f2)$ and Equation (8), we obtain:

$$\lim_{n\to\infty} g(gz_n) = g(\lim_{n\to\infty} gz_n) = g\bar{z}. \tag{12}$$

By the definition of $\{gz_n\}$, we have $\{Sz_n\}$ is also $\mathcal{R}^{\neq}$-preserving (i.e., $(Sz_n, Sz_{n+1}) \in \mathcal{R}^{\neq}$, for all n), so using assumption $(f2)$ and Equation (11), we get:

$$\lim_{n\to\infty} g(Sz_n) = g(\lim_{n\to\infty} Sz_n) = g\bar{z}. \tag{13}$$

By using Equation (8) and $\mathcal{R}^{\neq}$-continuity of S, we obtain:

$$\lim_{n\to\infty} S(gz_n) = S(\lim_{n\to\infty} gz_n) = S\bar{z}. \tag{14}$$

As $\{Sz_n\}$ and $\{gz_n\}$ are $\mathcal{R}^{\neq}$-preserving and $\lim_{n\to\infty} Sz_n = \lim_{n\to\infty} gz_n = \bar{z}$, by the condition $(f1)$, we have:

$$\lim_{n\to\infty} d_\rho(g(Sz_n), S(gz_n)) = 0. \tag{15}$$

Now, from Equations (13)–(15) and continuity of d_ρ, it follows that:

$$d_\rho(g\bar{z}, S\bar{z}) = d_\rho(\lim_{n\to\infty} g(Sz_n), \lim_{n\to\infty} S(gz_n))$$
$$= \lim_{n\to\infty} d_\rho(g(Sz_n), S(gz_n))$$
$$= 0,$$

i.e., $g\bar{z} = S\bar{z}$ and we are done. Secondly, suppose that (f^*) is satisfied. Then, by (f^*1), there exists some $z \in M$ such that $\bar{z} = gz$. Hence, Equations (8) and (11) respectively reduce to:

$$\lim_{n\to\infty} gz_n = gz, \tag{16}$$

and:

$$\lim_{n\to\infty} Sz_n = gz. \tag{17}$$

Next, to accomplish that z is a coincidence point of S and g, we utilize (f^*2). Thus, suppose that S is $\mathcal{R}^{\neq}$-g-continuous, then using Equation (16), we obtain:

$$\lim_{n\to\infty} Sz_n = Sz. \tag{18}$$

Now, by virtue of uniqueness of limit, Equations (17) and (18) give $Sz = gz$.

Next, assume that S and g are continuous. Then owing to Lemma 3, there exists $D \subseteq M$ such that $g(D) = g(M)$ and $g : D \to M$ is injective. Now, define a mapping $\tilde{S} : g(D) \to g(M)$ by:

$$\tilde{S}(gt) = St, \; \forall gt \in g(D). \tag{19}$$

As $g : D \to M$ is injective and $S(M) \subseteq g(M)$, $\tilde{S}$ is well-defined. Further, due to the continuity of S and g, $\tilde{S}$ is continuous. The fact that $g(D) = g(M)$, assumptions (c) and (f^*1) imply that:

$$S(M) \subseteq g(D) \cap N \text{ and } N \subseteq g(D).$$

Thus, without loss of generality, we can construct $\{z_n\} \subseteq D$, satisfying Equation (16) with $z \in D$. On using Equations (16), (17), and (19) with continuity of $\tilde{S}$, we obtain:

$$Sz = \tilde{S}(gz) = \tilde{S}(\lim_{n \to \infty} gz_n) = \lim_{n \to \infty} \tilde{S}(gz_n) = \lim_{n \to \infty} Sz_n = gz,$$

and we are done. Alternatively, if $\mathcal{R}^{\neq}|_N$ is ρ-self closed, then for any $\mathcal{R}^{\neq}$-preserving sequence $\{gz_n\}$ in N with $\{gz_n\} \to gz$, there exists a subsequence $\{gz_{n_k}\}$ of $\{gz_n\}$ such that $[gz_{n_k}, gz] \in \mathcal{R}$, for all $k \in \mathbb{N}_0$. Suppose $\rho(gz, Sz) > 0$, then we have:

$$\mathcal{M}_{\rho,g}(z_{n_k}, z) = \max\left\{\rho(gz_{n_k}, gz), \rho(gz_{n_k}, Sz_{n_k}), \rho(gz, Sz), \frac{\rho(gz_{n_k}, Sz) + \rho(gz, Sz_{n_k})}{2}\right\}.$$

Letting $k \to \infty$ and using Equation (8), we get:

$$\lim_{k \to \infty} \mathcal{M}_{\rho,g}(z_{n_k}, z) = \rho(gz, Sz). \qquad (20)$$

Now, applying $z = z_{n_k}$ and $w = z$, condition (e) gives:

$$\rho(Sz_{n_k}, Sz) \leq \mathcal{M}_{\rho,g}(z_{n_k}, z) - \phi(\rho(Sz_{n_k}, Sz)),$$

which, on letting $n \to \infty$ and using Equations (8) and (20) and Lemma 2, yields that:

$$\rho(gz, Sz) \leq \rho(gz, Sz) - \liminf_{k \to \infty} \phi(\rho(gz_{n_k+1}, Sz)),$$

a contradiction. Hence $\rho(gz, Sz) = 0$, i.e., $gz = Sz$. This completes the proof. □

Now, we present a corresponding uniqueness result.

Theorem 2. *In addition to the assumptions of Theorem 1, if we assume that the following condition is satisfied:*

(g) $S(M)$ is $(S, g, \mathcal{R}^s)$-connected,

then S and g have a unique point of coincidence. Moreover, if:

(h) S and g are weakly compatible,

then S and g have a unique common fixed point.

Proof. Firstly, Theorem 1 ensures that $Coin(S, g) \neq \emptyset$. Let $\bar{z}, \bar{w} \in Coin(S, g)$. Then, there exists $z, w \in M$ such that $\bar{z} = Sz = gz$ and $\bar{w} = Sw = gw$. Our claim is that $\bar{z} = \bar{w}$. Now, owing to hypothesis (g), there exists a path, say $\{gp_0, gp_1, gp_2, ..., gp_l\} \subseteq M$ of some finite length l in $\mathcal{R}|^s_{g(M)}$ from Sz to Sw with:

$$gp_0 = Sz, \; gp_l = Sw \text{ and } [gp_i, gp_{i+1}] \in \mathcal{R}, \text{ for each } 0 \leq i \leq l-1 \qquad (21)$$

and:

$$[gp_i, Sp_i] \in \mathcal{R}, \text{ for each } 1 \leq i \leq l-1. \qquad (22)$$

Define constant sequences $\{p_n^0 = z\}$ and $\{p_n^l = w\}$, then we have $gp_{n+1}^0 = Sp_n^0 = Sz = \bar{z}$ and $gp_{n+1}^l = Sp_n^l = Sw = \bar{w}$, for all $n \in \mathbb{N}_0$. Further, set $p_0^i = p_i$, for each $0 \leq i \leq l$ and define sequences $\{p_n^1\}, \{p_n^2\}, ..., \{p_n^{k-1}\}$ by:

$$gp_{n+1}^i = Sp_n^i, \; \forall n \in \mathbb{N}_0 \text{ and for each } 1 \leq i \leq l-1.$$

Hence:

$$gp_{n+1}^i = Sp_n^i, \; \forall n \in \mathbb{N}_0 \text{ and for each } 0 \leq i \leq l.$$

By mathematical induction, we will prove that:

$$[gp_n^i, gp_n^{i+1}] \in \mathcal{R}, \; \forall n \in \mathbb{N}_0 \text{ and for each } 0 \leq i \leq l-1.$$

In view of Equation (21), the result holds for $n = 0$. Now, suppose it holds for $n = k > 0$, i.e.:

$$[gp_k^i, gp_k^{i+1}] \in \mathcal{R}, \text{ for each } 0 \leq i \leq l-1.$$

By (S, g)-closedness of $\mathcal{R}$ and Proposition 2, we have:

$$[Sp_k^i, Sp_k^{i+1}] = [gp_{k+1}^i, gp_{k+1}^{i+1}] \in \mathcal{R}, \text{ for each } 0 \leq i \leq l-1,$$

i.e., the result holds for $n = k+1$ and hence, it holds for all $n \in \mathbb{N}_0$. Also from Equation (22), we have $[gp_0^i, gp_1^i] \in \mathcal{R}$ and $\mathcal{R}$ is (S, g)-closed, so by Proposition 2 and Equation (4), we have:

$$\lim_{n \to \infty} \rho(gp_n^i, gp_{n+1}^i) = 0. \tag{23}$$

Now, for all $n \in \mathbb{N}_0$ and for each $0 \leq i \leq l-1$, define $f_n^i = \rho(gp_n^i, gp_n^{i+1})$. Our claim is that:

$$\lim_{n \to \infty} f_n^i = 0.$$

Suppose, by contrast, that $\lim_{n \to \infty} f_n^i = f > 0$. Since $[gp_n^i, gp_n^{i+1}] \in \mathcal{R}$, $(gp_n^i, gp_n^{i+1}) \in \mathcal{R}$ or $(gp_n^{i+1}, gp_n^i) \in \mathcal{R}$, for all $n \in \mathbb{N}_0$ and for each $0 \leq i \leq l-1$. Making use of Equation (1), we have:

$$\rho(Sp_n^i, Sp_n^{i+1}) \leq \mathcal{M}_{\rho,g}(p_n^i, p_n^{i+1}) - \phi(\rho(Sp_n^i, Sp_n^{i+1}))$$

or:

$$\rho(gp_{n+1}^i, Sp_{n+1}^{i+1}) \leq \mathcal{M}_{\rho,g}(p_n^i, p_n^{i+1}) - \phi(\rho(gp_{n+1}^i, gp_{n+1}^{i+1})), \tag{24}$$

where:

$$\mathcal{M}_{\rho,g}(p_n^i, p_n^{i+1}) = \max\Big\{\rho(gp_n^i, gp_n^{i+1}), \rho(gp_n^i, Sp_n^i), \rho(gp_n^{i+1}, Sp_n^{i+1}),$$
$$\frac{\rho(gp_n^i, Sp_n^{i+1}) + \rho(gp_n^{i+1}, Sp_n^i)}{2}\Big\}$$
$$= \max\Big\{\rho(gp_n^i, gp_n^{i+1}), \rho(gp_n^i, gp_{n+1}^i), \rho(gp_n^{i+1}, gp_{n+1}^{i+1}),$$
$$\frac{\rho(gp_n^i, gp_{n+1}^{i+1}) + \rho(gp_n^{i+1}, gp_{n+1}^i)}{2}\Big\}$$
$$\leq \max\Big\{\rho(gp_n^i, gp_n^{i+1}), \rho(gp_n^i, gp_{n+1}^i), \rho(gp_n^{i+1}, gp_{n+1}^{i+1}),$$
$$\frac{\rho(gp_n^i, gp_{n+1}^i) + \rho(gp_{n+1}^i, gp_{n+1}^{i+1}) + \rho(gp_n^{i+1}, gp_n^i) + \rho(gp_n^i, gp_{n+1}^i)}{2}\Big\}.$$

Now, letting $n \to \infty$ and using Equation (23), we obtain:

$$\lim_{n \to \infty} \mathcal{M}_{\rho,g}(p_n^i, p_n^{i+1}) = f,$$

which, on applying Equation (24) after taking limit, yields that:

$$f \leq f - \liminf_{n \to \infty} \phi(\rho(p_{n+1}^i, p_{n+1}^{i+1})),$$

a contradiction. Therefore, $\lim_{n\to\infty} f_n^i = 0$.

Next, we have:

$$\rho(\bar{z}, \bar{w}) = \rho(gp_n^0, gp_n^l) \leq \sum_{i=0}^{k-1} \rho(gp_n^i, gp_n^{i+1}) - \sum_{i=1}^{k-1} \rho(gp_n^i, gp_n^{i+1})$$

$$\leq \sum_{i=0}^{k-1} \rho(gp_n^i, gp_n^{i+1})$$

$$= \sum_{i=0}^{k-1} f_n^i \to 0 \text{ (as } n \to \infty).$$

Hence, $\bar{z} = \bar{w}$, i.e., $Sz = Sw$. Thus, S and g have a unique point of coincidence.

Secondly, to justify the existence of a unique common fixed point, we consider $z \in Coin(S, g)$, i.e., $Sz = gz = \bar{z}$, for some $\bar{z} \in M$. By the condition (h), S and g commute at their coincidence points, i.e.,

$$S(gz) = g(Sz) = g(gz), \qquad (25)$$

thereby yielding $S\bar{z} = g\bar{z}$, i.e., $\bar{z} \in Coin(S, g)$. Thus, by the uniqueness of point of the coincidence point, we have:

$$\bar{z} = g\bar{z} = S\bar{z}.$$

The uniqueness of the common fixed point is a direct consequence of the uniqueness of the coincidence point. This finishes the proof. □

We present the following example to support our result.

Example 1. *Let $M = [0, \infty)$ with partial metric $\rho : M \times M \to [0, \infty)$ defined by:*

$$\rho(z_1, z_2) = \max\{z_1, z_2\}.$$

Define a binary relation $\mathcal{R} = \{(z_1, z_2) \in M \times M : z_1 \geq z_2\}$. Clearly, (M, ρ) is a complete partial metric space. Define $S, g : M \to M$ by:

$$Sz = \frac{z}{3} \text{ and } gz = \frac{z}{2}, \forall z \in M.$$

It is clear that $\mathcal{R}$ is (S,g)-closed and S and g are continuous. Next, define $\phi : [0, \infty) \to [0, \infty)$ by:

$$\phi(t) = \frac{t}{6}, \forall t \in [0, \infty).$$

Clearly, $\phi \in \Phi$. Observe that all the conditions of Theorems 1 and 2 are fulfilled (with $N = M$). Hence, S and g have a unique common fixed point (namely 0).

Next, we present the following corollaries.

Corollary 1. *The conclusion of Theorem 2 remains valid if we replace the condition (g) by any one of the following:*

(g1) $\mathcal{R}|_{g(M)}$ *is complete;*
(g2) $S(M)$ *is $(S, g, \mathcal{R}^s)$-directed.*

Proof. If (g1) holds true, then for any $z_1, z_2 \in S(M)$, we have $z_1 = gw_1$ and $z_2 = gw_2$, for some $w_1, w_2 \in M$ (as $S(M) \subseteq g(M)$). In view of (g1), we have $[gw_1, gw_2] \in \mathcal{R}|_{g(M)}$, i.e., $\{gw_1, gw_2\}$ is a path of length 1 in $\mathcal{R}|_{g(M)}^s$ from z_1 to z_2. Hence, condition (g) of Theorem 2 is fulfilled and the result is concluded by Theorem 2.

On the other hand, if condition (g2) holds, then for each $z_1, z_2 \in S(M)$ (such that $z_1 = gw_1$ and $z_2 = gw_2$, for $w_1, w_2 \in M$), there exists $w_3 \in M$ such that $[gw_1, gw_3], [gw_2, gw_3] \in \mathcal{R}|_{g(M)}$, i.e., $\{gw_1, gw_3, gw_2\}$ is a path of length 2 in $\mathcal{R}|^s_{g(M)}$ from z_1 to z_2 and $[gw_3, Sw_3] \in \mathcal{R}|_{g(M)}$. Hence, condition (g) of Theorem 2 is fulfilled and again by Theorem 2, the conclusion follows. □

Corollary 2. *The conclusions of Theorems 1 and 2 remain true if we replace assumption (e) by the following one:*

(e1)   *S satisfies*
$$\rho(Sz, Sw) \leq \rho(gz, gw) - \phi(\rho(Sz, Sw)), \tag{26}$$

for all $z, w \in M$ with $(gz, gw) \in \mathcal{R}^{\neq}$ and $\phi \in \Phi$.

Proof. As $\rho(gz, gw) \leq M_{\rho,g}(z, w)$, we have:
$$\rho(Sz, Sw) \leq \rho(gz, gw) - \phi(\rho(Sz, Sw)) \implies \rho(Sz, Sw) \leq M_{\rho,g}(z, w) - \phi(\rho(Sz, Sw)),$$

for all $z, w \in M$ with $(gz, gw) \in \mathcal{R}^{\neq}$. Thus, all the assumptions of Theorems 1 and 2 are satisfied and the conclusions hold. □

Following Reference [32], it can be easily seen that in a partial metric space (M, ρ), for all $(gz, gw) \in \mathcal{R}^{\neq}$, the conditions:
$$\rho(Sz, Sw) \leq \rho(gz, gw) - \phi(\rho(Sz, Sw)), \tag{27}$$

and:
$$\rho(Sz, Sw) \leq M_{\rho,g}(z, w) - \phi(\rho(Sz, Sw)), \tag{28}$$

are more weaker than:
$$\rho(Sz, Sw) \leq \rho(gz, gw) - \phi(\rho(gz, gw)), \tag{29}$$

and:
$$\rho(Sz, Sw) \leq M_{\rho,g}(z, w) - \phi(M_{\rho,g}(z, w)), \tag{30}$$

respectively. However, the converse need not be true in general (even the above assertion is true for any $z, w \in M$). This leads us to our next corollary.

Corollary 3. *The conclusions of Theorems 1 and 2 remain true if we replace assumption (e) by the following one:*

(e2)   *S satisfies:*
$$\rho(Sz, Sw) \leq \rho(gz, gw) - \phi(\rho(gz, gw)), \tag{31}$$

or:
$$\rho(Sz, Sw) \leq M_{\rho,g}(z, w) - \phi(M_{\rho,g}(z, w)), \tag{32}$$

for all $z, w \in M$ with $(gz, gw) \in \mathcal{R}^{\neq}$ and $\phi \in \Phi$.

By setting $\phi(t) = (1 - k)t$, with $k \in [0, 1)$ and $t \in [0, \infty)$ in Corollary 3, we deduce the following corollaries:

Corollary 4. *The conclusions of Theorems 1 and 2 remain true if we replace assumption (e) with the following one:*

(e3)   *there exists $k \in [0, 1)$ such that:*
$$\rho(Sz, Sw) \leq k\rho(gz, gw),$$

for all $z, w \in M$ with $(gz, gw) \in \mathcal{R}^{\neq}$ and $\phi \in \Phi$.

We see that the above corollary is a relatively new and somewhat refined version of Alam and Imdad [31] type result in partial metric space with some refinement, e.g.:

- We use $\mathcal{R}^{\neq}$-precompleteness of subspace $N \subseteq M$ in place of $\mathcal{R}$-completeness.
- We use $\mathcal{R}^{\neq}$-analogous of compatibility, continuity, closedness and ρ-self closedness instead of their $\mathcal{R}$-analogous.

Corollary 5. *The conclusions of Theorems 1 and 2 remain true if we replace assumption (e) with the following one:*

(e4)   *S satisfies:*
$$\rho(Sz, Sw) \leq k M_{\rho,g}(z, w), \tag{33}$$
for all $z, w \in M$ with $(gz, gw) \in \mathcal{R}^{\neq}$ and $\phi \in \Phi$.

By considering $g = I_M$, the following fixed point result can be deduced easily from Theorems 1 and 2.

Corollary 6. *Let (M, ρ) be a partial metric space equipped with a binary relation $\mathcal{R}$, $N \subseteq M$ an $\mathcal{R}^{\neq}$-precomplete subspace in M and $S : M \to M$. Assume that the following assumptions are satisfied:*

(a)   *There exists $z_0 \in M$ such that $(z_0, Sz_0) \in \mathcal{R}$;*
(b)   *$\mathcal{R}$ is S-closed;*
(c)   *$S(M) \subseteq N$;*
(d)   *$\mathcal{R}$ is locally S-transitive;*
(e)   *S satisfies generalized Ćirić-type weak $(\phi, \mathcal{R})$-contraction, i.e.:*

$$\rho(Sz, Sw) \leq M(z, w) - \phi(\rho(Sz, Sw)),$$

for all $z, w \in M$ with $(z, w) \in \mathcal{R}^{\neq}$ and $\phi \in \Phi$, where:

$$M(z, w) = \max \left\{ \rho(z, w), \rho(z, Sz), \rho(w, Sw), \frac{\rho(z, Sw) + \rho(w, Sz)}{2} \right\};$$

(f)   *either S is $\mathcal{R}^{\neq}$-continuous or $\mathcal{R}^{\neq}|_N$ is ρ-self closed.*

Then, S has a fixed point. In addition, if:

(g)   *N is $(S, \mathcal{R}^s)$-connected,*

then the fixed point is unique.

In place of $\mathcal{R}^{\neq}$-precomplete of N, if we use the $\mathcal{R}^{\neq}$-completeness of the whole space M, then we find a particular version of Theorem 1.

Corollary 7. *Let $(M, \rho, \mathcal{R})$ be an $\mathcal{R}^{\neq}$-complete partial metric space and $S, g : M \to M$ satisfy the following assumptions:*

(a) $M(g, S, \mathcal{R}) \neq \emptyset$;
(b) $\mathcal{R}$ is (S, g)-closed;
(c) $S(M) \subseteq g(M)$;
(d) $\mathcal{R}$ is locally S-transitive;
(e)   *S satisfies generalized Ćirić-type weak $(\phi_g, \mathcal{R})$-contraction, i.e.,:*

$$\rho(Sz, Sw) \leq M_{\rho,g}(z, w) - \phi(\rho(Sz, Sw)), \tag{34}$$

for all $z, w \in M$ with $(gz, gw) \in \mathcal{R}^{\neq}$ and $\phi \in \Phi$, where:

$$\mathcal{M}_{\rho,g}(z,w) = \max\left\{\rho(gz,gw), \rho(gz,Sz), \rho(gw,Sw), \frac{\rho(gz,Sw)+\rho(gw,Sz)}{2}\right\};$$

(f)   (f1) S and g are $\mathcal{R}^{\neq}$-compatible;
        (f2) S and g are $\mathcal{R}^{\neq}$-continuous;
        or alternatively:

(f*)   (f*1) there exists an $\mathcal{R}^{\neq}$-closed subspace N of M such that $S(M) \subseteq N \subseteq g(M)$;
         (f*2) either S is $\mathcal{R}^{\neq}$-g-continuous or S and g are continuous or $\mathcal{R}^{\neq}|_N$ is ρ-self closed.

Then, S and g have a coincidence point.

Proof. The result follows by Proposition 3 and Remark 3. □

Moreover, in Corollary 7, if we assume g to be surjective, then assumption (c) as well as assumption (f*1) can be removed trivially since $N = g(M) = M$.

5. Consequences

5.1. Results in Abstract Spaces

By considering $\mathcal{R}$ to be the universal relation, i.e., $\mathcal{R} = M \times M$, we deduce the following results from Theorems 1 and 2.

Corollary 8. Let (M, ρ) be a partial metric space and $S, g : M \to M$. Assume that the following conditions are satisfied:

(a) $S(M) \subseteq g(M) \cap N$;
(b)   S satisfies:

$$\rho(Sz, Sw) \leq \mathcal{M}_{\rho,g}(z,w) - \phi(\rho(Sz,Sw)),$$

for all $z, w \in M$ with $gz \neq gw$ and $\phi \in \Phi$;

(c)   (c_1) S and g are compatible;
        (c_2) S and g are continuous;
        or alternatively:

(c*) $N \subseteq g(M)$.

Then, S and g have a coincidence point.

Corollary 9. Moreover, if S and g are weakly compatible, then S and g have a unique common fixed point.

In view of Corollary 4 under $\mathcal{R} = M \times M$, it can be easily seen that Corollary 8 is a more generalized and sharpened version of Goebel and Jungck type results in partial metric spaces.

5.2. Results in Ordered Partial Metric Spaces via Increasing Mappings

The idea under consideration was initiated by Turinici [17], which was later generalized by several authors, e.g., Ran and Reurings [18], Nieto and Rodríguez-López [19], and some others, e.g., the authors of [34–37]. In this section, from now on, $\preceq$ denotes a partial order on a non-empty set M, $(M, \preceq)$ denotes a partially ordered set, and $(M, \rho, \preceq)$ stands for a partial metric space with partial order $\preceq$, which we call ordered partial metric space.

Now, we recall the following notions which are needed in the sequel.

Definition 19. [38] A mapping $S : M \to M$ is said to be g-increasing if $Sz_1 \preceq Sz_2$, for any $z_1, z_2 \in M$ with $gz_1 \preceq gz_2$.

Remark 5. *Notice that S is g-increasing and the notion $\preceq$ is (S,g)-closed in our sense coincide with each other.*

Definition 20. *[39] Let $\{z_n\}$ be a sequence in an ordered set $(M, \preceq)$. Then:*

(a) $\{z_n\}$ *is said to be increasing if for all $m, n \in \mathbb{N}_0$:*

$$m \leq n \implies z_m \preceq z_n.$$

(b) $\{z_n\}$ *is said to be decreasing if for all $m, n \in \mathbb{N}_0$:*

$$m \leq n \implies z_n \preceq z_m.$$

(c) $\{z_n\}$ *is said to be monotone if it is either increasing or decreasing.*

Now, we introduce the notion of increasing-convergence-upper bound (ICU) property in the setting of ordered partial metric spaces.

Definition 21. *Let $(M, \rho, \preceq)$ be an ordered partial metric space. We say that $(M, \rho, \preceq)$ has ICU (increasing-convergence-upper bound) property if every increasing sequence $\{z_n\} \subseteq M$ such that $\{z_n\} \to z$ is bounded above by limit, i.e., $z_n \preceq z$, for all $n \in \mathbb{N}$.*

Remark 6. *It is observed that $(M, \rho, \preceq)$ has ICU property is equivalent to $\preceq$ is ρ-self closed.*

Notice that Alam et al. [40] defined ICU property in the setting of ordered metric spaces.

Definition 22. *In an ordered partial metric space $(M, \rho, \preceq)$, we define the following:*

(a) *$(M, \rho, \preceq)$ is said to be $\overline{O}$-complete (resp. $\underline{O}$-complete, O-complete) if every increasing (resp. decreasing, monotone) Cauchy sequence in M converges in M.*
(b) *a self-mapping S on M is said to be $(g, \overline{O})$-continuous (resp. $(g, \underline{O})$-continuous, (g, O)-continuous) at $z \in M$, if for any increasing (resp. decreasing, monotone) sequence $\{z_n\} \subseteq M$ such that $\{z_n\} \to z$, we have $\{Sz_n\} \to Sz$.*
S is $(g, \overline{O})$-continuous (resp. $(g, \underline{O})$-continuous, (g, O)-continuous) on M if it is $(g, \overline{O})$-continuous (resp. $(g, \underline{O})$-continuous, (g, O)-continuous) at every $z \in M$.
(c) *two self-mappings S and g are said to be $\overline{O}$-compatible (resp. $\underline{O}$-compatible, O-compatible) if for any sequence $\{z_n\}$ and $z \in M$ such that $\{Sz_n\}$ and $\{gz_n\}$ are increasing (resp. decreasing and monotone) and $\lim_{n\to\infty} Sz_n = \lim_{n\to\infty} gz_n = z$, we have:*

$$\lim_{n\to\infty} \rho(S(gz_n), g(Sz_n)) = 0.$$

Remark 7. *Notice that for $g = I$, $(g, \overline{O})$-continuity reduces to $\overline{O}$-continuity, and the same happens to the others.*

The above notions were defined by Kutbi et al. [41] in the setting of ordered metric spaces. Now, we introduce the following notion.

Definition 23. *A subset N of an ordered partial metric space $(M, \rho, \preceq)$ is said to be $\overline{O}$-precomplete (resp. $\underline{O}$-precomplete, O-precomplete) if every increasing (resp. decreasing, monotone) Cauchy sequence in N converges to a point of M.*

Under consideration of Remarks 5 and 6 and $\mathcal{R} = \preceq$, we obtained the below result from Theorem 1, which is new for the existing literature.

Corollary 10. Let $(M, \rho, \preceq)$ be an ordered partial metric space, $N \subseteq M$ an $\tilde{O}$-precomplete subspace in M and $S, g : M \to M$. Assume that the following assumptions are satisfied:

(a) There exists $z_0 \in M$ such that $gz_0 \preceq Sz_0$;
(b) S is g-increasing;
(c) $S(M) \subseteq g(M) \cap N$;
(d) S satisfies generalized Ćirić-type weak $(\phi_g, \preceq)$-contraction, i.e.,

$$\rho(Sz, Sw) \leq \mathcal{M}_{\rho,g}(z, w) - \phi(\rho(Sz, Sw)), \tag{35}$$

for all $z, w \in M$ with $gz \preceq gw$ and $\phi \in \Phi$, where:

$$\mathcal{M}_{\rho,g}(z, w) = \max\left\{\rho(gz, gw), \rho(gz, Sx), \rho(gw, Sw), \frac{\rho(gz, Sw) + \rho(gw, Sz)}{2}\right\};$$

(e) (e1) S and g are $\overline{O}$-compatible;
(e2) S and g are $\overline{O}$-continuous;
or alternatively:
(e*) (e*1) $N \subseteq g(M)$;
(e*2) either S is $(g, \overline{O})$-continuous or S and g are continuous or $(N, \rho, \preceq)$ has ICU property.

Then, S and g have a coincidence point.

5.3. Results in Ordered Partial Metric Spaces via Comparable Mappings

Definition 24. [42] For $S, g : M \to M$, S is said to be g-comparable if for all $z_1, z_2 \in M$ such that $gz_1 \prec\succ gz_2$, we have $Sz_1 \prec\succ Sz_2$.

Remark 8. Observe that the notion S is g-comparable is equivalent to saying that $\prec\succ$ is (S, g)-closed.

Definition 25. [43] Let $(M, \preceq)$ be an ordered set and $\{z_n\}$ a sequence in M. Then:

(a) $\{z_n\}$ is said to be termwise bounded if there is an element $z \in M$ such that each term of $\{z_n\}$ is comparable with z, i.e., $z_n \prec\succ z$, for all $n \in \mathbb{N}_0$ and z is a c-bound of $\{z_n\}$.
(b) $\{z_n\}$ is said to be termwise monotone if consecutive terms of $\{z_n\}$ are comparable, i.e., $z_n \prec\succ z_{n+1}$, for all $n \in \mathbb{N}_0$.

Now, we introduce TCC property in the setting of ordered partial metric spaces.

Definition 26. We say that an ordered partial metric space $(M, \rho, \preceq)$ has TCC property if every termwise monotone convergent sequence $\{z_n\}$ in M has a subsequence, which is termwise bounded by the limit (of the sequence) as a c-bound, i.e.:

$$z_n \updownarrow z \implies \text{there exists a subsequence } \{z_{n_k}\} \text{ of } \{z_n\} \text{ with } z_{n_k} \prec\succ z, \forall k \in \mathbb{N}_0.$$

Remark 9. It is observed that $(M, \rho, \preceq)$ has TCC property which is equivalent to $\prec\succ$, which is ρ-self closed.

In view of Remarks 8 and 9 and using $\mathcal{R} = \prec\succ$ in Theorem 1, we again obtained a new result for the existing literature.

Corollary 11. Let $(M, \rho, \preceq)$ be an ordered partial metric space, $N \subseteq M$, an O-precomplete subspace in M and $S, g : M \to M$. Assume that the following assumptions are satisfied:

(a) There exists $z_0 \in M$ such that $gz_0 \prec\succ Sz_0$;

(b) S is g-increasing;
(c) $S(M) \subseteq g(M) \cap N$;
(d) S satisfies generalized Ćirić-type weak $(\phi_g, \mathcal{R})$-contraction, i.e.:

$$\rho(Sz, Sw) \leq \mathcal{M}_{\rho,g}(z,w) - \phi(\rho(Sz, Sw)), \tag{36}$$

for all $z, w \in M$ with $gz \prec\!\!\succ gy$ and $\phi \in \Phi$, where;

$$\mathcal{M}_{\rho,g}(z,w) = \max\left\{\rho(gz,gw), \rho(gz,Sz), \rho(gw,Sw), \frac{\rho(gz,Sw)+\rho(gw,Sz)}{2}\right\};$$

(e) (e1) S and g are O-compatible;
 (e2) S and g are O-continuous;
 or alternatively:
(e*) (e*1) $N \subseteq g(M)$;
 (e*2) either S is (g, O)-continuous or S and g are continuous or $(N, \rho, \preceq)$ has TCC property.

Then, S and g have a coincidence point.

6. Application

Let us consider the following system of equations:

$$\begin{cases} z(t) = \int_0^T K_1(t, \tau, z(\tau)) d\tau + a(t); \\ z(t) = \int_0^T K_2(t, \tau, z(\tau)) d\tau + a(t), \end{cases} \tag{37}$$

for all $t \in \Omega = [0, T]$, $T > 0$, where $K_1, K_2 : \Omega \times \Omega \times \mathbb{R}^n \to \mathbb{R}^n$ and $a : \Omega \to \mathbb{R}^n$.

Our aim is to provide an existence theorem in order to find the solution of the above system of integral equations using Theorem 1.

Let $\mathcal{R}$ be an arbitrary transitive binary relation on $\mathbb{R}^n$ and $M = \mathcal{C}(\Omega, \mathbb{R}^n)$, set of all continuous mappings from $\Omega \to \mathbb{R}^n$, with sup norm $\|z\|_M = \max_{t \in \Omega} \|z(t)\|$, $z \in M$. Consider a binary relation $\mathcal{R}_M$ on M as:

$$(z_1, z_2) \in \mathcal{R}_M \iff (z_1(t), z_2(t)) \in \mathcal{R}, \ \forall t \in \Omega.$$

For any $\mathcal{R}_M$-preserving sequence $\{z_n\}$ in M converging to $z \in M$, we have $(z_n(t), z(t)) \in \mathcal{R}$, for all $t \in \Omega$. Further, define $S, g : M \to M$ by:

$$Sz(t) = \int_0^T K_1(t, \tau, z(\tau)) d\tau + a(t) \text{ and } gz(t) = \int_0^T K_2(t, \tau, z(\tau)) d\tau + a(t),$$

for all $t \in \Omega$, where g is surjective.

Theorem 3. *Suppose the following conditions are satisfied:*

(A) $K_1, K_2 : \Omega \times \Omega \times \mathbb{R}^n \to \mathbb{R}^n$ and $a : \Omega \to \mathbb{R}^n$ are continuous;
(B) There exists some $z_0 \in M$ such that:

$$\left(\int_0^T K_2(t, \tau, z_0(\tau)) d\tau + a(t), \int_0^T K_1(t, \tau, z_0(\tau)) d\tau + a(t)\right) \in \mathcal{R}, \ \forall t \in \Omega;$$

(C) $(gz(t), gw(t)) \in \mathcal{R} \implies (Sz(t), Sw(t)) \in \mathcal{R}, \ \forall t \in \Omega$;
(D) For each $z, w \in M$ such that $(z, w) \in \mathcal{R}^{\neq}$ and $t, \tau \in \Omega$, there exists a number $\lambda \in [0, \frac{1}{T}]$ such that:

$$\|K_1(t, \tau, z(\tau)) - K_1(t, \tau, w(\tau))\| \leq \lambda \|gz(t) - gw(t)\|.$$

Then, Equation (37) has a solution in M.

Proof. Define $\rho : M \times M \to [0, \infty)$ as:

$$\rho(z, w) = \|z - w\|_M, \ \forall z, w \in M.$$

Now, for $(z, w) \in \mathcal{R}^{\neq}$, we have:

$$\rho(Sz, Sw) = \max_{t \in \Omega} \left\| \int_0^T (K_1(t, \tau, z(\tau)) - K_1(t, \tau, w(\tau))) d\tau \right\|$$

$$\leq \max_{t \in \Omega} \int_0^T \|K_1(t, \tau, z(\tau)) - K_1(t, \tau, w(\tau))\| d\tau$$

$$\leq \lambda \max_{t \in \Omega} \|gz(t) - gw(t)\| \int_0^T d\tau$$

$$= \lambda T \|gz - gw\|_M$$

$$= \lambda_1 \rho(gz, gw),$$

where $\lambda_1 = \lambda T$. Now, define $\phi : [0, \infty) \to [0, \infty)$ as $\phi(t) = (1 - \lambda_1)t$, $\lambda_1 \in [0, 1)$. It can be easily seen that $\phi \in \Phi$. Applying it in the above inequality, we obtain:

$$\rho(Sz, Sw) \leq \rho(gz, gw) - \phi(\rho(gz, gw))$$
$$\leq \rho(gz, gw) - \phi(\rho(Sz, Sw))$$
$$\leq \mathcal{M}_{\rho, g}(z, w) - \phi(\rho(Sz, Sw)),$$

where $\mathcal{M}_{\rho, g}$ is as defined in Theorem 1. By choosing $N = M$, it is also clear that $S(M) \subseteq M = g(M)$. Hence, by fulfilling all the necessary requirements of Theorem 1, S and g have a coincidence point. Hence, the system (Equation (37)) has a solution. This completes the proof. □

7. Conclusions

Essentially, inspired by Alam and Imdad [21] and Zhiqun Xue [32], we introduced a new contraction condition and used the same to prove some new fixed point results in the setting of partial metric space. To establish our claim, we deduced some corollaries which are still new and refined versions of earlier known results in literature. Finally, by presenting an application, we exhibited the usability of our main result.

Author Contributions: All the authors have contributed equally in the preparation of this manuscript.

Funding: This research received no external funding.

Conflicts of Interest: The authors declare no conflict of interest.

References

1. Matthews, S.G. Partial metric topology. *Ann. N. Y. Acad. Sci.* **1994**, *728*, 183–197. [CrossRef]
2. Abdeljawad, T.; Karapınar, E.; Taş, K. Existence and uniqueness of a common fixed point on partial metric spaces. *Appl. Math. Lett.* **2011**, *24*, 1900–1904. [CrossRef]
3. Altun, I.; Erduran, A. Fixed point theorems for monotone mappings on partial metric spaces. *Fixed Point Theory Appl.* **2011**, *2011*, 508730. [CrossRef]
4. Amini-Harandi, A. Metric-like spaces, partial metric spaces and fixed points. *Fixed Point Theory Appl.* **2012**, *2012*, 204. [CrossRef]
5. Aydi, H. Some coupled fixed point results on partial metric spaces. *Int. J. Math. Math. Sci.* **2011**, *2011*, 647091. [CrossRef]

6. Bejenaru, A.; Pitea, A. Fixed point and best proximity point theorems on partial metric spaces. *J. Math. Anal.* **2016**, *7*, 25–44.
7. Chi, K.P.; Karapınar, E.; Thanh, T.D. Generalized contraction principle in partial metric spaces. *Math. Comput. Model.* **2012**, *55*, 1673–1681. [CrossRef]
8. Chi, K.P.; Karapınar, E.; Thanh, T.D. On the fixed point theorems for generalized weakly contractive mappings on partial metric spaces. *Bull. Iran. Math. Soc.* **2013**, *39*, 269–381.
9. Karapınar, E. Generalizations of Caristi Kirk's theorem on partial metric spaces. *Fixed Point Theory Appl.* **2011**. [CrossRef]
10. Karapınar, E. A note on common fixed point theorems in partial metric spaces. *Miscolc Math. Notes* **2011**, *12*, 185–191. [CrossRef]
11. Karapınar, E.; Sadarangani, K. Fixed point theory for cyclic (ϕ-ψ)-contractions. *Fixed Point Theory Appl.* **2011**. [CrossRef]
12. Karapınar, E.; Erhan, I.M. Cyclic contractions and fixed point theorems. *FILOMAT* **2012**, *26*, 777–782. [CrossRef]
13. Karapınar, E.; Shatanawi, W.; Tas, K. Fixed point theorem on partial metric spaces involving rational expressions. *Miskolc Math. Notes* **2013**, *14*, 135–142. [CrossRef]
14. Shatanawi, W.; Postolache, M. Coincidence and fixed point results for generalized weak contractions in the sense of Berinde on partial metric spaces. *Fixed Point Theory Appl.* **2013**, *2013*, 54. [CrossRef]
15. Valero, O. On Banach fixed point theorems for partial metric spaces. *Appl. Gen. Topol.* **2005**, *2*, 229–240. [CrossRef]
16. Karapınar, E.; Romaguera, S. Nonunique fixed point theorems in partial metric spaces. *FILOMAT* **2013**, *27*, 1305–1314. [CrossRef]
17. Turinici, M. Fixed points for monotone iteratively local contractions. *Demonstr. Math.* **1986**, *19*, 171–180.
18. Ran, A.C.M.; Reurings, M.C.B. A fixed point theorem in partially ordered sets and some applications to matrix equations. *Proc. Am. Math. Soc.* **2004**, *132*, 1435–1443. [CrossRef]
19. Nieto, J.J.; Rodríguez-López, R. Contractive mapping theorems in partially ordered sets and applications to ordinary differential equations. *Order* **2005**, *22*, 223–239. [CrossRef]
20. Nieto, J.J.; Rodríguez-López, R. Existence and uniqueness of fixed point in partially ordered sets and applications to ordinary differential equations. *Acta Math. Sin. Engl. Ser.* **2007**, *23*, 2205–2212. [CrossRef]
21. Alam, A.; Imdad, M. Relation-theoretic contraction principle. *J. Fixed Point Theory Appl.* **2015**, *17*, 693–702. [CrossRef]
22. Ahmadullah, M.; Ali, J.; Imdad, M. Unified relation-theoretic metrical fixed point theorems under an implicit contractive condition with an application. *Fixed Point Theory Appl.* **2016**, *2016*, 42. [CrossRef]
23. Alam, A.; Imdad, M. Nonlinear contractions in metric spaces under locally T-transitive binary relations. *Fixed Point Theory* **2018**, *19*, 13–24. [CrossRef]
24. Imdad, M.; Khan, Q.H.; Alfaqih, W.M.; Gubran, R. A relation theoretic $(F, \mathcal{R})$-contraction principle with applications to matrix equations. *Bull. Math. Anal. Appl.* **2018**, *10*, 1–12.
25. Sawangsup, K.; Sintunavarat, W. Fixed point and multidimensional fixed point theorems with applications to nonlinear matrix equations in terms of weak altering distance functions. *Open Math.* **2017**, *15*, 111–125. [CrossRef]
26. Maddux, R.D. *Relation Algebras*; Elsevier Science Limited: Amsterdam, The Netherlands, 2006; Volume 150.
27. Roldán, A.; Shahzad, N. Fixed point theorems by combining Jleli and Samet's and Branciari's inequalities. *J. Nonlinear Sci. Appl.* **2016**, *9*, 3822–3849.
28. Shahzad, N.; Roldán-López-de-Hierro, A.F.; Khojasteh, F. Some new fixed point theorems under $(\mathcal{A}, \mathcal{S})$-contractivity conditions. *RACSAM* **2017**, *111*, 307–324. [CrossRef]
29. Samet, B.; Turinici, M. Fixed point theorems on a metric space endowed with an arbitrary binary relation and applications. *Commun. Math. Anal.* **2012**, *13*, 82–97.
30. Kolman, B.; Busby, R.C.; Ross, S.C. *Discrete Mathematical Structures*; Prentice-Hall, Inc.: Upper Saddle River, NJ, USA, 2003.
31. Alam, A.; Imdad, M. Relation-theoretic metrical coincidence theorems. *FILOMAT* **2017**, *31*, 4421–4439. [CrossRef]
32. Xue, Z. The convergence of fixed point for a kind of weak contraction. *Nonlinear Funct. Anal. Appl.* **2016**, *21*, 497–500.

33. Popescu, O. Fixed points for (ψ, ϕ)-weak contractions. *Appl. Math. Lett.* **2011**, *24*, 1–4. [CrossRef]
34. Aydi, H. Some fixed point results in ordered partial metric spaces. *arXiv* **2011**, arXiv:1103.3680.
35. Aydi, H. Common fixed point results for mappings satisfying (ψ, ϕ)-weak contractions in ordered partial metric spaces. *Int. J. Math. Stat.* **2012**, *12*, 53–64.
36. Gülyaz, S.; Karapınar, E. A coupled fixed point result in partially ordered partial metric spaces through implicit function. *Hacet. J. Math. Stat.* **2013**, *42*, 347–357.
37. Roldán, A.; Martínez-Moreno, J.; Roldán, C.; Karapınar, E. Multidimensional fixed-point theorems in partially ordered complete partial metric spaces under (ψ, φ)-contractivity conditions. *Abstr. Appl. Anal.* **2013**, *2013*, 634371. [CrossRef]
38. Ćirić, L.B.; Cakić, N.; Rajović, M.; Ume, J.S. Monotone generalized nonlinear contractions in partially ordered metric spaces. *Fixed Point Theory Appl.* **2008**, *2008*, 131294. [CrossRef]
39. Turinici, M. Abstract comparison principles and multivariable Gronwall-Bellman inequalities. *J. Math. Anal. Appl.* **1986**, *117*, 100–127. [CrossRef]
40. Alam, A.; Khan, A.R.; Imdad, M. Some coincidence theorems for generalized nonlinear contractions in ordered metric spaces with applications. *Fixed Point Theory Appl.* **2014**, *2014*, 1–30. [CrossRef]
41. Kutbi, M.A.; Alam, A.; Imdad, M. Sharpening some core theorems of Nieto and Rodríguez-López with application to boundary value problem. *Fixed Point Theory Appl.* **2015**, *2015*, 198. [CrossRef]
42. Alam, A.; Imdad, M. Comparable linear contractions in ordered metric spaces. *Fixed Point Theory* **2017**, *18*, 415–432. [CrossRef]
43. Alam, A.; Imdad, M. Monotone generalized contractions in ordered metric spaces. *Bull. Korean Math. Soc.* **2016**, *53*, 61–81. [CrossRef]

 © 2019 by the authors. Licensee MDPI, Basel, Switzerland. This article is an open access article distributed under the terms and conditions of the Creative Commons Attribution (CC BY) license (http://creativecommons.org/licenses/by/4.0/).

Article
Applications of Square Roots of Diffeomorphisms

Yoshihiro Sugimoto

Research Institute for Mathematical Sciences, Kyoto University, Kyoto 606-8502, Japan; sugimoto@kurims.kyoto-u.ac.jp

Received: 09 March 2019; Accepted: 09 April 2019; Published: 11 April 2019

Abstract: In this paper, we prove that on any contact manifold (M,ξ) there exists an arbitrary C^∞-small contactomorphism which does not admit a square root. In particular, there exists an arbitrary C^∞-small contactomorphism which is not "autonomous". This paper is the first step to study the topology of $Cont_0(M,\xi)\backslash Aut(M,\xi)$. As an application, we also prove a similar result for the diffeomorphism group $Diff(M)$ for any smooth manifold M.

Keywords: diffeomorphism; contactomorphism; symplectomorphism

1. Introduction

For any closed manifold M, the set of diffeomorphisms $Diff(M)$ forms a group and any one-parameter subgroup $f : \mathbb{R} \to Diff(M)$ can be written in the following form

$$f(t) = \exp(tX).$$

Here, $X \in \Gamma(TM)$ is a vector field and $\exp : \Gamma(TM) \to Diff(M)$ is the time 1 flow of vector fields. From the inverse function theorem, one might expect that there exists an open neighborhood of the zero section $\mathcal{U} \subset \Gamma(TM)$ such that

$$\exp : \mathcal{U} \longrightarrow Diff(M)$$

is a diffeomorphism onto an open neighborhood of $Id \in Diff(M)$. However, this is far from true ([1], Warning 1.6). So one might expect that the set of "autonomous" diffeomorphisms

$$Aut(M) = \exp(\Gamma(TM))$$

is a small subset of $Diff(M)$.

For a symplectic manifold (M,ω), the set of Hamiltonian diffeomorphisms $Ham^c(M,\omega)$ contains "autonomous" subset $Aut(M,\omega)$ which is defined by

$$Aut(M,\omega) = \left\{ \exp(X) \,\middle|\, \begin{array}{c} X \text{ is a time-independent Hamiltonian vector field} \\ \text{whose support is compact} \end{array} \right\}.$$

In [2], Albers and Frauenfelder proved that on any symplectic manifold there exists an arbitrary C^∞-small Hamiltonian diffeomorphism not admitting a square root. In particular, there exists an arbitrary C^∞-small Hamiltonian diffeomorphism in $Ham^c(M,\omega)\backslash Aut(M,\omega)$.

Polterovich and Shelukhin used spectral spread of Floer homology and Conley conjecture to prove that $Ham^c(M,\omega)\backslash Aut(M,\omega) \subset Ham^c(M,\omega)$ is C^∞-dense and dense in the topology induced from Hofer's metric if (M,ω) is closed symplectically aspherical manifold ([3]). The author generalized this theorem to arbitrary closed symplectic manifolds and convex symplectic manifolds ([4]).

One might expect that "contact manifold" version of these theorems hold. In this paper, we prove that there exists an arbitrary C^∞-small contactomorphism not admitting a square root. In particular, there exists an arbitrary C^∞-small contactomorphism in $\mathrm{Cont}^c_0(M,\xi)\backslash\mathrm{Aut}(M,\xi)$. So, this paper is a contact manifold version of [2]. As an application, we prove that there exists an arbitrary C^∞-small diffeomorphism in $\mathrm{Diff}^c_0(M)$ not admitting a square root. This also implies that there exists an arbitrary C^∞-small diffeomorphism in $\mathrm{Diff}^c_0(M)\backslash\mathrm{Aut}(M)$.

2. Main Result

Let M be a smooth $(2n+1)$-dimensional manifold without boundary. A 1-form α on M is called contact if $(\alpha \wedge (d\alpha)^n)(p) \neq 0$ holds on any $p \in M$. A codimension 1 tangent distribution ξ on M is called contact structure if it is locally defined by $\ker(\alpha)$ for some (locally defined) contact form α. A diffeomorphism $\phi \in \mathrm{Diff}(M)$ is called contactomorphism if $\phi_*\xi = \xi$ holds (i.e., ϕ preserves the contact structure ξ). Let $\mathrm{Cont}^c_0(M,\xi)$ be the set of compactly supported contactomorphisms which are isotopic to Id through compactly supported contactomorphisms. In other words, $\mathrm{Cont}^c_0(M,\xi)$ is a connected component of compactly supported contactomorphisms ($\mathrm{Cont}^c(M,\xi)$) which contains Id.

$$\mathrm{Cont}^c_0(M,\xi) = \left\{\phi_1 \;\middle|\; \begin{array}{l} \phi_t\ (t \in [0,1])\text{ is an isotopy of contactomorphisms} \\ \phi_0 = \mathrm{Id},\ \cup_{t\in[0,1]}\mathrm{supp}(\phi_t)\text{ is compact} \end{array}\right\}$$

Let $X \in \Gamma^c(TM)$ be a compactly supported vector field on M. X is called contact vector field if the flow of X preserves the contact structure ξ (i.e., $\exp(X)_*\xi = \xi$ holds). Let $\Gamma^c_\xi(TM)$ be the set of compactly supported contact vector fields on M and let $\mathrm{Aut}(M,\xi)$ be their images

$$\mathrm{Aut}(M,\xi) = \{\exp(X) \mid X \in \Gamma^c_\xi(TM)\}.$$

We prove the following theorem.

Theorem 1. *Let (M,ξ) be a contact manifold without boundary. Let $\mathcal{W}$ be any C^∞-open neighborhood of $\mathrm{Id} \in \mathrm{Cont}^c_0(M,\xi)$. Then, there exists $\phi \in \mathcal{W}$ such that*

$$\phi \neq \psi^2$$

holds for any $\psi \in \mathrm{Cont}^c_0(M,\xi)$. In particular, $\mathcal{W}\backslash\mathrm{Aut}(M,\xi)$ is not empty.

Remark 1. *If ϕ is autonomous ($\phi = \exp(X)$), ϕ has a square root $\psi = \exp(\frac{1}{2}X)$.*

Corollary 1. *The exponential map $\exp : \Gamma^c_\xi(TM) \to \mathrm{Cont}^c_0(M,\xi)$ is not surjective.*

We also consider the diffeomorphism version of this theorem and corollary. Let M be a smooth manifold without boundary and let $\mathrm{Diff}^c(M)$ be the set of compactly supported diffeomorhisms

$$\mathrm{Diff}^c(M) = \{\phi \in \mathrm{Diff}(M) \mid \mathrm{supp}(\phi)\text{ is compact}\}.$$

Let $\mathrm{Diff}^c_0(M)$ be the connected component of $\mathrm{Diff}^c(M)$ (i.e., any element of $\mathrm{Diff}^c_0(M)$ is isotopic to Id). We define the set of autonomous diffeomorphisms by

$$\mathrm{Aut}(M) = \{\exp(X) \mid X \in \Gamma^c(TM)\}.$$

By combining the arguments in this paper and in [2], we can prove the following theorem.

Theorem 2. *Let M be a smooth manifold without boundary. Let $\mathcal{W}$ be any C^∞-open neighborhood of $\mathrm{Id} \in \mathrm{Diff}^c_0(M)$. Then, there exists $\phi \in \mathcal{W}$ such that*

$$\phi \neq \psi^2$$

holds for any $\psi \in \mathrm{Diff}^c(M)$. In particular, $\mathcal{W} \backslash \mathrm{Aut}(M)$ is not empty.

Corollary 2. *The exponential map* $\exp : \Gamma^c(TM) \to \mathrm{Diff}_0^c(M)$ *is not surjective.*

3. Milnor's Criterion

In [1], Milnor gave a criterion for the existence of a square root of a diffeomorphism. We use this criterion later. We fix $l \in \mathbb{N}_{\geq 2}$ and a diffeomorphism $\phi \in \mathrm{Diff}(M)$. Let $P^l(\phi)$ be the set of "l-periodic orbits" which is defined by

$$P^l(\phi) = \{(x_1, \cdots, x_l) \mid x_i \neq x_j (i \neq j), x_j = \phi^{j-1}(x_1), x_1 = \phi(x_l)\} / \sim .$$

This equivalence relation $\sim$ is given by the natural $\mathbb{Z}/l\mathbb{Z}$-action

$$(x_1, \cdots, x_l) \to (x_l, x_1, \cdots, x_{l-1}).$$

Proposition 1 (Milnor [1], Albers-Frauenfelder [2])**.** *Assume that* $\phi \in \mathrm{Diff}(M)$ *has a square root (i.e., there exists* $\psi \in \mathrm{Diff}(M)$ *such that* $\phi = \psi^2$ *holds). Then, there exists a free* $\mathbb{Z}/2\mathbb{Z}$*-action on* $P^{2k}(\phi)$ *(*$k \in \mathbb{N}$*). In particular,* $\sharp P^{2k}(\phi)$ *is even if* $\sharp P^{2k}(\phi)$ *is finite.*

4. Proof of Theorem 1

Proof. Before stating the proof of Theorem 1, we introduce the notion of a contact Hamiltonian function. Let M be a smooth manifold without boundary and let $\alpha \in \Omega^1(M)$ be a contact form on M ($\xi = \ker(\alpha)$). A Reeb vector field $R_\alpha \in \Gamma(TM)$ is the unique vector field which satisfies

$$\alpha(R_\alpha) = 1$$
$$d\alpha(R_\alpha, \cdot) = 0.$$

For any smooth function $h \in C_c^\infty(M)$, there exists only one contact vector field $X_h \in \Gamma_\xi^c(TM)$ which satisfies

$$X_h = h \cdot R_\alpha + Z \text{ where } Z \in \xi.$$

In fact, X_h is a contact vector field if and only if $\mathcal{L}_{X_h}(\alpha)|_\xi = 0$ holds ($\mathcal{L}$ is the Lie derivative). So,

$$\mathcal{L}_{X_h}(\alpha)(Y) = dh(Y) + d\alpha(X_h, Y) = dh(Y) + d\alpha(Z, Y) = 0$$

holds for any $Y \in \xi$. Because $d\alpha$ is non-degenerate on ξ, above equation determines $Z \in \xi$ uniquely. X_h is the contact vector field associated to the contact Hamiltonian function h. We denote the time t flow of X_h by ϕ_h^t and time 1 flow of X_h by ϕ_h.

Let (M, ξ) be a contact manifold without boundary. We fix a point $p \in (M, \xi)$ and a sufficiently small open neighborhood $U \subset M$ of p. Let $(x_1, y_1, \cdots, x_n, y_n, z)$ be a coordinate of $\mathbb{R}^{2n+1}$. Let $\alpha_0 \in \Omega^1(\mathbb{R}^{2n+1})$ be a contact form

$$\alpha_0 = \frac{1}{2} \sum_{1 \leq i \leq n} (x_i dy_i - y_i dx_i) + dz$$

on $\mathbb{R}^{2n+1}$. By using the famous Moser's arguments, we can assume that there exists an open neighborhood of the origin $V \subset \mathbb{R}^{2n+1}$ and a diffeomorphism

$$F : V \longrightarrow U \qquad (1)$$

which satisfies
$$\xi|_U = \ker((F^{-1})^*\alpha_0).$$

So, we first prove the theorem for $(V, \ker(\alpha_0))$ and apply this to (M, ξ).
We fix $k \in \mathbb{N}_{\geq 1}$ and $R > 0$ so that

$$\{(x_1, y_1, \cdots, z) \in \mathbb{R}^{2n+1} \mid |(x_1, \cdots, y_n)| < R, |z| < R\} \subset V$$

holds. Let $f \in C_c^\infty(V)$ be a contact Hamiltonian function. Then its contact Hamiltonian vector field X_f can be written in the following form

$$X_f(x_1, \cdots, z) = \sum_{1 \leq i \leq n} (-\frac{\partial f}{\partial y_i} + \frac{x_i}{2}\frac{\partial f}{\partial z})\frac{\partial}{\partial x_i}$$
$$+ \sum_{1 \leq i \leq n} (\frac{\partial f}{\partial x_i} + \frac{y_i}{2}\frac{\partial f}{\partial z})\frac{\partial}{\partial y_i}$$
$$+ (f - \sum_{1 \leq i \leq n} \frac{x_i}{2}\frac{\partial f}{\partial x_i} - \sum_{1 \leq i \leq n} \frac{y_i}{2}\frac{\partial f}{\partial y_i})\frac{\partial}{\partial z}.$$

Let $e : \mathbb{R}^{2n} \longrightarrow \mathbb{R}$ be a quadric function

$$e(x_1, y_1, \cdots, x_n, y_n) = x_1^2 + y_1^2 + \sum_{2 \leq i \leq n} \frac{x_i^2 + y_i^2}{2}.$$

We define a contact Hamiltonian function h on V by

$$h(x_1, y_1, \cdots, x_n, y_n, z) = \beta(z)\rho(e(x_1, y_1, \cdots, x_n, y_n)).$$

Here, $\beta : \mathbb{R} \to [0, 1]$ and $\rho : \mathbb{R}_{\geq 0} \to \mathbb{R}_{\geq 0}$ are smooth functions which satisfy the following five conditions.

1. $\mathrm{supp}(\rho) \subset [0, \frac{R^2}{2}]$
2. $\rho(r) \geq \rho'(r) \cdot r, -\frac{\pi}{2k} < \rho'(r) \leq \frac{\pi}{2k}$
3. There exists an unique $a \in [0, \frac{R^2}{2}]$ which satisfies the following conditions

$$\begin{cases} \rho'(r) = \frac{\pi}{2k} \Longleftrightarrow r = a \\ \rho(a) = \frac{\pi}{2k} \cdot a \end{cases}.$$

4. $\mathrm{supp}(\beta) \subset [-\frac{R}{2}, \frac{R}{2}]$
5. $\beta(0) = 1, \beta^{-1}(1) = 0$

Then, we can prove the following lemma.

Lemma 1. *Let $h \in C_c^\infty(V)$ be a contact Hamiltonian function as above. Then,*

$$[q, \phi_h(q), \cdots, \phi_h^{2k-1}(q)] \in P^{2k}(\phi_h)$$

holds if and only if

$$q \in \{(x_1, y_1, 0, \cdots, 0) \in V \mid x_1^2 + y_1^2 = a\} \stackrel{\mathrm{def.}}{=} S_a$$

holds.

Proof of Lemma 1. In order to prove this lemma, we first calculate the behavior of the function $z(\phi_h^t(q))$ for a fixed $q \in V$ (Here, z is the $(2n+1)$-th coordinate of $\mathbb{R}^{2n+1}$).

$$\frac{d}{dt}(z(\phi_h^t(q))) = h - \sum_{1 \le i \le n} \frac{x_i}{2}\frac{\partial h}{\partial x_i} - \sum_{1 \le i \le n} \frac{y_i}{2}\frac{\partial h}{\partial y_i}$$

$$= \beta(z)\{\rho(e) - \sum_{1 \le i \le n} \frac{x_i}{2}\frac{\partial}{\partial x_i}(\rho(e)) - \sum_{1 \le i \le n} \frac{y_i}{2}\frac{\partial}{\partial y_i}(\rho(e))\}$$

$$= \beta(z)\{\rho(e) - \rho'(e) \cdot e\} \ge 0$$

In the last inequality, we used the condition 2. So, this inequality implies that

$$\phi_h^{2k}(q) = q \implies \frac{d}{dt}(z(\phi_h^t(q))) = 0$$

holds.

Next, we study the behavior of $x_i(\phi_h^t(q))$ and $y_i(\phi_h^t(q))$. Let π_i be the projection

$$\pi_i : \mathbb{R}^{2n+1} \longrightarrow \mathbb{R}^2.$$
$$(x_1, y_1, \cdots, x_n, y_n, z) \mapsto (x_i, y_i)$$

Then, $Y_h^i = \pi_i(X_h)$ can be decomposed into the angular component $Y_h^{i,\theta}$ and the radius component $Y_h^{i,r}$ as follows

$$Y_h^{i,\theta}(x_1, y_1, \cdots, z) = -\frac{\partial h}{\partial y_i}\frac{\partial}{\partial x_i} + \frac{\partial h}{\partial x_i}\frac{\partial}{\partial y_i}$$

$$Y_h^{i,r}(x_1, y_1, \cdots, z) = (\frac{1}{2}\frac{\partial h}{\partial z})(x_i\frac{\partial}{\partial x_i} + y_i\frac{\partial}{\partial y_i}).$$

Let w_i be the complex coordinate of (x_i, y_i) ($w_i = x_i + \sqrt{-1}y_i$). Then, the angular component causes the following rotation on w_i, if we ignore the z-coordinate,

$$\arg(w_i) \longrightarrow \arg(w_i) + 2\rho'(e(x_1, \cdots, y_n))\beta(z)C_i t$$

$$C_i = \begin{cases} 1 & i = 1 \\ \frac{1}{2} & 2 \le i \le n \end{cases}.$$

By conditions 2, 3, and 5 in the definition of β and ρ, $|2\rho'(e(x_1, \cdots, y_n))\beta(z)C_i|$ is at most $\frac{2\pi}{2k}$ and the equality holds if and only if $(x_1, y_1, \cdots, x_n, y_n, z) \in S_a$ holds. On the circle S_a, ϕ_h is the $\frac{2\pi}{2k}$-rotation of the circle S_a. This implies that Lemma 1 holds. □

Next, we perturb the contactomorphism ϕ_h. Let (r, θ) be a coordinate of $(x_1, y_1) \in \mathbb{R}^2 \backslash (0,0)$ as follows

$$x_1 = r\cos\theta, \quad y_1 = r\sin\theta.$$

We fix $\epsilon_k > 0$. Then $\epsilon_k(1 - \cos(k\theta))$ is a contact Hamiltonian function on $\mathbb{R}^2\backslash(0,0) \times \mathbb{R}^{2n-1}$ and its contact Hamiltonian vector field can be written in the following form

$$X_{\epsilon_k(1-\cos(k\theta))} = -\frac{\epsilon_k k}{r}\sin(k\theta)\frac{\partial}{\partial r} + \epsilon_k(1 - \cos(k\theta))\frac{\partial}{\partial z}.$$

So $\phi_{\epsilon_k(1-\cos(k\theta))}$ only changes the r of (x_1, y_1)-coordinate and z-coordinate as follows

$$(r, \theta, x_2, y_2, \cdots, x_n, y_n, z) \mapsto (\sqrt{r^2 - 2\epsilon_k k\sin(k\theta)}, \theta, x_2, \cdots, y_n, z + \epsilon_k(1 - \cos(k\theta))).$$

We fix two small open neighborhoods of the circle S_a as follows

$$S_a \subset W_1 \subset W_2 \subset \mathbb{R}^2 \backslash (0,0) \times \mathbb{R}^{2n-1}$$
$$X_h(p) \neq 0 \text{ on } p \in W_2.$$

We also fix a cut-off function $\eta : \mathbb{R}^{2n+1} \to [0,1]$ which satisfies the following conditions

$$\eta((x_1, \cdots, z)) = 1 \quad ((x_1, \cdots, z) \in W_1)$$
$$\eta((x_1, \cdots, z)) = 0 \quad ((x_1, \cdots, z) \in \mathbb{R}^{2n+1} \backslash W_2)$$
$$\phi_h^j(\mathbb{R}^{2n+1} \backslash W_2) \cap \mathrm{supp}(\eta) = \varnothing \quad (1 \leq j \leq 2k).$$

We will use the last condition in the proof of Lemma 2. Then, $\eta(x_1, \cdots, z) \cdot \epsilon_k(1 - \cos(k\theta))$ is defined on $\mathbb{R}^{2n+1}$. We denote this contact Hamiltonian function by g_{ϵ_k}. We define $\phi_{\epsilon_k} \in \mathrm{Cont}_0^c(\mathbb{R}^{2n+1}, \ker(\alpha_0))$ by the composition $\phi_{g_{\epsilon_k}} \circ \phi_h$.

Lemma 2. *We take $\epsilon_k > 0$ sufficiently small. We define $2k$ points $\{a_i\}_{1 \leq i \leq 2k}$ by*

$$a_i = (\sqrt{a} \cos(\frac{i\pi}{k}), \sqrt{a} \sin(\frac{i\pi}{k}), 0, \cdots, 0)) \in S_a.$$

Then $P^{2k}(\phi_{\epsilon_k})$ has only one point $[a_1, a_2, \cdots, a_{2k}]$.

Proof of Lemma 2. The proof of this lemma is as follows. On W_1, $\phi_{g_{\epsilon_k}}$ only changes the r-coordinate of (x_1, y_1) and z-coordinate. So, ϕ_{ϵ_k} increases the angle of each (x_i, y_i) coordinate at most $\frac{2\pi}{2k}$ and the equality holds on only S_a. On the circle S_a, the fixed points of $\phi_{g_{\epsilon_k}}$ are $2k$ points $\{a_i\}$. From the arguments in the proof of Lemma 1, this implies that

$$[a_1, a_2, \cdots, a_{2k}] \in P^{2k}(\phi_{\epsilon_k})$$

holds and this is the only element of $P^{2k}(\phi_{\epsilon_k})$ on W_1. So, it suffices to prove that this is the only element in $P^{2k}(\phi_{\epsilon_k})$ if $\epsilon_k > 0$ is sufficiently small. We prove this by contradiction. Let $\{\epsilon_k^{(j)} > 0\}_{j \in \mathbb{N}}$ be a sequence which satisfies $\epsilon_k^{(j)} \to 0$. We assume that there exists a sequence

$$[b_1^{(j)}, \cdots, b_{2k}^{(j)}] \in P^{2k}(\phi_{\epsilon_k^{(j)}}) \backslash [a_1, a_2, \cdots, a_{2k}].$$

We may assume without loss of generality that $b_1^{(j)} \notin W_1$ holds because

$$(b_1^{(j)}, \cdots, b_{2k}^{(j)}) \notin W_1^{2k}$$

holds. We may assume that $b_1^{(j)}$ converges to a point $b \notin W_1$. Then, $\phi_h^{2k}(b) = b$ holds. If $X_h(b) \neq 0$, ϕ_h increases the angle of every (x_i, y_i) coordinate less than $\frac{2\pi}{2k}$ and this contradicts $\phi_h^{2k}(b) = b$. Thus $X_h(b) = 0$ holds. Because we assumed $X_h(p) \neq 0$ on $p \in W_2$, $X_h(b) = 0$ implies that $b \notin W_2$ holds. Let $N \in \mathbb{N}$ be a large integer so that $b_1^{(N)} \notin W_2$ holds. Then, $\phi_h^j(\mathbb{R}^{2n+1} \backslash W_2) \cap \mathrm{supp}(\eta) = \varnothing$ ($1 \leq j \leq 2k$) implies that $\phi_{\epsilon_k^{(N)}}^j(b_1^{(N)}) = \phi_h^j(b_1^{(N)})$ holds for $1 \leq j \leq 2k$ and $[b_1^{(N)}, \cdots, b_{2k}^{(N)}] \in P^{2k}(\phi_h)$ holds. This contradicts Lemma 1 because $b_1^{(N)} \notin S_a$. So, we proved Lemma 2. □

We assume that $\epsilon_k > 0$ is sufficiently small so that the conclusion of Lemma 2 holds and we define ϕ_k by $\phi_k = \phi_{\epsilon_k}$. Thus, we have constructed $\phi_k \in \mathrm{Cont}_0^c(V, \mathrm{Ker}(\alpha_0))$ which does not admit a square root for each $k \in \mathbb{N}$. Without loss of generality, we may assume that $\epsilon_k \to 0$ holds. Then ϕ_k converges to Id.

Finally, we prove Theorem 1. We define $\psi_k \in \mathrm{Cont}_0^c(M,\xi)$ for $k \in \mathbb{N}$ as follows. Recall that F is a diffeomorphism which was defined in Equation (1).

$$\psi_k(x) = \begin{cases} F \circ \phi_k \circ F^{-1}(x) & x \in U \\ x & x \in M \setminus U \end{cases}$$

Lemma 2 implies that

$$P^{2k}(\psi_k) = \{[F(a_1), \cdots, F(a_{2k})]\}$$

holds. Proposition 1 implies that ψ_k does not admit a square root. Because $p \in M$ is any point and U is any small open neighborhood of p, we proved Theorem 1. □

5. Proof of Theorem 2

Proof. Let M be a m-dimensional smooth manifold without boundary. We fix a point $p \in M$. Let U be an open neighborhood of p and let $V \subset \mathbb{R}^m$ be an open neighborhood of the origin such that there is a diffeomorphism

$$F : V \longrightarrow U.$$

In order to prove Theorem 2, it suffices to prove that there exists a sequence ψ_k ($k \in \mathbb{N}$) so that

- ψ_k does not admit a square root
- $\mathrm{supp}(\psi_k) \subset U$
- $\psi_k \longrightarrow \mathrm{Id}$ as $k \longrightarrow +\infty$

hold.

First, assume that m is odd ($m = 2n+1$). In this case, α_0 is a contact form on V. Let ϕ_k be a contactomorphism which we constructed in the proof of Theorem 1

- $\phi_k \in \mathrm{Cont}_0^c(V, \ker(\alpha_0))$
- $\sharp P^{2k}(\phi_k) = 1$.

We define $\psi_k \in \mathrm{Diff}_0^c(M)$ by

$$\psi_k(x) = \begin{cases} F \circ \phi_k \circ F^{-1}(x) & x \in U \\ x & x \in M \setminus U \end{cases}.$$

Then, $\sharp P^{2k}(\psi_k) = 1$ holds and this implies that ψ_k does not admit a square root and satisfies the above conditions. So, we proved Theorem 2 if m is odd.

Next, assume that m is even ($m = 2n$). Let ω_0 be a standard symplectic form on $(x_1, y_1, \cdots, x_n, y_n) \in \mathbb{R}^{2n}$ which is defined by

$$\omega_0 = \sum_{1 \le i \le n} dx_i \wedge dy_i.$$

By using the arguments in [2], we can construct a sequence $\phi_k \in \mathrm{Ham}^c(V, \omega_0)$ for $k \in \mathbb{N}$ which satisfies the following conditions

- $\sharp P^{2k}(\phi_k) = 1$
- $\phi_k \longrightarrow \mathrm{Id}$ as $k \longrightarrow +\infty$.

We define $\psi_k \in \mathrm{Diff}_0^c(M)$ by

$$\psi_k = \begin{cases} F \circ \phi_k \circ F^{-1} & x \in U \\ x & x \in M \setminus U \end{cases}.$$

Then $\sharp P^{2k}(\psi_k) = 1$ holds and this implies that ψ_k does not admit a square root and satisfies the above conditions. Hence, we have proved Theorem 2. □

Funding: This research received no external funding.

Acknowledgments: The author thanks Kaoru Ono and Urs Frauenfelder for many useful comments, discussion and encouragement.

Conflicts of Interest: The author declares no conflict of interest.

References

1. Milnor, J. Remarks on infinite-dimensional Lie groups. In *Relativity, Groups and Topology II*; Elsevier Science Ltd.: Amsterdam, The Netherlands, 1984.
2. Albers, P.; Frauenfelder, U. Square roots of Hamiltonian diffeomorphisms. *J. Symplectic Geom.* **2014**, *12*, 427–434. [CrossRef]
3. Polterovich, L.; Shelukhin, E. Autonomous Hamiltonian flows, Hofer's geometry and persistence modules. *Sel. Math.* **2016**, *22*, 227–296. [CrossRef]
4. Sugimoto, Y. Spectral spread and non-autonomous Hamiltonian diffeomorphisms. *Manuscr. Math.* **2018**. [CrossRef]

© 2019 by the author. Licensee MDPI, Basel, Switzerland. This article is an open access article distributed under the terms and conditions of the Creative Commons Attribution (CC BY) license (http://creativecommons.org/licenses/by/4.0/).

Article

Fixed Point Theorems for Geraghty Contraction Type Mappings in b-Metric Spaces and Applications

Hamid Faraji [1], Dragana Savić [2] and Stojan Radenović [3,4,*]

[1] Department of Mathematics, College of Technical and Engineering, Saveh Branch, Islamic Azad University, Saveh 39187/366, Iran; faraji@iau-saveh.ac.ir
[2] Primari School "Kneginja Milica", Beograd 11073, Serbia; gagasavic89@gmail.com
[3] Nonlinear Analysis Research Group, Ton Duc Thang University, Ho Chi Minh City 700000, Vietnam
[4] Faculty of Mathematics and Statistics, Ton Duc Thang University, Ho Chi Minh City 700000, Vietnam
* Correspondence: stojan.radenovic@tdtu.edu.vn

Received: 8 February 2019; Accepted: 12 March 2019; Published: 14 March 2019

Abstract: In this paper, some new results are given on fixed and common fixed points of Geraghty type contractive mappings defined in b-complete b-metric spaces. Moreover, two examples are represented to show the compatibility of our results. Some applications for nonlinear integral equations are also given.

Keywords: fixed point; Geraghty; b-metric space

1. Introduction

In 1989, Bakhtin [1] introduced b-metric spaces as a generalization of metric spaces. Since then, several papers have been published on the fixed point theory in such spaces. For further works and results in b-metric spaces, we refer readers to References [2–22].

Definition 1. *Let X be a (nonempty) set and $s \geq 1$ be a given real number. A function $d : X \times X \to [0, \infty)$ is called a b-metric on X if the following conditions hold for all $x, y, z \in X$:*

(i) $d(x,y) = 0$ if and only if $x = y$,
(ii) $d(x,y) = d(y,x)$,
(iii) $d(x,y) \leq s[d(x,z) + d(z,y)]$ (b-triangular inequality).

Then, the pair (X, d) is called a b-metric space with parameter s.

Example 1. *[14] Let (X, d) be a metric space and let $\beta > 1, \lambda \geq 0$ and $\mu > 0$. For $x, y \in X$, set $\rho(x,y) = \lambda d(x,y) + \mu d(x,y)^\beta$. Then (X, ρ) is a b-metric space with the parameter $s = 2^{\beta-1}$ and not a metric space on X.*

In 1973, Geraghty [23] introduced a class of functions to generalize the Banach contraction principle. Let S be the family of all functions $\alpha : [0, \infty) \to [0, 1)$ satisfying the property:

$$\lim_{n \to \infty} \alpha(t_n) = 1 \quad \text{implies} \quad \lim_{n \to \infty} t_n = 0.$$

Theorem 1. *[23] Let (X, d) be a complete metric space. Let $T : X \to X$ be given mapping satisfying:*

$$d(Tx, Ty) \leq \alpha(d(x,y))d(x,y), \quad x, y \in X,$$

where $\alpha \in S$. Then T has a unique fixed point.

In 2011, Dukic et al. [24] reconsidered Theorem 1 in the framework of b-metric spaces (see also Reference [25]).

Let (X,d) be a b-metric space with parameter $s \geq 1$ and S denote the set of all functions $\alpha : [0,\infty) \to [0, \frac{1}{s})$, satisfing the following condition:

$$\lim_{n\to\infty} \alpha(t_n) = \frac{1}{s} \Rightarrow \lim_{n\to\infty} t_n = 0.$$

Theorem 2. [24] *Let (X,d) be a b-complete b-metric space with parameter $s \geq 1$ and let $T : X \to X$ be a self-map. Suppose that there exists $\beta \in S$ such that:*

$$d(Tx, Ty) \leq \beta(d(x,y))d(x,y),$$

holds for all $x, y \in X$. Then T has a unique fixed point $x^* \in X$.

In recent years, many researchers have extended the result of Geraghty in the context of various metric spaces (e.g., see References [26–29]). In the present paper, we extended some fixed point theorems for Geraghty contractive mappings in b-metric spaces.

2. Results

Let $\mathcal{B}$ denote the set of all functions $\beta : [0,\infty) \to [0, \frac{1}{s})$ which satisfies the condition $\limsup_{n\to\infty} \beta(t_n) = \frac{1}{s}$ implies that $t_n \to 0$ as $n \to \infty$ [25].

Theorem 3. *Let (X,d) be a b-complete b-metric space with parameter $s \geq 1$. Let $T : X \to X$ be a self-mapping satisfying:*

$$d(Tx, Ty) \leq \beta(M(x,y))M(x,y), \quad x, y \in X, \tag{1}$$

where:

$$M(x,y) = \max\{d(x,y), d(x, Tx), d(y, Ty), \frac{1}{2s}(d(x, Ty) + d(y, Tx))\},$$

and $\beta \in \mathcal{B}$. Then T has a unique fixed point.

Proof of Theorem 3. Let $x_0 \in X$ be arbitrary. Consider the sequence $\{x_n\}$ where:

$$x_n = Tx_{n-1} = T^n x_0, \quad n \in \mathbb{N}.$$

If there exists $n \in \mathbb{N}$ such that $x_{n+1} = x_n$, then x_n is a fixed point of T and the proof is finished. Otherwise, we have $d(x_{n+1}, x_n) > 0$ for all $n \in \mathbb{N}$. By Condition (1), for all $n \in \mathbb{N}$ we have:

$$d(x_n, x_{n+1}) = d(Tx_{n-1}, Tx_n) \leq \beta(M(x_{n-1}, x_n))M(x_{n-1}, x_n), \tag{2}$$

where:

$$\begin{aligned}
M(x_{n-1}, x_n) &= \max\{d(x_{n-1}, x_n), d(x_{n-1}, Tx_{n-1}), d(x_n, Tx_n), \frac{d(x_{n-1}, Tx_n) + d(x_n, Tx_{n-1})}{2s}\} \\
&= \max\{d(x_{n-1}, x_n), d(x_{n-1}, x_n), d(x_n, x_{n+1}), \frac{d(x_{n-1}, x_{n+1}) + d(x_n, x_n)}{2s}\} \\
&\leq \max\{d(x_{n-1}, x_n), d(x_n, x_{n+1}), \frac{s(d(x_{n-1}, x_n) + d(x_n, x_{n+1}))}{2s}\} \\
&= \max\{d(x_{n-1}, x_n), d(x_n, x_{n+1})\}.
\end{aligned}$$

If $d(x_{n-1}, x_n) \leq d(x_n, x_{n+1})$, then $M(x_{n-1}, x_n) = d(x_n, x_{n+1})$. From Condition (2), we have:

$$d(x_n, x_{n+1}) \leq \beta(M(x_{n-1}, x_n))M(x_{n-1}, x_n)$$
$$\leq \frac{1}{s} d(x_n, x_{n+1}) \quad n \in \mathbb{N}.$$

This is a contradiction. Thus, we have:

$$M(x_{n-1}, x_n) = d(x_n, x_{n-1})$$

Then, from Condition (2), we get:

$$d(x_n, x_{n+1}) \leq \beta(M(x_{n-1}, x_n))d(x_{n-1}, x_n) \tag{3}$$
$$< d(x_{n-1}, x_n), \quad n \in \mathbb{N}.$$

So $\{d(x_{n-1}, x_n)\}$ is a decreasing sequence of non-negative reals. Hence, there exists $r \geq 0$ such that $d(x_{n-1}, x_n) \to r$ as $n \to \infty$. We claimed that $r = 0$. Suppose on the contrary that $r > 0$, then from Condition (3), we have:

$$r \leq \limsup_{n \to \infty} \beta(M(x_{n-1}, x_n))r.$$

Then,

$$\frac{1}{s} \leq 1 \leq \limsup_{n \to \infty} \beta(M(x_{n-1}, x_n)) \leq \frac{1}{s}.$$

Since $\beta \in \mathcal{B}$, then $\lim_{n \to \infty} M(x_{n-1}, x_n) = 0$. So $\lim_{n \to \infty} d(x_{n-1}, x_n) = 0$, which is a contradiction, that is, $r = 0$. Now we show that $\{x_n\}$ is a b-Cauchy sequence. Suppose on the contrary that $\{x_n\}$ is not a b-Cauchy sequence. Then there exists $\varepsilon > 0$ for which we can find subsequences $\{x_{m(k)}\}$ and $\{x_{n(k)}\}$ of $\{x_n\}$ such that $n(k)$ is the smallest index for which $n(k) > m(k) > k$,

$$d(x_{m(k)}, x_{n(k)}) \geq \varepsilon, \tag{4}$$

and

$$d(x_{m(k)}, x_{n(k)-1}) < \varepsilon. \tag{5}$$

From Condition (5) and using the b-triangular inequality, we have:

$$\varepsilon \leq d(x_{m(k)}, x_{n(k)}) \leq s(d(x_{m(k)}, x_{m(k)+1}) + d(x_{m(k)+1}, x_{n(k)})).$$

Then, we get:

$$\frac{\varepsilon}{s} \leq \limsup_{k \to \infty} d(x_{m(k)+1}, x_{n(k)}). \tag{6}$$

Therefore,

$$\begin{aligned}
\limsup_{k\to\infty} M(x_{m(k)}, x_{n(k)-1}) &= \limsup_{k\to\infty} \max\{d(x_{m(k)}, x_{n(k)-1}), d(x_{m(k)}, Tx_{m(k)}),\\
&\quad d(x_{n(k)-1}, Tx_{n(k)-1}), \frac{d(x_{m(k)}, Tx_{n(k)-1}) + d(x_{n(k)-1}, Tx_{m(k)})}{2s}\}\\
&= \limsup_{k\to\infty} \max\{d(x_{m(k)}, x_{n(k)-1}), d(x_{m(k)}, x_{m(k)+1}),\\
&\quad d(x_{n(k)-1}, x_{n(k)}), \frac{d(x_{m(k)}, x_{n(k)}) + d(x_{n(k)-1}, x_{m(k)+1})}{2s}\}\\
&\leq \limsup_{k\to\infty} \max\{d(x_{m(k)}, x_{n(k)-1}), d(x_{m(k)}, x_{m(k)+1}), d(x_{n(k)-1}, x_{n(k)}),\\
&\quad \frac{sd(x_{m(k)}, x_{n(k)-1}) + sd(x_{n(k)}, x_{n(k)-1})}{2s}\\
&\quad + \frac{sd(x_{n(k)-1}, x_{m(k)}) + sd(x_{m(k)}, x_{m(k)+1})}{2s}\}\\
&\leq \varepsilon.
\end{aligned}$$

From Condition (6) and Condition (1), we have:

$$\begin{aligned}
\frac{\varepsilon}{s} &\leq \limsup d(x_{m(k)+1}, x_{n(k)})\\
&\leq \limsup_{k\to\infty} \beta(M(x_{m(k)}, x_{n(k)-1})) M(x_{m(k)}, x_{n(k)-1})\\
&\leq \varepsilon \limsup_{k\to\infty} \beta(M(x_{m(k)}, x_{n(k)-1})).
\end{aligned}$$

Then $\frac{1}{s} \leq \limsup_{k\to\infty} \beta(M(x_{m(k)}, x_{n(k)-1})) \leq \frac{1}{s}$. Since $\beta \in \mathcal{B}$, so $M(x_{m(k)}, x_{n(k)-1}) \to 0$, as a result, $d(x_{m(k)}, x_{n(k)-1}) \to 0$. From Condition (4) and using the b-triangular inequality, we have:

$$\varepsilon \leq d(x_{m(k)}, x_{n(k)}) \leq s(d(x_{m(k)}, x_{n(k)-1}) + d(x_{n(k)-1}, x_{n(k)})).$$

Therefore, $\lim_{k\to\infty} d(x_{m(k)}, x_{n(k)}) = 0$. This contradicts with Condition (4). Hence, $\{x_n\}$ is a b-Cauchy sequence. The completeness of X implies that there exists $u \in X$ such that $x_n \to u$. We showed that u is a fixed point of T. By b-triangular inequality and Condition (1), we have:

$$\begin{aligned}
d(u, Tu) &\leq s(d(u, Tx_n) + d(Tx_n, Tu))\\
&\leq sd(u, Tx_n) + s\beta(M(x_n, u)) M(x_n, u).
\end{aligned}$$

Letting $n \to \infty$ in the above inequality, we obtain:

$$d(u, Tu) \leq s \limsup_{n\to\infty} d(u, x_{n+1}) \qquad (7)$$
$$+ s \limsup_{n\to\infty} \beta(M(x_n, u)) \limsup_{n\to\infty} M(x_n, u),$$

where:

$$\begin{aligned}
\limsup_{n\to\infty} M(x_n, u) &= \limsup_{n\to\infty} \max\{d(x_n, u), d(x_n, Tx_n), d(u, Tu), \frac{1}{2s}(d(x_n, Tu) + d(u, Tx_n))\}\\
&\leq \limsup_{n\to\infty} \max\{d(x_n, u), d(x_n, x_{n+1}), d(u, Tu), \frac{1}{2s}(sd(x_n, u) + sd(u, Tu) + d(u, x_{n+1}))\}\\
&\leq d(u, Tu).
\end{aligned}$$

Hence, from Condition (7), we have:

$$d(u, Tu) \leq s \limsup \beta(M(x_n, u)) d(u, Tu).$$

Consequently, $\frac{1}{s} \leq \limsup_{n \to \infty} \beta(M(x_n, u)) \leq \frac{1}{s}$. Since $\beta \in \mathcal{B}$, we concluded $\lim_{n \to \infty} M(x_n, u) = 0$. Therefore, $Tu = u$. To see that the fixed point $u \in X$ is unique, suppose there is $v \neq u$ in X such that $Tv = v$. From Condition (1), we get:

$$d(u,v) = d(Tu, Tv) \leq \beta(M(u,v))M(u,v),$$

where:

$$M(u,v) = \max\{d(u,v), d(u, Tu), d(v, Tv), \frac{1}{2s}(d(u, Tv) + d(v, Tu))\}$$
$$\leq d(u,v).$$

Therefore, we have $d(u,v) < \frac{1}{s}d(u,v)$. Then $u = v$, which is a contradiction. □

Example 2. Let $X = \{1, 2, 3\}$ and $d : X \times X \to [0, \infty)$ be defined as follows:

(i) $d(1,2) = d(2,1) = 1$,
(ii) $d(1,3) = d(3,1) = \frac{1}{9}$,
(iii) $d(2,3) = d(3,2) = \frac{6}{9}$.
(iv) $d(1,1) = d(2,2) = d(3,3) = 0$.

It is easy to check that (X, d) is a b-metric space with constant $s = \frac{3}{2}$. Set $T1 = T3 = 1, T2 = 3$ and $\beta(t) = \frac{2}{3}e^{-t}, t > 0$ and $\beta(0) \in [0, \frac{2}{3})$. Then we have:

$$d(T1, T2) = d(1, 3) = \frac{1}{9} \leq \frac{2}{3}e^{-1} = \beta(M(1,2))M(1,2),$$

$$d(T1, T3) = d(1, 1) = 0 \leq \beta(M(1,3))M(1,3),$$

$$d(T2, T3) = d(3, 1) = \frac{1}{9} \leq \frac{2}{3}e^{-\frac{6}{9}}(\frac{6}{9}) = \beta(M(2,3))M(2,3).$$

Therefore, the conditions of Theorem 3 are satisfied.

Theorem 4. Let (X, d) be a b-complete b-metric space with parameter $s \geq 1$. Let T, S be self-mappings on X which satisfy:

$$sd(Tx, Sy) \leq \beta(M(x,y))M(x,y), \quad x, y \in X, \qquad (8)$$

where $M(x,y) = \max\{d(x,y), d(x, Tx), d(y, Sy)\}$ and $\beta \in \mathcal{B}$. If T or S are continuous, then T and S have a unique common fixed point.

Proof of Theorem 4. Let x_0 be arbitrary. Define the sequence $\{x_n\}$ in X by $x_{2n+1} = Tx_{2n}$ and $x_{2n+2} = Sx_{2n+1}$ for all $n = 0, 1, \ldots$. From Condition (8), for all $n = 0, 1, 2, \ldots$, we have:

$$sd(x_{2n+1}, x_{2n+2}) = sd(Tx_{2n}, Sx_{2n+1}) \qquad (9)$$
$$\leq \beta(M(x_{2n}, x_{2n+1}))M(x_{2n}, x_{2n+1}),$$

where:

$$M(x_{2n}, x_{2n+1}) = \max\{d(x_{2n}, x_{2n+1}), d(x_{2n}, Tx_{2n}), d(x_{2n+1}, Sx_{2n+1})\}$$
$$= \max\{d(x_{2n}, x_{2n+1}), d(x_{2n+1}, x_{2n+2})\}, \quad n = 0, 1, 2, \ldots.$$

If $M(x_{2n}, x_{2n+1}) = d(x_{2n+1}, x_{2n+2})$, then:

$$sd(x_{2n+1}, x_{2n+2}) \leq \beta(M(x_{2n}, x_{2n+1}))d(x_{2n+1}, x_{2n+2}) < \frac{1}{s}d(x_{2n+1}, x_{2n+2}),$$

which is a contradiction. Hence, we have $M(x_{2n}, x_{2n+1}) = d(x_{2n}, x_{2n+1})$. From Condition (9), we have:

$$d(x_{2n+1}, x_{2n+2}) \leq \beta(M(x_{2n}, x_{2n+1}))d(x_{2n}, x_{2n+1}) \tag{10}$$
$$\leq \frac{1}{s}d(x_{2n}, x_{2n+1}).$$

Then, we get $d(x_{2n+1}, x_{2n+2}) \leq d(x_{2n}, x_{2n+1})$. Similarly, $d(x_{2n+3}, x_{2n+2}) \leq d(x_{2n+2}, x_{2n+1})$. So, we have $d(x_n, x_{n+1}) \leq d(x_{n-1}, x_n)$. Thus $\{d(x_n, x_{n+1})\}$ is a nonincreasing sequence, hence there exists $r \geq 0$ such that $d(x_n, x_{n+1}) \to r$ as $n \to \infty$. We showed that $r = 0$. Suppose on the contrary that $r > 0$. Letting $n \to \infty$ in (10), we obtain:

$$r \leq \limsup_{n \to \infty} \beta(M(x_{2n}, x_{2n+1}))r.$$

Then, we have:

$$\frac{1}{s} \leq 1 \leq \limsup_{n \to \infty} \beta(M(x_{2n}, x_{2n+1})) \leq \frac{1}{s}.$$

Since $\beta \in \mathcal{B}$, we have:

$$\lim_{n \to \infty} M(x_{2n}, x_{2n+1}) = 0.$$

Hence,

$$r = \lim_{n \to \infty} d(x_{2n}, x_{2n+1}) = 0,$$

which is a contradiction. Now, we show that $\{x_{2n}\}$ is a b-Cauchy sequence. Suppose that $\{x_{2n}\}$ is not a b-Cauchy sequence. Then there exists $\varepsilon > 0$ for which we can find subsequences $\{x_{2m(k)}\}$ and $\{x_{2n(k)}\}$ of $\{x_{2n}\}$ such that $n(k)$ is the smallest index for which $n(k) > m(k) > k$,

$$d(x_{2n(k)}, x_{2m(k)}) \geq \varepsilon, \tag{11}$$

and

$$d(x_{2n(k)}, x_{2m(k)-2}) < \varepsilon. \tag{12}$$

From Condition (8) and Condition (11) and the b-triangular inequality, we have:

$$\begin{aligned}
\varepsilon &\leq d(x_{2n(k)}, x_{2m(k)}) \\
&\leq sd(x_{2n(k)}, x_{2n(k)+1}) + sd(x_{2n(k)+1}, x_{2m(k)}) \\
&= sd(x_{2n(k)}, x_{2n(k)+1}) + sd(Tx_{2n(k)}, Sx_{2m(k)-1}) \\
&\leq sd(x_{2n(k)}, x_{2n(k)+1}) \\
&\quad + \beta(M(x_{2n(k)}, x_{2m(k)-1}))M(x_{2n(k)}, x_{2m(k)-1}),
\end{aligned} \tag{13}$$

where:

$$M(x_{2n(k)}, x_{2m(k)-1}) = \max\{d(x_{2n(k)}, x_{2m(k)-1}), d(x_{2n(k)}, Tx_{2n(k)}), d(x_{2m(k)-1}, Sx_{2m(k)-1})\}.$$

Letting $k \to \infty$, we have:

$$\limsup_{k \to \infty} M(x_{2n(k)}, x_{2m(k)-1}) = \limsup_{k \to \infty} d(x_{2n(k)}, x_{2m(k)-1}).$$

From the b-triangular inequality, we have:

$$d(x_{2n(k)}, x_{2m(k)-1}) \leq s(d(x_{2n(k)}, x_{2m(k)-2}) + d(x_{2m(k)-2}, x_{2m(k)-1})).$$

Letting again $k \to \infty$ in the above inequality, we get:

$$\limsup_{k \to \infty} d(x_{2n(k)}, x_{2m(k)-1}) \leq s\varepsilon. \tag{14}$$

From Condition (13) and Condition (14), we obtain:

$$\begin{aligned} \varepsilon &\leq \limsup_{k \to \infty} \left(\beta(M(x_{2n(k)}, x_{2m(k)-1})) M(x_{2n(k)}, x_{2m(k)-1}) \right) \\ &= \limsup_{k \to \infty} \beta(M(x_{2n(k)}, x_{2m(k)-1})) \limsup_{k \to \infty} d(x_{2n(k)}, x_{2m(k)-1}) \\ &\leq s\varepsilon \limsup_{k \to \infty} \beta(M(x_{2n(k)}, x_{2m(k)-1})). \end{aligned}$$

Therefore,

$$\frac{1}{s} \leq \limsup_{k \to \infty} \beta(M(x_{2n(k)}, x_{2m(k)-1})) \leq \frac{1}{s}.$$

Since $\beta \in \mathcal{B}$, it follows that:

$$\lim_{k \to \infty} M(x_{2n(k)}, x_{2m(k)-1}) = 0.$$

Consequently,

$$\lim_{k \to \infty} d(x_{2n(k)}, x_{2m(k)-1}) = 0. \tag{15}$$

From Condition (11) and using the b-triangular inequality, we get:

$$\varepsilon \leq d(x_{2n(k)}, x_{2m(k)}) \leq s(d(x_{2n(k)}, x_{2m(k)-1}) + d(x_{2m(k)-1}, x_{2m(k)})).$$

Letting $k \to \infty$ in the above inequality and using Condition (15), we obtain:

$$\limsup_{k \to \infty} d(x_{2n(k)}, x_{2m(k)}) = 0.$$

This contradicts Condition (11). This implies that $\{x_{2n}\}$ is a b-Cauchy sequence and so is $\{x_n\}$. There exists $x^* \in X$ such that $\lim_{n \to \infty} x_n = x^*$. If T is continuous, we have:

$$Tx^* = \lim_{n \to \infty} Tx_{2n} = \lim_{n \to \infty} x_{2n+1} = x^*.$$

From Condition (8), we have:

$$sd(x^*, Sx^*) = sd(Tx^*, Sx^*) \leq \beta(M(x^*, x^*))M(x^*, x^*),$$

where:

$$\begin{aligned} M(x^*, x^*) &= \max\{d(x^*, x^*), d(x^*, Tx^*), d(x^*, Sx^*)\} \\ &= d(x^*, Sx^*). \end{aligned}$$

Since $\beta \in \mathcal{B}$, we have,

$$sd(x^*, Sx^*) \leq \beta(M(x^*, x^*)) d(x^*, Sx^*) \leq \frac{1}{s} d(x^*, Sx^*).$$

Hence, $Sx^* = x^*$. If S is continuous, then, by a similar argument to that of above, one can show that T, S have a common fixed point. Now, we prove the uniqueness of the common fixed point. Let $y = Ty = Sy$, is another common fixed point for T and S. From Condition (8), we obtain:

$$sd(x^*, y) = sd(Tx^*, Sy) \leq \beta(M(x^*, y))M(x^*, y),$$

where:

$$M(x^*, y) = max\{d(x^*, y), d(x^*, Tx^*), d(y, Sy)\} = d(x^*, y).$$

Therefore, $x^* = y$ and the common fixed point T and f is unique. □

In Theorem 4, if $T = S$, we get the following result.

Corollary 1. *Let (X, d) be a b-complete b-metric space with parameter $s \geq 1$ and T be self-mapping on X which satisfy:*

$$sd(Tx, Ty) \leq \beta(M(x, y))M(x, y), \quad x, y \in X, \tag{16}$$

where $M(x, y) = max\{d(x, y), d(x, Tx), d(y, Ty)\}$ and T is continuous. Then T has a unique fixed point.

Example 3. Let $X = [0, 1]$ and $d : X \times X \to [0, \infty)$ be defined by $d(x, y) = |x - y|^2$, for all $x, y \in [0, 1]$. It is easy to check that (X, d) is a b-metric space with parameter $s = 2$. Set $Tx = \frac{x}{4}$ for all $x \in X$ and $\beta(t) = \frac{1}{4}$ for all $t > 0$. Then,

$$\begin{aligned}
2d(Tx, Ty) &= 2|\frac{x}{4} - \frac{y}{4}|^2 \\
&\leq \frac{1}{4}|x - y|^2 \\
&\leq \beta(M(x, y))M(x, y).
\end{aligned}$$

Then, the conditions of Corollary 1 are satisfied.

3. Applications to Nonlinear Integral Equations

In this section, we studied the existence of solutions for nonlinear integral equations, as an application to the fixed point theorems proved in the previous section.

Let $X = C[0, l]$ be the set of all real continuous functions on $[0, l]$ and $d : X \times X \to [0, \infty)$ be defined by:

$$d(u, v) = max_{0 \leq t \leq l} |u(t) - v(t)|^2, \quad u, v \in X.$$

Obviously, (X, d) is a complete b-metric space with parameter $s = 2$. First, consider the integral equation:

$$u(t) = h(t) + \int_0^l G(t, s) k(t, s, u(s))\, ds, \tag{17}$$

where $l > 0$ and $h : [0, l] \to \mathbb{R}, G : [0, l] \times [0, l] \to \mathbb{R}$ and $k : [0, l] \times [0, l] \times \mathbb{R} \to \mathbb{R}$ are continuous functions.

Theorem 5. *Suppose that the following hypotheses hold:*
(1) for all $t, s \in [0, l]$ and $u, v \in X$, we have:

$$|k(t, s, u(s)) - k(t, s, v(s))| \leq \frac{\sqrt{e^{-M(u,v)} M(u, v)}}{2},$$

(2) for all $t, s \in [0, l]$, we have:
$$\max \int_0^l G(t,s)^2 \, ds \leq \frac{1}{l}.$$

Then, the integral equation (see Condition (17)) has a unique solution $u \in X$.

Proof of Theorem 5. Let $T : X \to X$ be a mapping defined by:
$$Tu(t) = h(t) + \int_0^l G(t,s) k(t,s,u(s)) \, ds, \quad u \in X, t, s \in [0, l].$$

From Condition (1) and Condition (2), we can write:

$$\begin{aligned}
d(Tu, Tv) &= \max_{t \in [0,l]} |Tu(t) - Tv(t)|^2 \\
&= \max_{t \in [0,l]} \{|h(t) + \int_0^l G(t,s) k(t,s,u(s)) \, ds - h(t) - \int_0^l G(t,s) k(t,s,v(s)) \, ds|^2\} \\
&= \max_{t \in [0,l]} \{|\int_0^l G(t,s)(k(t,s,u(s)) - k(t,s,v(s))) \, ds|^2\} \\
&\leq \max_{t \in [0,l]} \{\int_0^l G(t,s)^2 \, ds \int_0^l |k(t,s,u(s)) - k(t,s,v(s))|^2 \, ds\} \\
&\leq \frac{1}{l} \int_0^l |\frac{\sqrt{e^{-M(u,v)} M(u,v)}}{2}|^2 \, ds \\
&\leq \frac{e^{-M(u,v)}}{2} M(u,v).
\end{aligned}$$

So, we get:
$$d(Tu, Tv) \leq \beta(M(u,v)) M(u,v).$$

Thus, all conditions in Theorem 3 for $\beta(t) = \frac{e^{-t}}{2}, t > 0$ and $\beta(0) \in [0, \frac{1}{2})$ are satisfied and hence T has a fixed point. □

Let $X = C[a,b]$ be the set of all real continuous functions on $[a,b]$ and X equipped with the b-metric below,
$$d(u,v) = \max_{a \leq t \leq b} \{(|u(t) - v(t)|)^p\}, \quad p > 1, u, v \in X.$$

Then (X, d) is a complete b-metric space with parameter $s = 2^{p-1}$. Now, consider the integral equations:
$$u(t) = \int_a^b G(t,s) k_1(t,s,u(s)) \, ds, \tag{18}$$

and
$$u(t) = \int_a^b G(t,s) k_2(t,s,u(s)) \, ds, \tag{19}$$

where $G : [a,b] \times [a,b] \to \mathbb{R}$ and $k_1, k_2 : [a,b] \times [a,b] \times \mathbb{R} \to \mathbb{R}$ are continuous functions.

Theorem 6. *Suppose that:*
(1) For all $t, s \in [a,b]$ and $u, v \in X$, we have:
$$|k_1(t,s,u(s)) - k_2(t,s,v(s))| \leq \left(\frac{\ln(1 + (|u(s) - v(s)|)^p)}{2^{2p-1}}\right)^{\frac{1}{p}}.$$

(2) For all $t, s \in [a, b]$, we have:

$$max_{a \leq t \leq b} \int_a^b G(t,s)^q \, ds \leq \frac{1}{(b-a)^{\frac{q}{p}}}, \quad \frac{1}{p} + \frac{1}{q} = 1.$$

Then the integral equations (Condition (18) and Condition (19)) have a unique common solution.

Proof of Theorem 6. Let $T, S : X \to X$ be mappings defined by:

$$Tu(t) = \int_a^b G(t,s) k_1(t,s,u(s)) \, ds, \qquad (20)$$

and

$$Su(t) = \int_a^b G(t,s) k_2(t,s,u(s)) \, ds. \qquad (21)$$

From Condition (1) and Condition (2), we have:

$$\begin{aligned}
d(Tu, Tv) &= max_{a \leq t \leq b}\{(|Tu(t) - Sv(t)|)^p\} \\
&\leq max_{a \leq t \leq b}\{(|\int_a^b G(t,s)k_1(t,s,u(s))\,ds - \int_a^b G(t,s)k_2(t,s,v(s))\,ds|)^p\} \\
&\leq max_{a \leq t \leq b}\{(\int_a^b |G(t,s)|(|k_1(t,s,u(s)) - k_2(t,s,v(s))|)\,ds)^p\} \\
&\leq max_{a \leq t \leq b}\{((\int_a^b |G(t,s)^q|\,ds)^{\frac{1}{q}}(\int_a^b (|k_1(t,s,u(s)) - k_2(t,s,v(s))|)^p\,ds)^{\frac{1}{p}})^p\} \\
&\leq max_{a \leq t \leq b}\{(\int_a^b |G(t,s)^q|\,ds)^{\frac{p}{q}}(\int_a^b (|k_1(t,s,u(s)) - k_2(t,s,v(s))|)^p\,ds)\} \\
&\leq max_{a \leq t \leq b}\{(\frac{1}{(b-a)^{\frac{q}{p}}})^{\frac{p}{q}}(\int_a^b (\frac{\ln(1+(|u(s)-v(s)|)^p)}{2^{2p-1}})\,ds)\} \\
&\leq \frac{1}{(b-a)}(\int_a^b (\frac{\ln(1+d(u,v))}{2^{2p-1}})\,ds \\
&\leq \frac{\ln(1+M(u,v))}{2^{2p-1}}.
\end{aligned}$$

Therefore, we get the following result:

$$\begin{aligned}
2^{p-1} d(Tu, Tv) &\leq \frac{M(u,v)}{2^p}. \\
&\leq \beta(M(u,v)) M(u,v).
\end{aligned}$$

Hence, all of the hypotheses of Theorem 4 for $s = 2^{p-1}$ and $\beta(t) = \frac{1}{2^p}$ are satisfied. Then T and S have a common fixed point $u \in X$. □

Author Contributions: H.F. contributed in conceptualization, methodology, analysis, data curation, original draft writing and editing. D.S. contributed in analysis, data curation. S.R. contributed in conceptualization, methodology, analysis, data curation, writing, review and editing the revision manuscript.

Funding: This research received no external funding.

Conflicts of Interest: The authors declare no conflict of interest.

References

1. Bakhtin, I.A. The contraction mapping principle in almost metric space. In *Functional Analysis*; Ulýanovsk Gos. Ped. Inst.: Ulýanovsk, Russia, 1989; pp. 26–37. (In Russian)
2. Aghajani, A.; Abbas, M.; Roshan, J.R. Common fixed point of generalized weak contractive mappings in partially ordered b-metric spaces. *Math. Slovaca* **2014**, *64*, 941–960. [CrossRef]
3. Aleksić, S.; Huang, H.; Mitrović, Z.; Radenović, S. Remarks on some fixed point results in b-metric space. *J. Fixed Point Theory Appl.* **2018**, *20*, 147. [CrossRef]
4. Czerwik, S. Nonlinear Set-valued contraction mappings in b-metric spaces. *Atti Sem. Mat. Fis. Univ. Modena* **1998**, *46*, 263–276.
5. Ding, H.S.; Imdad, M.; Radenoví'c, S.; Vujakoví'c, J. On some fixed point results in b-metric, rectangular and b-rectangular metric spaces. *Arab J. Math. Sci* **2016**, *22*, 151–164.
6. Dosenoví'c, T.; Pavloví'c, M.; Radenoví'c, S. Contractive conditions in b-metric spaces. *Vojnotehnicki Glasnik/Mil. Tech. Cour.* **2017**, *65*, 851–865. [CrossRef]
7. Faraji, H.; Nourouzi, K. A generalization of Kannan and Chatterjea fixed point theorems on complete b-metric spaces. *Sahand Commun. Math. Anal.* **2017**, *6*, 77–86.
8. Faraji, H.; Nourouzi, K. Fixed and common fixed points for (ψ, φ)-weakly contractive mappings in b-metric spaces. *Sahand Commun. Math. Anal.* **2017**, *7*, 49–62.
9. Faraji, H.; Nourouzi, K.; O'Regan, D. A fixed point theorem in uniform spaces generated by a family of b-pseudometrics. *Fixed Point Theory.* **2019**, *20*, 177–183. [CrossRef]
10. Hussain, N.; Mitrović, Z.D.; Radenović, S. A common fixed point theorem of Fisher in b-metric spaces. *RACSAM* **2018**. [CrossRef]
11. Ilchev, A.; Zlatanov, B. *On Fixed Points for Reich Maps in B-Metric Spaces*; Annual of Konstantin Preslavski University of Shumen, Faculty of Mathematics and Computer Science VI: Shumen, Bulgaria, 2016; pp. 77–88.
12. Jovanović, M.; Kadelburg, Z.; Radenović, S. Common fixed point results in metric-type spaces. *Fixed Point Theory Appl.* **2010**, *2010*, 978121. [CrossRef]
13. Khamsi, M.A.; Hussain, N. KKM mappings in metric type spaces. *Nonlinear Anal.* **2010**, *73*, 3123–3129. [CrossRef]
14. Kirk, W.; Shahzad, N. *Fixed Point Theory in Distance Spaces*; Springer: Berlin, Germany, 2014.
15. Koleva, R.; Zlatanov, B. On fixed points for Chatterjeas maps in b-metric spaces. *Turk. J. Anal. Number Theory* **2016**, *4*, 31–34.
16. Miculsecu, R.; Mihail, A. New fixed point theorems for set-valued contractions in b-metric spaces. *J. Fixed Point Theory Appl.* **2017**, *19*, 2153–2163. [CrossRef]
17. Mitrović, Z.D. A note on the results of Suzuki, Miculescu and Mihail. *J. Fixed Point Theory Appl.* **2019**, *21*, 24. [CrossRef]
18. Roshan, J.R.; Parvaneh, V.; Altun, I. Some coincidence point results in ordered b-metric spaces and applications in a system of integral equations. *Appl. Math. Comput.* **2014**, *226*, 725–737.
19. Sing, S.L.; Czerwik, S.; Krol, K.; Singh, A. Coincidences and fixed points of hybrid contractions. *Tamsui Oxf. J. Math. Sci.* **2008**, *24*, 401–416.
20. Suzuki, T. Basic inequality on a b-metric space and its applications. *J. Inequal. Appl.* **2017**, *2017*, 256. [CrossRef] [PubMed]
21. Zoto, K.; Rhoades, B.; Radenović, S. Common fixed point theorems for a class of (s, q)-contractive mappings in b-metric-like spaces and applications to integral equations. *Math. Slovaca* **2019**, *69*, 233–247. [CrossRef]
22. Mitrović, Z.D.; Hussain, N. On weak quasicontractions in b-metric spaces. *Publ. Math. Debrecen* **2019**, *94*, 29.
23. Geraghty, M.A. On contractive mappings. *Proc. Am. Math. Soc.* **1973**, *40*, 604–608. [CrossRef]
24. Dukic, D.; Kadelburg, Z.; Radenović, S. Fixed points of Geraghty-type mappings in various generalized metric spaces. *Abstr. Appl. Anal.* **2011**, *2011*, 561245. [CrossRef]
25. Shahkoohi, R.J.; Razani, A. Some fixed point theorems for rational Geraghty contractive mappings in ordered b-metric spaces. *J. Inequal. Appl.* **2014**, *2014*, 373. [CrossRef]
26. Huang, H.; Paunović, L.; Radenović, S. On some fixed point results for rational Geraghty contractive mappings in ordered b-metric spaces. *J. Nonlinear Sci. Appl.* **2015**, *8*, 800–807. [CrossRef]
27. Hussain, N.; Roshan, J.R.; Parvaneh, V.; Abbas, M. Common fixed point results for weak contractive mappings in ordered b-dislocated metric spaces with applications. *J. Inequal. Appl.* **2013**, *2013*, 486. [CrossRef]

28. Ishak, A.; Kishin, S. Generalized Geraghty type mappings on partial metric spaces and fixed point results. *Arab. J. Math.* **2013**, *2*, 247–253.
29. Zabihi, F.; Razani, A. Fixed point theorems for Hybrid Rational Geraghty Contractive Mappings in orderd *b*-Metric spaces. *J. Appl. Math.* **2014**, *2014*, 929821. [CrossRef]

© 2019 by the authors. Licensee MDPI, Basel, Switzerland. This article is an open access article distributed under the terms and conditions of the Creative Commons Attribution (CC BY) license (http://creativecommons.org/licenses/by/4.0/).

Article

Best Proximity Point Theorems on Rectangular Metric Spaces Endowed with a Graph

Nizar Souayah [1,2,*,†], Hassen Aydi [3,†], Thabet Abdeljawad [4,†] and Nabil Mlaiki [4]

1. Department of Natural Sciences, Community College Al-Riyadh, King Saud University, Riyadh 4545, Saudi Arabia
2. Ecole Supérieure des Sciences Economiques et Commerciales de Tunis, University of Tunis, Montfleury, Tunis 1089, Tunisia
3. Department of Mathematics, College of Education in Jubail, Imam Abdulrahman Bin Faisal University, P.O. Box 12020, Industrial Jubail 31961, Saudi Arabia; hmaydi@iau.edu.sa or hassen.aydi@isima.rnu.tn
4. Department of Mathematics and General Sciences, Prince Sultan University, Riyadh 11586, Saudi Arabia; tabdeljawad@psu.edu.sa (T.A.); nmlaiki@psu.edu.sa (N.M.)
* Correspondence: nizar.souayah@yahoo.fr or nsouayah@ksu.edu.sa; Tel.: +966-540373511
† These authors contributed equally to this work.

Received: 3 January 2019; Accepted: 28 January 2019; Published: 1 February 2019

Abstract: In this paper, we ensure the existence and uniqueness of a best proximity point in rectangular metric spaces endowed with a graph structure.

Keywords: proximity point; rectangular metric; *G*-contraction; graph

1. Introduction

Over the last decades, many researchers have focused on fixed point theory since it plays a very important role in the resolution of several mathematical models in various fields, see References [1–14]. One of the tools used is the well-known Banach contraction principle, which states that if (X, d) is a complete metric space and $f : X \longrightarrow X$ is a contraction self-mapping, then f has a unique fixed point in X. On the other hand, if f is a non-self mapping, that is, $f : A \longrightarrow B$, where A and B are two subsets of X, then f might not necessary have a fixed point, which leads one to think of an approximate solution x of $fx = x$ such that x is closet to fx: thereby, best proximity point theory appeared. We recall the definition of a *best proximity point*.

Definition 1. *Let (X, d) be a metric space, A and B two subsets of X and a mapping $f : A \longrightarrow B$. We denote by $d(A, B)$ the distance between A and B as follows*

$$d(A, B) = \min\{d(x, y) : x \in A, \ b \in B\}.$$

An element $u \in A$ is called a best proximity point of the mapping f if

$$d(u, fu) = d(A, B). \tag{1}$$

There are many variants and extensions of results for the existence of a best proximity point. For more details, we refer to References [15–29].

One of the generalized metric spaces is the rectangular metric space introduced first by Branciari [30]. Metric spaces endowed with a graph were introduced by Jachymski [31], which is an extension of metric spaces with partial order structures. In this paper, we consider rectangular metric spaces with the additional structure of a graph. Our contribution is that of proving the existence of a unique best proximity point for mappings satisfying different contractive conditions.

2. Preliminaries

In this section, we present some useful preliminary definitions and results related to our study. First, we remind the reader of the definition of rectangular metric spaces along with the topology.

Definition 2. *[30] Let X be a nonempty set. If the function $d : X^2 \longrightarrow [0, \infty)$ satisfies the following conditions for all $x, y, z \in X$:*

(r_1) $x = y$ if and only if $d(x, y) = 0$;
(r_2) $d(x, y) = d(y, x)$;
(r_3) $d(x, y) \leq d(x, u) + d(u, v) + d(v, y)$ for all different $u, v \in X \setminus \{x, y\}$,

then the pair (X, d) is called a rectangular metric space.

Definition 3. *[30] Let (X, d) be a rectangular metric space. Then,*

1. *a sequence $\{x_n\}$ in X converges to a point x if and only if $\lim_{n \to \infty} d(x_n, x) = 0$.*
2. *a sequence $\{x_n\}$ in X is called Cauchy if $\lim_{n,m \to \infty} d(x_n, x_m) = 0$.*
3. *(X, d) is said to be complete if every Cauchy sequence $\{x_n\}$ in X converges to a point $x \in X$.*
4. *Let $B_r(x_0, \delta) = \{y \in X$ such that $d(x_0, y) < \delta\}$ be an open ball in (X, d). A mapping $f : X \longrightarrow X$ is continuous at $x_0 \in X$ if for each $\epsilon > 0$, there exists $\delta > 0$ so that $f(B_r(x_0, \delta)) \subset B_r(fx_0, \epsilon)$.*

Now, we present the definition of a best proximity point in the rectangular metric spaces (X, d).

Definition 4. *Let A, B be nonempty subsets of (X, d) and $f : A \longrightarrow B$ be a given mapping. We denote by $d(A, B) = \inf\{d(a, b) : a \in A, b \in B\}$. An element $u \in A$ is called a best proximity point for the mapping f if $d(u, fu) = d(A, B)$. We denote by A_0 and B_0 the following sets:*

$$A_0 = \{x \in A : d(x, y) = d(A, B) \text{ for some } y \in B\} \tag{2}$$
$$B_0 = \{y \in B : d(x, y) = d(A, B) \text{ for some } x \in A\}. \tag{3}$$

The concept of P-property was defined by Raj in Reference [32].

Definition 5. *[32] Let (A, B) be a pair of non-empty subsets of (X, d) such that $A_0 \neq \emptyset$. We say that the pair (A, B) has the P-property if and only if for $x_1, x_2 \in A_0$ and $y_1, y_2 \in B_0$*
$$\left.\begin{array}{l} d(x_1, y_1) = d(A, B) \\ d(x_2, y_2) = d(A, B) \end{array}\right\} \Longrightarrow d(x_1, x_2) = d(y_1, y_2).$$

Here, let us recall some preliminaries from graph theory. Let X be a nonempty set and $\Delta = \{(x, x) \in X \times X, x \in X\}$. A graph G is a pair (V, E) where $V = V(G)$ is a set of vertices coinciding with X and $E = E(G)$ the set of its edges such that $\Delta \subset E(G)$. Furthermore, throughout this paper, we assume that the graph G has no parallel edges, that is, we do not allow it to get two or more edges that are incident to the same two vertices. By reversing the direction of edges in G, we get the graph denoted G^{-1} where its set of edges and vertices are defined as follows:

$$E(G^{-1}) = \{(x, y) \in X^2 : (y, x) \in E(G)\} \text{ and } V(G^{-1}) = V(G).$$

Consider the graph $\tilde{G}$ consisting of all vertices and edges of G and G^{-1}, that is,

$$E(\tilde{G}) = E(G) \cup E(G^{-1}). \tag{4}$$

We denote by $\tilde{G}$ the undirected graph obtained by ignoring the direction of edges of G.

Definition 6. *[31] A subgraph is a graph which consists of a subset of a graph's edges and associated vertices.*

Definition 7. [31] Let x and y be two vertices in a graph G. A path in G from x to y of length n ($n \in \mathbb{N} \cup \{0\}$) is a sequence $(x_i)_{i=0}^n$ of $n+1$ distinct vertices such that $x_0 = x$, $x_n = y$ and $(x_i, x_{i+1}) \in E(G)$ for $i = 1, 2, ..., n$.

Definition 8. [31] A graph G is said to be connected if there is a path between any two vertices of G and it is weakly connected if $\tilde{G}$ is connected.

Definition 9. [31] A path is called elementary if no vertices appear more than once in it. For more details see Figures 1 and 2.

Let (X, d) be a rectangular metric space. The graph G may be converted to a weighted graph by assigning to each edge the distance given by the rectangular metric between its vertices. In order to later apply the rectangular inequality to the vertices of the graph, we need to consider a graph of length bigger than 2, which means that between two vertices, we can find a path through at least two other vertices.

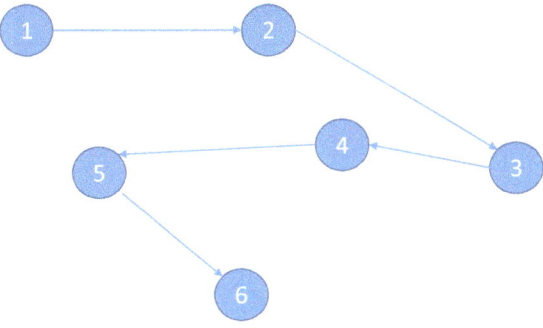

Figure 1. Elementary path.

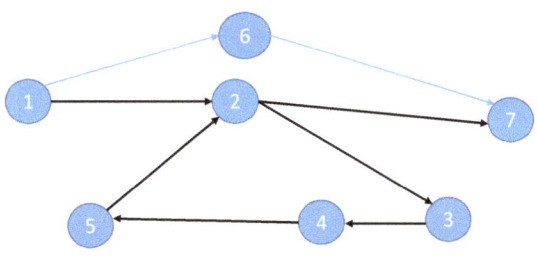

Figure 2. Non Elementary path.

3. Main Results

First, let (X, d) be a rectangular metric space and G be a directed graph without parallel edges such that $V(G) = X$.

Definition 10. Let A and B be two nonempty subsets of (X, d). A mapping $f : A \longrightarrow B$ is said to be a G-contraction mapping if for all $x, y \in A$, $x \neq y$ with $(x, y) \in E(G)$:

(i) $d(fx, fy) \leq \alpha d(x, y)$, for some $\alpha \in [0, 1)$,

(ii) $\left. \begin{array}{l} d(x_1, fx) = d(A,B) \\ d(y_1, fy) = d(A,B) \end{array} \right\} \implies (x_1, y_1) \in E(G), \ \forall x_1, y_1 \in A.$

Our first main result is as follows:

Theorem 1. *Let (X,d) be a complete rectangular metric space, A and B be two nonempty closed subsets of (X,d) such that (A,B) has the P-property. Let $f : A \longrightarrow B$ be a continuous G-contractive mapping such that $f(A_0) \subseteq B_0$ and $A_0 \neq \emptyset$. Assume that d is continuous and the following condition (C_1) holds: there exist x_0 and x_1 in A_0 such that there is an elementary path in A_0 between them and $d(x_1, fx_0) = d(A,B)$.*
Then, there exists a sequence $\{x_n\}_{n \in \mathbb{N}}$ with $d(x_{n+1}, fx_n) = d(A,B)$ for $n \in \mathbb{N}$. Moreover, if there exists a path $(y^i)_{i=0}^s \subseteq A_0$ in G between any two elements x and y, then f has a unique best proximity point.

Proof. From the condition (C_1), there exist two points x_0 and x_1 in A_0 such that $d(x_1, fx_0) = d(A,B)$ and a path $(t_0^i)_{i=0}^N$ in G between them such that the sequence $\{t_0^i\}_{i=0}^N$ containing points of A_0. Consequently, $t_0^0 = x_0$, $t_0^N = x_1$ and $(t_0^i, t_0^{i+1}) \in E(G) \ \forall 0 \leq i \leq N$.

Given that $t_0^i \in A_0$, $f(A_0) \subseteq B_0$ and from the definition of A_0, there exists $t_1^i \in A_0$ such that $d(t_1^1, ft_0^1) = d(A,B)$. Similarly, for $i = 2, ..., N$, there exists $t_1^i \in A_0$ such that $d(t_1^i, ft_0^i) = d(A,B)$.

As $(t_0^i)_{i=0}^N$ is a path in G then $(t_0^0, t_0^1) = (x_0, t_0^1) \in E(G)$. From the above, we have $d(x_1, fx_0) = d(A,B)$ and $d(t_1^1, ft_0^1) = d(A,B)$. Therefore, as f is a G-contraction, it follows that $(x_1, t_1^1) \in E(G)$. In a similar manner, it follows that

$$(t_1^{i-1}, t_1^i) \in E(G) \text{ for } i = 2, ..., N. \tag{5}$$

Let $x_2 = t_1^N$. Then, $(t_1^i)_{i=0}^N$ is a path from $x_1 = t_1^0$ to $x_2 = t_1^N$. For each $i = 2, ..., N$, as $t_1^i \in A_0$ and $ft_1^i \in f(A_0) \subseteq B_0$, then by the definition of B_0, there exists $t_2^i \in A_0$ such that $d(t_2^i, ft_1^i) = d(A,B)$. In addition, we have $d(x_2, fx_1) = d(A,B)$. As above mentioned, we obtain

$$(x_2, t_2^1) \in E(G) \text{ and } (t_2^{i-1}, t_2^i) \in E(G) \forall i = 1, 2, ..., N. \tag{6}$$

Let $x_3 = t_2^N$. Then, $(t_2^i)_{i=0}^N$ is a path from $t_2^0 = x_2$ and $t_2^N = x_3$.

Continuing in this process, for all $n \in \mathbb{N}$, we generate a path $(t_n^i)_{i=0}^N$ from $x_n = t_n^0$ and $x_{n+1} = t_n^N$. As a consequence, we build a sequence $\{x_n\}_{n \in \mathbb{N}}$ where $x_{n+1} \in [x_n]_G^N$ and $d(x_{n+1}, fx_n) = d(A,B)$ such that

$$d(t_{n+1}^i, ft_n^i) = d(A,B) \ \forall \ i = 0, ..., N. \tag{7}$$

From the P-property of (A,B) and (7), it follows for each $n \in \mathbb{N}$,

$$d(t_n^{i-1}, t_n^i) = d(ft_{n-1}^{i-1}, ft_{n-1}^i) \forall i = 1, ..., N. \tag{8}$$

Next, we claim that $d(x_n, x_{n+1}) \leq \alpha^n C$, where C is a constant. To prove the claim, we need to consider the following two cases where $(t_n^i)_{i=0,...,N}$ is a path from x_n to x_{n+1}.

Note that for all $i = 0, ..., N$, $(t_n^i)_{i=0,...,N}$ are different owing to the fact that the considered path (t_n^i) is elementary. Then, we can apply the triangular inequality (r_3).

Case 1: $N = 2k + 1$ (N is odd).

For any positive integer n, we get

$$\begin{aligned} d(x_n, x_{n+1}) &= d(t_n^0, t_n^N) = d(t_n^0, t_n^{2k+1}) \\ &\leq d(t_n^0, t_n^1) + d(t_n^1, t_n^2) + d(t_n^2, t_n^{2k+1}) \\ &\leq d(t_n^0, t_n^1) + d(t_n^1, t_n^2) + \ldots + d(t_n^{2k}, t_n^{2k+1}) \\ &\leq \sum_{i=1}^{2k+1} d(t_n^{i-1}, t_n^i) \\ &= \sum_{i=1}^{N} d(ft_{n-1}^{i-1}, ft_{n-1}^i). \end{aligned} \qquad (9)$$

Knowing that $(t_{n-1}^{i-1}, t_{n-1}^i) \in E(G)$ for all $n \in \mathbb{N}$, and f is a G-contraction, we obtain from (9)

$$d(x_n, x_{n+1}) \leq \alpha \sum_{i=1}^{N} d(t_{n-1}^{i-1}, t_{n-1}^i) \quad \forall n \in \mathbb{N}. \qquad (10)$$

By induction, it follows that for all $n \in \mathbb{N}$

$$d(x_n, x_{n+1}) \leq \alpha^n \sum_{i=1}^{N} d(t_0^{i-1}, t_0^i) = C\alpha^n \qquad (11)$$

where $C = \sum_{i=1}^{N} d(t_0^{i-1}, t_0^i)$.

Case 2: $N = 2k$ (N is even).

$$\begin{aligned} d(x_n, x_{n+1}) &= d(t_n^0, t_n^N) = d(t_n^0, t_n^{2k}) \\ &\leq d(t_n^0, t_n^1) + d(t_n^1, t_n^2) + d(t_n^2, t_n^{2k}) \\ &\leq d(t_n^0, t_n^1) + d(t_n^1, t_n^2) + \ldots + d(t_n^{2k-3}, t_n^{2k-2}) + d(t_n^{2k-2}, t_n^{2k}) \\ &= \sum_{i=1}^{2k} d(t_n^{i-1}, t_n^i) - d(t_n^{2k-2}, t_n^{2k-1}) - d(t_n^{2k-1}, t_n^{2k}) + d(t_n^{2k-2}, t_n^{2k}) \\ &\leq \sum_{i=1}^{2k} d(t_n^{i-1}, t_n^i) + d(t_n^{2k-2}, t_n^{2k}) \\ &\leq \sum_{i=1}^{2k} d(ft_{n-1}^{i-1}, ft_{n-1}^i) + d(t_n^{2k-2}, t_n^{2k}). \end{aligned}$$

By the same arguments used in Case 1, we deduce that $\sum_{i=1}^{2k} d(ft_{n-1}^{i-1}, ft_{n-1}^i) \leq \alpha^n \sum_{i=1}^{N} d(t_0^{i-1}, t_0^i)$.

On the other hand, $d(t_n^{2k-2}, t_n^{2k}) \leq \alpha^n d(t_0^{2k-2}, t_0^{2k})$. Indeed, from (7), we have $d(t_n^{2k-2}, ft_{n-1}^{2k-2}) = d(A, B)$ and $d(t_n^{2k}, ft_{n-1}^{2k}) = d(A, B)$ and using the P-property, we get

$$\begin{aligned} d(t_n^{2k-2}, t_n^{2k}) &= d(ft_{n-1}^{2k-2}, ft_{n-1}^{2k}) \\ &\leq \alpha d(t_{n-1}^{2k-2}, t_{n-1}^{2k}) \\ &\leq \alpha^n d(t_0^{2k-2}, t_0^{2k}). \end{aligned} \qquad (12)$$

Then, we conclude that $d(x_n, x_{n+1}) \leq \alpha^n C$ where $C = \sum_{i=1}^{N} d(t_0^{i-1}, t_0^i) + d(t_0^{2k-2}, t_0^{2k})$.

Let us prove that $\{x_n\}$ is a Cauchy sequence. Let $n, m \in \mathbb{N}$ such that $m \geq n$. We suppose w.l.o.g that m is odd ($m = 2k+1$) since the case $m = 2k$ is similar. Note that $x_n = t_0^n$, $x_{n+1} = t_n^N$ and $t_0^n \neq t_n^N$

for all n since the path $(t_n^i)_{i=0,\dots,N}$ is elementary. Then, using the triangular inequality of the rectangular metric, we obtain

$$\begin{aligned}
d(x_n, x_m) &\leq d(x_n, x_{n+1}) + d(x_{n+1}, x_{n+2}) + \dots + d(x_{m-1}, x_m) \\
&\leq d(x_n, x_{n+1}) + d(x_{n+1}, x_{n+2}) + \dots + d(x_{m-1}, x_m) \\
&\leq C\alpha^n + C\alpha^{n+1} + \dots + C\alpha^{m-1} \\
&= C\alpha^n(1 + \alpha + \dots + \alpha^{m-n-1}) \\
&\leq C\frac{\alpha^n}{1-\alpha}.
\end{aligned}$$

As $\alpha < 1$, then $\lim_{n,m \to \infty} d(x_n, x_m) = 0$. Therefore, $\{x_n\}_{n \in \mathbb{N}}$ is a Cauchy sequence and there exists $u \in A$ such that $x_n \longrightarrow u$ as $n \longrightarrow \infty$.

Using the continuity of f, we get $fx_n \longrightarrow fu$ as $n \longrightarrow \infty$. Now, using the continuity of the rectangular metric function, we obtain $d(x_{n+1}, fx_n)$ converges to $d(u, fu)$ as $n \longrightarrow \infty$.

Since $d(x_{n+1}, fx_n) = d(A, B)$, the sequence $\{d(x_{n+1}, fx_n)\}_n$ is constant. Consequently, $d(u, fu) = d(A, B)$. Then, u is a best proximity point of f.

In order to prove the uniqueness of the best proximity point u, we assume that there exist u and u' such that

$$d(u, fu) = d(A, B) \tag{13}$$
$$d(u', fu') = d(A, B). \tag{14}$$

Knowing that the pair (A, B) has the P-property, from (13) and (14), we get $d(u, u') = d(fu, fu')$. Since f is a G-contraction, we obtain $d(u, u') = d(fu, fu') \leq \alpha d(u, u')$, which holds unless

$$d(u, u') = 0, \text{ then } u = u'.$$

□

Definition 11. *Let $f : A \longrightarrow B$ be a mapping. Define $X_f(G_{A_0})$ as*

$$X_f(G_{A_0}) := \{x \in A_0 : \exists y \in A_0 \text{ for which } d(y, fx) = \text{dist}(A, B) \text{ and } (x, y) \in E(G)\}. \tag{15}$$

Definition 12. *Let A and B be two non-empty subsets of (X, d). A mapping $f : A \longrightarrow B$ is said to be a G-weakly contractive mapping if for all $x, y \in A$, $x \neq y$ with $(x, y) \in E(G)$:*

(i) $d(fx, fy) \leq d(x, y) - \psi(d(x, y))$, where $\psi : [0, \infty) \longrightarrow [0, \infty)$ is a continuous and nondecreasing function such that ψ is positive on $(0, \infty)$, $\psi(0) = 0$ and $\lim_{t \to \infty} \psi(t) = \infty$. If A is bounded, then the infinity condition can be omitted.

(ii) $\left.\begin{array}{l} d(x_1, fx) = d(A, B) \\ d(y_1, fy) = d(A, B) \end{array}\right\} \Longrightarrow (x_1, y_1) \in E(G), \ \forall x_1, y_1 \in A.$

Our second main result is as follows:

Theorem 2. *Let (X, d) be a complete rectangular metric space endowed with a directed graph, A and B be two nonempty closed subsets of (X, d) such that (A, B) has the P-property. Let $f : A \longrightarrow B$ be a continuous G-weakly contractive mapping such that $f(A_0) \subseteq B_0$. Assume that d is continuous and A_0 is a closed nonempty set. Then, there exists a sequence $\{x_n\}_{n \in \mathbb{N}}$ in A_0 such that $d(x_{n+1}, fx_n) = d(A, B)$ for $n \in \mathbb{N}$. Moreover, f has a unique best proximity point.*

Proof. It follows from the definition of A_0 and B_0 that for every $x \in A_0$, there exists $y \in B_0$ such that $d(x,y) = dist(A,B)$. Conversely, for every $y' \in B_0$ there exists $x' \in A_0$ such that $d(x',y') = dist(A,B)$. Since $f(A_0) \subset B_0$, for every $x \in A_0$ there exists $y \in A_0$ such that $d(y, fx) = dist(A,B)$.

Let $x_0 \in X_f(G_{A_0})$, then there exists $x_1 \in A_0$ such that $(x_0, x_1) \in E(G)$ and $d(x_1, fx_0) = dist(A,B)$. On the other hand, since $x_1 \in A_0$ and $f(A_0) \subset B_0$, there exists $x_2 \in A_0$ such that $d(x_2, fx_1) = dist(A,B)$ and because f is a G-weakly contractive mapping, we get $(x_1, x_2) \in E(G)$. We repeat this process in a similar way, we build a sequence $\{x_n\}$ in A_0 such that

$$(x_n, x_{n+1}) \in E(G) \tag{16}$$
$$d(x_{n+1}, fx_n) = dist(A,B) \forall n \in \mathbb{N}. \tag{17}$$

Since the pair (A, B) has the P-property, we conclude that $d(x_n, x_{n+1}) = d(fx_{n-1}, fx_n)$ for all $n \in \mathbb{N}$. Then, for any positive integer n

$$\begin{aligned} d(x_n, x_{n+1}) &= d(fx_{n-1}, fx_n) \\ &\leq d(x_{n-1}, x_n) - \psi(d(x_{n-1}, x_n)) \\ &\leq d(x_{n-1}, x_n). \end{aligned} \tag{18}$$

If we denote by $v_n = d(x_n, x_{n+1})$, from (18), $\{v_n\}$ is a nonnegative decreasing sequence. Hence, $\{v_n\}$ converges to some real number $v \geq 0$. Suppose that $v > 0$. As ψ is increasing, for any positive integer n, we have

$$\begin{aligned} v_n = d(x_n, x_{n+1}) &\leq d(x_{n-1}, x_n) - \psi(d(x_{n-1}, x_n)) \\ &= v_{n-1} - \psi(v_{n-1}) \\ &\leq v_{n-1} - \psi(v). \end{aligned}$$

At the limit, $v \leq v - \psi(v) < v$, which is a contradiction, so $v = 0$, that is,

$$d(x_n, x_{n+1}) \longrightarrow 0 \text{ as } n \longrightarrow \infty. \tag{19}$$

Similarly, we find that

$$d(x_n, x_{n+2}) \longrightarrow 0 \text{ as } n \longrightarrow \infty. \tag{20}$$

Now, let us prove that $\{x_n\}$ is a Cauchy sequence.
For any $\epsilon > 0$, choose N such that

$$d(x_N, x_{N+1}) < \min\{\tfrac{\epsilon}{8}, \psi(\tfrac{\epsilon}{8})\} \tag{21}$$
$$d(x_N, x_{N+2}) < \min\{\tfrac{\epsilon}{8}, \psi(\tfrac{\epsilon}{8})\}. \tag{22}$$

Let $B[x_N, \epsilon] := \{x \in X : d(x_N, x) < \epsilon\}$ be a closed ball with center x_N and radius ϵ. We claim that $f(B[x_N, \epsilon]) \subseteq B[fx_{N-1}, \epsilon]$.

Using the P-property, we obtain
$$\left. \begin{aligned} d(x_N, fx_{N-1}) &= dist(A, B) \\ d(x_{N+1}, fx_N) &= dist(A, B) \end{aligned} \right\} \Longrightarrow$$

$$d(x_N, x_{N+1}) = d(fx_{N-1}, fx_N). \tag{23}$$

Consider $x \in B[x_N, \epsilon]$, i.e., $d(x_N, x) \leq \epsilon$. We distinguish two cases $d(x_N, x) \leq \dfrac{\epsilon}{2}$ and $d(x_N, x) > \dfrac{\epsilon}{2}$.

Case 1: $d(x_N, x) \leq \dfrac{\epsilon}{2}$.

Using the rectangular inequality, we distinguish the following two subcases:

- If $fx_{N-1} = fx_{N+1}$, $fx_{N+2} = fx$ and $fx_{N+1} \neq fx_{N+2}$, we have

$$\begin{aligned}
d(fx_{N-1}, fx) &= d(fx_{N+1}, fx_{N+2}) \\
&\leq d(x_{N+1}, x_{N+2}) - \psi(d(x_{N+1}, x_{N+2})) \\
&\leq d(x_{N+1}, x_{N+2}) \\
&= d(fx_N, fx_{N+1}) \\
&\leq d(x_N, x_{N+1}) \\
&\leq \frac{\epsilon}{8}.
\end{aligned}$$

In the case where $fx_{N+1} = fx_{N+2}$, we obtain $d(fx_{N-1}, fx) = 0$.

- If $fx_{N-1} \neq fx_{N+1}$, $fx_{N+2} \neq fx$ and $fx_{N+1} \neq fx_{N+2}$, we have

$$\begin{aligned}
d(fx_{N-1}, fx) &\leq d(fx_{N-1}, fx_{N+1}) + d(fx_{N+1}, fx_{N+2}) + d(fx_{N+2}, fx) \\
&= d(x_N, x_{N+2}) + d(fx_{N+1}, fx_{N+2}) + d(fx_{N+2}, fx) \\
&\leq d(x_N, x_{N+2}) + d(x_{N+1}, x_{N+2}) - \psi(d(x_{N+1}, x_{N+2})) + d(x_{N+2}, x) - \psi(d(x_{N+2}, x)) \\
&\leq d(x_N, x_{N+2}) + d(x_{N+1}, x_{N+2}) + d(x_{N+2}, x) \\
&\leq d(x_N, x_{N+2}) + d(x_{N+1}, x_{N+2}) + d(x_{N+2}, x_{N+1}) + d(x_{N+1}, x_N) + d(x_N, x) \\
&\leq d(x_N, x_{N+2}) + 2d(x_{N+1}, x_{N+2}) + d(x_{N+1}, x_N) + d(x_N, x) \\
&\leq d(x_N, x_{N+2}) + 2d(x_N, x_{N+1}) - 2\psi(d(x_N, x_{N+1})) + d(x_{N+1}, x_N) + d(x_N, x) \\
&\leq d(x_N, x_{N+2}) + 3d(x_N, x_{N+1}) + d(x_N, x) \\
&\leq \frac{\epsilon}{8} + 3 \times \frac{\epsilon}{8} + \frac{\epsilon}{2} = \epsilon
\end{aligned}$$

which implies that $fx \in B[fx_{N-1}, \epsilon]$.

Case 2: $\frac{\epsilon}{2} < d(x_N, x) \leq \epsilon$.

- If $fx_{N-1} = fx_{N+1}$, $fx_N = fx$ and $fx_{N+1} = fx_N$, we get

$$\begin{aligned}
d(fx_{N-1}, fx) &\leq d(fx_{N+1}, fx_N) \\
&\leq d(x_{N+1}, x_N) - \psi(d(x_{N+1}, x_N)) \\
&\leq d(x_{N+1}, x_N) \\
&\leq \frac{\epsilon}{8}.
\end{aligned}$$

- If $fx_{N-1} \neq fx_{N+1}$, $fx_N \neq fx$ and $fx_{N+1} \neq fx_N$, we have

$$\begin{aligned}
d(fx_{N-1}, fx) &\leq d(fx_{N-1}, fx_{N+1}) + d(fx_{N+1}, fx_N) + d(fx_N, fx) \\
&\leq d(x_N, x_{N+2}) + d(x_{N+1}, x_N) - \psi(d(x_{N+1}, x_N)) + d(x_N, x) - \psi(d(x_N, x)) \\
&\leq d(x_N, x_{N+2}) + d(x_{N+1}, x_N) + d(x_N, x) - \psi(d(x_N, x)) \\
&\leq \frac{\epsilon}{8} + \frac{\epsilon}{8} + \epsilon - \psi(\frac{\epsilon}{2}) \\
&= \frac{\epsilon}{4} + \epsilon - \psi(\frac{\epsilon}{2}) \\
&\leq \frac{\epsilon}{2} + \epsilon - \psi(\frac{\epsilon}{2}) \\
&\leq \psi(\frac{\epsilon}{2}) + \epsilon - \psi(\frac{\epsilon}{2}) = \epsilon. \text{(since } \psi \text{ is increasing)}.
\end{aligned}$$

Then, $d(fx_{N-1}, fx) \leq \epsilon$, which gives that $fx \in B[fx_{N-1}, \epsilon]$. Thus, we obtain that

$$f(B[x_N, \epsilon]) \subseteq B[fx_{N-1}, \epsilon]. \tag{24}$$

Claim: If $y \in B[fx_{N-1}, \epsilon]$ with $d(x, y) = dist(A, B)$ for some $x \in A_0$, then $x \in B[x_N, \epsilon]$.

Let $y \in B[fx_{N-1}, \epsilon]$. Then,

$$d(fx_{N-1}, y) \leq \epsilon. \tag{25}$$

Assume that there exists $x \in A_0$ such that $d(x, y) = dist(A, B)$. From (17), we get $d(x_N, fx_{N-1}) = dist(A, B)$ which gives us using the P-property,

$$d(x, x_N) = d(y, fx_{N-1}). \tag{26}$$

From (25) and (26), we obtain that $d(x, x_N) \leq \epsilon$, i.e., $x \in B[x_N, \epsilon]$ and the claim is proved.
From (21) and (23), we have $x_{N+1} \in B[x_N, \epsilon]$. Then, using (24), we get $fx_{N+1} \in B[fx_{N-1}, \epsilon]$, i.e.,

$$d(fx_{N+1}, fx_{N-1}) \leq \epsilon. \tag{27}$$

Since $d(x_{N+2}, fx_{N+1}) = dist(A, B)$, by the precedent claim $d(x_{N+2}, fx_N) \leq \epsilon$. Again, from (24), $d(x_{N+2}, fx_{N-1}) \leq \epsilon$ and from the claim $d(x_{N+3}, fx_N) \leq \epsilon$. In this way, we obtain

$$d(x_{N+m}, x_N) \leq \epsilon \ \forall m \in \mathbb{N}. \tag{28}$$

Thus, the sequence $\{x_n\}$ is Cauchy. Since A is a closed subset of the complete rectangular metric space, there exists $x^* \in A$ such that

$$\lim_{n \to \infty} x_N = x^*. \tag{29}$$

From the continuity of f, we obtain

$$\lim_{n \to \infty} fx_N = fx^*. \tag{30}$$

Then, using the continuity of the rectangular metric, we obtain

$$d(x_{N+1}, fx_N) \longrightarrow d(x^*, fx^*) \text{ as } N \longrightarrow \infty. \tag{31}$$

From (17), $d(x_{N+1}, x_N) = dist(A, B)$, we conclude that $\{d(x_{N+1}, x_N)\}_N$ is a constant sequence equal to $dist(A, B)$. Therefore, from (31), $d(x^*, fx^*) = dist(A, B)$. Thereby, x^* is a best proximity point of f.

Let us prove the uniqueness of the best proximity point. Consider x_1, x_2 two different best proximity points. Then, $d(x_1, fx_1) = d(x_2, fx_2) = dist(A, B)$. From the P-property, we obtain $d(x_1, x_2) = d(fx_1, fx_2)$. Using that f is weakly G-contractive, we get

$$0 < d(x_1, x_2) = d(fx_1, fx_2) \leq d(x_1, x_2) - \psi(d(x_1, x_2)) < d(x_1, x_2), \tag{32}$$

which is a contradiction. Therefore, $x_1 = x_2$. □

Definition 13. *Let (X, d) be a rectangular metric space and G be a directed graph. Let A, B be two nonempty subsets of X. A non-self mapping $T : A \longrightarrow B$ is said to be*

- *a G- proximal Kannan mapping if for $x, y, u, v \in A$, there exists $b \in [0, \frac{1}{2})$ such that*
$$\left.\begin{array}{ll} (x, y) & \in E(G) \\ d(u, Tx) & = d(A, B) \\ d(v, Ty) & = d(A, B) \end{array}\right\} \implies d(u, v) \leq b[d(x, v) + d(y, u)].$$

- *proximally G-edge preserving if for each $x, y, u, v \in A$*

$$\left.\begin{array}{l}(x,y) \in E(G) \\ d(u, Tx) = d(A, B) \\ d(v, Ty) = d(A, B)\end{array}\right\} \implies (u, v) \in E(G).$$

Our third main result is as follows:

Theorem 3. *Let (X, d) be a rectangular metric space and G a directed graph. Let A, B be two nonempty closed subsets of X. Assume that A_0 is nonempty and d is continuous. Let $T : A \longrightarrow B$ be a continuous non-self mapping satisfying the following properties:*

- *T is proximal G-edge preserving and a G-proximal Kannan mapping such that $T(A_0) \subseteq B_0$.*
- *There exist $x_0, x_1 \in A_0$ such that*

$$d(x_1, Tx_0) = d(A, B) \text{ and } (x_0, x_1) \in E(G). \tag{33}$$

Then, T has a best proximity point x^* in A. Furthermore, the sequence $\{x_n\}$ defined by $d(x_n, Tx_{n-1}) = d(A, B)$ for all $n \in \mathbb{N}$ converges to x^*. Moreover, if there exists a path in G between any two points of A, then the best proximity point is unique.

Proof. From (33), there exist $x_0, x_1 \in A_0$ such that

$$d(x_1, Tx_0) = d(A, B) \text{ and } (x_0, x_1) \in E(G). \tag{34}$$

Since $T(A_0) \subseteq B_0$, we have $Tx_1 \in B_0$ and there exists $x_2 \in A_0$ such that

$$d(x_2, Tx_1) = d(A, B). \tag{35}$$

Using the proximally G-edge preserving of T, (34) and (35), we get $(x_1, x_2) \in E(G)$. By continuing this process, we obtain the sequence $\{x_n\}$ in A_0 such that

$$d(x_n, Tx_{n-1}) = d(A, B) \tag{36}$$

with $(x_n, Tx_{n-1}) \in E(G) \; \forall n \in \mathbb{N}. \tag{37}$

Now, let us prove that $\{x_n\}$ is a Cauchy sequence in A. Note that if there exists $n_0 \in \mathbb{N}$ such that $x_{n_0} = x_{n_0+1}$, from (36), we get that x_{n_0} is a best proximity point of T. Therefore, we may assume that $x_{n-1} \neq x_n$ for all $n \in \mathbb{N}$.

Since T is a G-proximal Kannan mapping for each $n \in \mathbb{N}$, we obtain $(x_{n-1}, x_n) \in E(G)$, $d(x_n, Tx_{n-1}) = d(A, B)$ and $d(x_{n+1}, Tx_n) = d(A, B)$ which imply that

$$d(x_n, x_{n+1}) \leq b[d(x_{n-1}, x_{n+1}) + d(x_n, x_n)] \leq bd(x_{n-1}, x_{n+1}).$$

By induction, we obtain

$$d(x_n, x_{n+1}) \leq b^n d(x_0, x_2) = b^n C \; \forall n \in \mathbb{N}. \tag{38}$$

As $b < \dfrac{1}{2}$, then $d(x_n, x_{n+1}) \longrightarrow 0$ as $n \longrightarrow \infty$. Let $p \geq 1$.

Case 1:

$$\begin{aligned}
d(x_n, x_{n+(2p+1)}) &= d(x_n, x_{n+1}) + d(x_{n+1}, x_{n+2}) + \ldots + d(x_{n+2p}, x_{n+2p+1}) \\
&\leq b^n C + b^{n+1} C + \ldots + b^{n+2p} C \\
&= (b^n + b^{n+1} + \ldots + b^{n+2p}) C \longrightarrow 0 \text{ as } n, p \longrightarrow \infty.
\end{aligned} \quad (39)$$

Case 2:

$$\begin{aligned}
d(x_n, x_{n+(2p)}) &= d(x_n, x_{n+1}) + d(x_{n+1}, x_{n+2}) + \ldots + d(x_{n+2p-2}, x_{n+2p}) \\
&\leq d(x_n, x_{n+1}) + d(x_{n+1}, x_{n+2}) + \ldots + d(x_{n+2p-2}, x_{n+2p-1}) + d(x_{n+2p-1}, x_{n+2p}) \\
&\quad + d(x_{n+2p-2}, x_{n+2p}) \\
&\leq \sum_{k=n}^{n+2p-1} Cb^k + d(x_{n+2p-2}, x_{n+2p}).
\end{aligned} \quad (40)$$

Knowing that $\sum_{k=n}^{n+2p-1} Cb^k \longrightarrow 0$ as $n, p \longrightarrow \infty$, we shall prove that $d(x_{n+2p-2}, x_{n+2p}) \longrightarrow 0$ as $n, p \longrightarrow \infty$. From (36), we can conclude that

$$d(x_{n+2p-2}, Tx_{n+2p-1}) = d(A, B) \quad (41)$$
$$d(x_{n+2p}, Tx_{n+2p+1}) = d(A, B). \quad (42)$$

On the other hand, from (37) we get $(x_{n+2p-1}, x_{n+2p}) \in E(G)$ and $(x_{n+2p}, x_{n+2p+1}) \in E(G)$. Then, since G is a connected graph, there exists a path between x_{n+2p-1} and x_{n+2p+1} in G. Therefore,

$$(x_{n+2p-1}, x_{n+2p+1}) \in E(G). \quad (43)$$

Knowing that T is a G-proximal Kannan mapping and from (41)–(43), we obtain

$$\begin{aligned}
d(x_{n+2p-2}, x_{n+2p}) &\leq b[d(x_{n+2p-1}, x_{n+2p}) + d(x_{n+2p+1}, x_{n+2p-2})] \\
&\leq b[d(x_{n+2p-1}, x_{n+2p}) + d(x_{n+2p-2}, x_{n+2p-1}) + d(x_{n+2p-1}, x_{n+2p}) + d(x_{n+2p}, x_{n+2p+1})] \\
&= b[2d(x_{n+2p-1}, x_{n+2p}) + d(x_{n+2p-2}, x_{n+2p-1}) + d(x_{n+2p}, x_{n+2p+1})] \\
&\leq b[2Cb^{n+2p-1} + Cb^{n+2p-2} + Cb^{n+2p}] \longrightarrow 0 \text{ as } n, p \to \infty.
\end{aligned} \quad (44)$$

Therefore, from (40), we conclude that $d(x_n, x_{n+2p}) \longrightarrow 0$ as $n, p \longrightarrow \infty$. It follows that $\{x_n\}$ is a Cauchy sequence in A. Since A is closed, there exists $x^* \in A$ such that $x_n \longrightarrow x^*$ as $n \longrightarrow \infty$. By the continuity of T, we obtain $Tx_n \longrightarrow Tx^*$ as $n \longrightarrow \infty$. Since d is assumed to be continuous, we get $d(x_{n+1}, Tx_n) \longrightarrow d(x^*, Tx^*)$ as $n \longrightarrow \infty$. By (36), we conclude that

$$d(x^*, Tx^*) = d(A, B).$$

Thus, x^* is a best proximity point of T and the sequence $\{x_n\}$ defined by $d(x_{n+1}, Tx_n) = d(A, B)$ converges to x^* for all $n \in \mathbb{N}$.

Let us prove the uniqueness of the best proximity point x^*. Suppose that x_1^* and x_2^* are two best proximity points. Then, we obtain $d(x_1^*, Tx_1^*) = d(A, B)$, $d(x_2^*, Tx_2^*) = d(A, B)$ and $(x_1^*, x_2^*) \in E(G)$, which gives $d(x_1^*, x_2^*) \leq b[d(x_1^*, x_2^*) + d(x_1^*, x_2^*)] = 2bd(x_1^*, x_2^*)$. Therefore, we get $(1 - 2b)d(x_1^*, x_2^*) \leq 0$, which implies that $1 - 2b \leq 0 \Longrightarrow b \geq \frac{1}{2}$. It is a contradiction with respect to $b < \frac{1}{2}$. Then, $d(x_1^*, x_2^*) = 0$, that is, $x_1^* = x_2^*$ and so the uniqueness of the best proximity point follows. □

4. Conclusions and Perspectives

In Theorems 1–3, we assumed that the rectangular metric space is continuous, which is a strong hypothesis and does not hold in general. To our knowledge, our work is the first attempt to prove best proximity point results not only in the setting of rectangular metric spaces, but with the addition of a graph theory structure. Finally, an open question, how does one prove the above three theorems when omitting the continuity of the rectangular metric?

Author Contributions: All authors read and approved the manuscript.

Funding: This research was funded by the research group Nonlinear Analysis Methods in Applied Mathematics (NAMAM) group number RG-DES-2017-01-17.

Acknowledgments: The third and fourth authors would like to thank Prince Sultan University for funding this work through the research group Nonlinear Analysis Methods in Applied Mathematics (NAMAM) group number RG-DES-2017-01-17.

Conflicts of Interest: The authors declare no conflict of interest.

References

1. Abdeljawad, T.; Alzabut, J.O.; Mukheimer, E.; Zaidan, Y. Banach contraction principle for cyclical mappings on partial metric spaces. In *Fixed Point Theory and Applications*; Springer International Publishing: Cham, Switzerland, 2012.
2. Mlaiki, M.; Abodayeh, K.; Aydi, H.; Abdeljawad, T.; Abuloha, M. Rectangular Metric-Like Type Spaces and Related Fixed Points. *J. Math.* **2018**, *2018*, 3581768. [CrossRef]
3. Abodayeh, K.; Bataihah, A.; Shatanawi, W. Generalized w-distance mappings and some fixed point theorems. *UPB Sci. Bull. Ser. A Appl. Math. Phys.* **2017**, *79*, 223–232.
4. Afshari, H.; Marasi, H.R.; Aydi, H. Existence and uniqueness of positive solutions for boundary value problems of fractional differential equations. *Filomat* **2017**, *31*, 2675–2682. [CrossRef]
5. Afshari, H.; Aydi, H. Some results about Krosnoselski-Mann iteration process. *J. Nonlinear Sci. Appl.* **2016**, *9*, 4852–4859. [CrossRef]
6. Ameer, E.; Aydi, H.; Arshad, M.; Alsamir, H.; Noorani, M.S. Hybrid multivalued type contraction mappings in α_K-complete partial b-metric spaces and applications. *Symmetry* **2019**, *11*, 86. [CrossRef]
7. Espinola, R.; Kirk, W.A. Fixed point theorems in R-trees with applications to graph theory. *Topol. Appl.* **2006**, *153*, 215–218. [CrossRef]
8. Felhi, A.; Aydi, H. New fixed point results for multi-valued maps via manageable functions and an application on a boundary value problem. *UPB Sci. Bull. Ser. A* **2018**, *80*, 81–92.
9. Felhi, A.; Aydi, H.; Zhang, D. Fixed points for α-admissible contractive mappings via simulation functions. *J. Nonlinear Sci. Appl.* **2016**, *9*, 5544–5560. [CrossRef]
10. Marasi, H.R.; Piri, H.; Aydi, H. Existence and multiplicity of solutions for nonlinear fractional differential equations. *J. Nonlinear Sci. Appl.* **2016**, *9*, 4639–4646. [CrossRef]
11. Radenović, S.; Chandok, S.; Shatanawi, W. Some cyclic fixed point results for contractive mappings. *Univ. Thought Publ. Nat. Sci.* **2016**, *6*, 38–40. [CrossRef]
12. Shatanawi, W.; Abodaye, K.; Bataihah, A. Fixed point theorem through ω-distance of Suzuki type contraction condition. *Gazi Univ. J. Sci.* **2016**, *29*, 129–133.
13. Souayah, N.; Mlaiki, N.; Mrad, M. The G_M−Contraction Principle for Mappings on M−Metric Spaces Endowed with a Graph and Fixed Point Theorems. *IEEE Access* **2018**, *6*, 25178–25184. [CrossRef]
14. Souayah, N.; Mlaiki, N. A fixed point theorem in S_b-metric spaces. *J. Math. Comput. Sci.* **2016**, *16*, 131–139. [CrossRef]
15. Abdeljawad, T.; Alzabut, J.O.; Mukheimer, E.; Zaidan, Y. Best proximity points for cyclical contraction mappings with boundedly compact decompositions in partial metric spaces. *J. Comput. Anal. Appl.* **2013**, *15*, 678–685.
16. Aydi, H.; Felhi, A. On best proximity points for various α-proximal contractions on metric-like spaces. *J. Nonlinear Sci. Appl.* **2016**, *9*, 5202–5218. [CrossRef]
17. Aydi, H.; Felhi, A.; Karapinar, E. On common best proximity points for generalized $\alpha - \psi$-proximal contractions. *J. Nonlinear Sci. Appl.* **2016**, *9*, 2658–2670. [CrossRef]

18. Aydi, H.; Felhi, A. Best proximity points and stability results for controlled proximal contractive set valued mappings. *Fixed Point Theory Appl.* **2016**, *2016*, 22.
19. di Bari, C.; Suzuki, T.; Vetro, C. Best proximity points for cyclic Meir-Keeler contractions. *Nonlinear Anal.* **2008**, *69*, 3790–3794. [CrossRef]
20. Basha, S.S.; Veeramani, P.; Pai, D.V. Best proximity pair theorems. *Indian J. Pure Appl. Math.* **2001**, *32*, 1237–1246.
21. Basha, S.S.; Veeramani, P. Best proximity pair theorems for multifunctions with open fibres. *J. Approx. Theory* **2000**, *103*, 119–129. [CrossRef]
22. Choudhury, B.S.; Metiya, N.; Postolache, M.; Konar, P. A discussion on best proximity point and coupled best proximity point in partially ordered metric spaces. *Fixed Point Theory Appl.* **2015**, *2015*, 170. [CrossRef]
23. Eldred, A.A.; Veeramani, P. On best proximity pair solutions with applications to differential equations. *J. Indian Math. Soc.* **2008**, *1907*, 51–62.
24. Eldred, A.A.; Kirk, W.A.; Veeramani, P. Proximal normal structure and relatively nonexpansive mappings. *Studia Math.* **2005**, *171*, 283–293. [CrossRef]
25. Jacob, G.K.; Postolache, M.; Marudai, M.; Raja, V. Norm convergence iterations for best proximity points of non-self non-expansive mappings. *UPB Sci. Bull. Ser. A Appl. Math. Phys.* **2017**, *79*, 49–56.
26. Karpagam, S.; Agarwal, S. Best proximity point theorems for p-cyclic Meir-Keeler contractions. *Fixed Point Theory Appl.* **2009**, *2009*, 9. [CrossRef]
27. Kirk, W.A.; Reich, S.; Veeramani, P. Proximinal retracts and best proximity pair theorems. *Numer. Funct. Anal. Optim.* **2003**, *24*, 851–862. [CrossRef]
28. Shatanawi, W. Best proximity point on nonlinear contractive condition. *J. Phys. Conf. Seri.* **2013**, *435*, 012006. [CrossRef]
29. Shatanawi, W.; Pitea, A. Best Proximity Point and Best Proximity Coupled Point in a Complete Metric Space with (P)-Property. *Filomat* **2015**, *29*, 63–74. [CrossRef]
30. Branciari, A. A fixed point theorem of Banach-Caccioppoli type on a class of generalized metric spaces. *Publ. Math. Debr.* **2000**, *57*, 31–37.
31. Jachymski, J. The contraction principle for mappings on a metric space with a graph. *Proc. Am. Math. Soc.* **2008**, *136*, 1359–1373. [CrossRef]
32. Raj, V.S. Best proximity point theorems for non-self mappings. *Fixed Point Theory* **2013**, *14*, 447–454.

© 2019 by the authors. Licensee MDPI, Basel, Switzerland. This article is an open access article distributed under the terms and conditions of the Creative Commons Attribution (CC BY) license (http://creativecommons.org/licenses/by/4.0/).

MDPI
St. Alban-Anlage 66
4052 Basel
Switzerland
Tel. +41 61 683 77 34
Fax +41 61 302 89 18
www.mdpi.com

Axioms Editorial Office
E-mail: axioms@mdpi.com
www.mdpi.com/journal/axioms

www.ingramcontent.com/pod-product-compliance
Lightning Source LLC
LaVergne TN
LVHW071943080526
838202LV00064B/6665